SOURCE
The Prentice Hall
ENGINEERING SOURCE

Introduction
to Engineering Analysis

Second Edition

Kirk D. Hagen

Weber State University, Ogden, UT

PEARSON

Prentice
Hall

Upper Saddle River, NJ 07458

Library of Congress Cataloging-in-Publication Data

Hagen, Kirk D., 1953-
    Introduction to engineering analysis / Kirk D. Hagen.-- 2nd ed.
       p. cm. -- (The Prentice Hall engineering source)
    Includes bibliographical reference and index.
    ISBN 0-13-145332-7
    1. Engineering mathematics. I. Title. II. ESource--the Prentice Hall engineering source.
TA330.H34 2005
620'.001'51--dc22   2003069012

Vice President and Editorial Director, ECS: *Marcia J. Horton*
Executive Editor: *Eric Svendsen*
Associate Editor: *Dee Bernhard*
Vice President and Director of Production and Manufacturing, ESM: *David W. Riccardi*
Executive Managing Editor: *Vince O'Brien*
Managing Editor: *David A. George*
Production Editor: *Barbara A. Till*
Art Director: *Jayne Conte*
Cover Designer: *Bruce Kenselaar*
Art Editor: *Greg Dulles*
Manufacturing Manager: *Trudy Pisciotti*
Manufacturing Buyer: *Lisa McDowell*
Marketing Manager: *Holly Stark*

© 2005, 2000 Pearson Education, Inc.
Pearson Prentice Hall
Pearson Education, Inc.
Upper Saddle River, NJ 07458

Printed in the United States of America

10 9 8 7 6 5 4 3 2 1

0-13-145332-7

Pearson Education Ltd., *London*
Pearson Education Australia Pty. Ltd., *Sydney*
Pearson Education Singapore, Pte. Ltd.
Pearson Education North Asia Ltd., *Hong Kong*
Pearson Education Canada, Inc., *Toronto*
Pearson Educación de Mexico, S.A. de C.V.
Pearson Education—Japan, *Tokyo*
Pearson Education Malaysia, Pte. Ltd.
Pearson Education, Inc., *Upper Saddle River, New Jersey*

About ESource

ESource—The Prentice Hall Engineering Source—
www.prenhall.com/esource

ESource—The Prentice Hall Engineering Source gives professors the power to harness the full potential of their text and their first-year engineering course. More than just a collection of books, ESource is a unique publishing system revolving around the ESource website—www.prenhall.com/esource. ESource enables you to put your stamp on your book just as you do your course. It lets you:

Control You choose exactly which chapters are in your book and in what order they appear. Of course, you can choose the entire book if you'd like and stay with the authors' original order.

Optimize Get the most from your book and your course. ESource lets you produce the optimal text for your students needs.

Customize You can add your own material anywhere in your text's presentation, and your final product will arrive at your bookstore as a professionally formatted text. Of course, all titles in this series are available as stand-alone texts, or as bundles of two or more books sold at a discount. Contact your PH sales rep for discount information.

ESource ACCESS

Professors who choose to bundle two or more texts from the ESource series for their class, or use an ESource custom book will be providing their students with an on-line library of intro engineering content—ESource Access. We've designed ESource ACCESS to provide students a flexible, searchable, on-line resource. Free access codes come in bundles and custom books are valid for one year after initial log-on. Contact your PH sales rep for more information.

ESource Content

All the content in ESource was written by educators specifically for freshman/first-year students. Authors tried to strike a balanced level of presentation, an approach that was neither formulaic nor trivial, and one that did not focus too heavily on advanced topics that most introductory students do not encounter until later classes. Because many professors do not have extensive time to cover these topics in the classroom, authors prepared each text with the idea that many students would use it for self-instruction and independent study. Students should be able to use this content to learn the software tool or subject on their own.

While authors had the freedom to write texts in a style appropriate to their particular subject, all followed certain guidelines created to promote a consistency that makes students comfortable. Namely, every chapter opens with a clear set of **Objectives**, includes **Practice Boxes** throughout the chapter, and ends with a number of **Problems**, and a list of **Key Terms**. **Applications Boxes** are spread throughout the book with the intent of giving students a real-world perspective of engineering. **Success Boxes** provide the student with advice about college study skills, and help students avoid the common pitfalls of first-year students. In addition, this series contains an entire book titled *Engineering Success* by Peter Schiavone of the University of Alberta intended to expose students quickly to what it takes to be an engineering student.

Creating Your Book

Using ESource is simple. You preview the content either on-line or through examination copies of the books you can request on-line, from your PH sales rep, or by calling 1-800-526-0485. Create an on-line outline of the content you want, in the order you want, using ESource's simple interface. Insert your own material into the text flow. If you are not ready to order, ESource will save your work. You can come back at any time and change, re-arrange, or add more material to your creation. Once you're finished you'll automatically receive an ISBN. Give it to your bookstore and your book will arrive on their shelves four to six weeks after they order. Your custom desk copies with their instructor supplements will arrive at your address at the same time.

To learn more about this new system for creating the perfect textbook, go to www.prenhall.com/esource. You can either go through the on-line walkthrough of how to create a book, or experiment yourself.

Supplements

Adopters of ESource receive an instructor's CD that contains professor and student resources and **350 PowerPoint transparencies** created by Jack Leifer of University of Kentucky–Paducah for various books in the series. Professors can either follow these transparencies as pre-prepared lectures or use them as the basis for their own custom presentations.

Titles in the ESource Series

Design Concepts for Engineers, 2/e
0-13-093430-5
Mark Horenstein

Engineering Success, 2/e
0-13-041827-7
Peter Schiavone

Engineering Design and Problem Solving, 2E
0-13-093399-6
Steven K. Howell

Exploring Engineering
0-13-093442-9
Joe King

Engineering Ethics
0-13-784224-4
Charles B. Fleddermann

Introduction to Engineering Analysis, 2/e
0-13-145332-7
Kirk D. Hagen

Introduction to Engineering Communication
0-13-146102-8
Hillary Hart

Introduction to Engineering Experimentation
0-13-032835-9
Ronald W. Larsen, John T. Sears, and Royce Wilkinson

Introduction to Mechanical Engineering
0-13-019640-1
Robert Rizza

Introduction to Electrical and Computer Engineering
0-13-033363-8
Charles B. Fleddermann and Martin Bradshaw

Introduction to MATLAB 6—Update
0-13-140918-2
Delores Etter and David C. Kuncicky, with Douglas W. Hull

MATLAB Programming
0-13-035127-X
David C. Kuncicky

Introduction to Mathcad 2000
0-13-020007-7
Ronald W. Larsen

Introduction to Mathcad 11
0-13-008177-9
Ronald W. Larsen

Introduction to Maple 8
0-13-032844-8
David I. Schwartz

Mathematics Review
0-13-011501-0
Peter Schiavone

Power Programming with VBA/Excel
0-13-047377-4
Steven C. Chapra

Introduction to Excel 2002
0-13-008175-2
David C. Kuncicky

Introduction to Excel, 2/e
0-13-016881-5
David C. Kuncicky

About the Authors

No project could ever come to pass without a group of authors who have the vision and the courage to turn a stack of blank paper into a book. The authors in this series, who worked diligently to produce their books, provide the building blocks of the series.

Martin D. Bradshaw was born in Pittsburg, KS in 1936, grew up in Kansas and the surrounding states of Arkansas and Missouri, graduating from Newton High School, Newton, KS in 1954. He received the B.S.E.E. and M.S.E.E. degrees from the University of Wichita in 1958 and 1961, respectively. A Ford Foundation fellowship at Carnegie Institute of Technology followed from 1961 to 1963 and he received the Ph.D. degree in electrical engineering in 1964. He spent his entire academic career with the Department of Electrical and Computer Engineering at the University of New Mexico (1961-1963 and 1991-1996). He served as the Assistant Dean for Special Programs with the UNM College of Engineering from 1974 to 1976 and as the Associate Chairman for the EECE Department from 1993 to 1996. During the period 1987-1991 he was a consultant with his own company, EE Problem Solvers. During 1978 he spent a sabbatical year with the State Electricity Commission of Victoria, Melbourne, Australia. From 1979 to 1981 he served an IPA assignment as a Project Officer at the U.S. Air Force Weapons Laboratory, Kirkland AFB, Albuquerque, NM. He has won numerous local, regional, and national teaching awards, including the George Westinghouse Award from the ASEE in 1973. He was awarded the IEEE Centennial Medal in 2000.

Acknowledgments: Dr. Bradshaw would like to acknowledge his late mother, who gave him a great love of reading and learning, and his father, who taught him to persist until the job is finished. The encouragement of his wife, Jo, and his six children is a never-ending inspiration.

Stephen J. Chapman received a B.S. degree in Electrical Engineering from Louisiana State University (1975), the M.S.E. degree in Electrical Engineering from the University of Central Florida (1979), and pursued further graduate studies at Rice University. Mr. Chapman is currently Manager of Technical Systems for British Aerospace Australia, in Melbourne, Australia. In this position, he provides technical direction and design authority for the work of younger engineers within the company. He also continues to teach at local universities on a part-time basis.

Mr. Chapman is a Senior Member of the Institute of Electrical and Electronics Engineers (and several of its component societies). He is also a member of the Association for Computing Machinery and the Institution of Engineers (Australia).

Steven C. Chapra presently holds the Louis Berger Chair for Computing and Engineering in the Civil and Environmental Engineering Department at Tufts University. Dr. Chapra received engineering degrees from Manhattan College and the University of Michigan. Before joining the faculty at Tufts, he taught at Texas A&M University, the University of Colorado, and Imperial College, London. His research interests focus on surface water-quality modeling and advanced computer applications in environmental engineering. He has published over 50 refereed journal articles, 20 software packages and 6 books. He has received a number of awards including the 1987 ASEE Merriam/Wiley Distinguished Author Award, the 1993 Rudolph Hering Medal, and teaching awards from Texas A&M, the University of Colorado, and the Association of Environmental Engineering and Science Professors.

Acknowledgments: To the Berger Family for their many contributions to engineering education. I would also like to thank David Clough for his friendship and insights, John Walkenbach for his wonderful books, and my colleague Lee Minardi and my students Kenny William, Robert Viesca and Jennifer Edelmann for their suggestions.

 **Mark Dix** began working with AutoCAD in 1985 as a programmer for CAD Support Associates, Inc. He helped design a system for creating estimates and bills of material directly from AutoCAD drawing databases for use in the automated conveyor industry. This system became the basis for systems still widely in use today. In 1986 he began collaborating with Paul Riley to create AutoCAD training materials, combining Riley's background in industrial design and training with Dix's background in writing, curriculum development, and programming. Mr. Dix received the M.S. degree in education from the University of Massachusetts. He is currently the Director of Dearborn Academy High School in Arlington, Massachusetts.

 **Delores M. Etter** is a Professor of Electrical and Computer Engineering at the University of Colorado. Dr. Etter was a faculty member at the University of New Mexico and also a Visiting Professor at Stanford University. Dr. Etter was responsible for the Freshman Engineering Program at the University of New Mexico and is active in the Integrated Teaching Laboratory at the University of Colorado. She was elected a Fellow of the Institute of Electrical and Electronics Engineers for her contributions to education and for her technical leadership in digital signal processing.

 **Charles B. Fleddermann** is a professor in the Department of Electrical and Computer Engineering at the University of New Mexico in Albuquerque, New Mexico. All of his degrees are in electrical engineering: his Bachelor's degree from the University of Notre Dame, and the Master's and Ph.D. from the University of Illinois at Urbana-Champaign. Prof. Fleddermann developed an engineering ethics course for his department in response to the ABET requirement to incorporate ethics topics into the undergraduate engineering curriculum. *Engineering Ethics* was written as a vehicle for presenting ethical theory,

analysis, and problem solving to engineering undergraduates in a concise and readily accessible way.

Acknowledgments: I would like to thank Profs. Charles Harris and Michael Rabins of Texas A & M University whose NSF sponsored workshops on engineering ethics got me started thinking in this field. Special thanks to my wife Liz, who proofread the manuscript for this book, provided many useful suggestions, and who helped me learn how to teach "soft" topics to engineers.

 **Kirk D. Hagen** is a professor at Weber State University in Ogden, Utah. He has taught introductory-level engineering courses and upper-division thermal science courses at WSU since 1993. He received his B.S. degree in physics from Weber State College and his M.S. degree in mechanical engineering from Utah State University, after which he worked as a thermal designer/analyst in the aerospace and electronics industries. After several years of engineering practice, he resumed his formal education, earning his Ph.D. in mechanical engineering at the University of Utah. Hagen is the author of an undergraduate heat transfer text.

 **Mark N. Horenstein** is a Professor in the Department of Electrical and Computer Engineering at Boston University. He has degrees in Electrical Engineering from M.I.T. and U.C. Berkeley and has been involved in teaching engineering design for the greater part of his academic career. He devised and developed the senior design project class taken by all electrical and computer engineering students at Boston University. In this class, the students work for a virtual engineering company developing products and systems for real-world engineering and social-service clients.

Acknowledgments: I would like to thank Prof. James Bethune, the architect of the Peak Performance event at Boston University, for his permission to highlight the competition in my text. Several of the ideas relating to brainstorming and teamwork were derived from a workshop on engineering design offered by Prof. Charles Lovas of Southern Methodist University. The principles of estimation were derived in part from a freshman engineering problem posed by Prof. Thomas Kincaid of Boston University.

 **Steven Howell** is the Chairman and a Professor of Mechanical Engineering at Lawrence Technological University. Prior to joining LTU in 2001, Dr. Howell led a knowledge-based engineering project for Visteon Automotive Systems and taught computer-aided design classes for Ford Motor Company engineers. Dr. Howell also has a total of 15 years experience as an engineering faculty member at Northern Arizona University, the University of the Pacific, and the University of Zimbabwe. While at Northern Arizona University, he helped develop and implement an award-winning interdisciplinary series of design courses simulating a corporate engineering-design environment.

 **Douglas W. Hull** is a graduate student in the Department of Mechanical Engineering at Carnegie Mellon University in Pittsburgh, Pennsylvania. He is the author of *Mastering Mechanics I Using Matlab 5*, and contributed to *Mechanics of Materials* by Bedford and Liechti. His research in the Sensor Based Planning lab involves motion planning for hyper-redundant manipulators, also known as serpentine robots.

 **Scott D. James** is a staff lecturer at Kettering University (formerly GMI Engineering & Management Institute) in Flint, Michigan. He is currently pursuing a Ph.D. in Systems Engineering with an emphasis on software engineering and computer-integrated manufacturing. He chose teaching as a profession after several years in the computer industry. "I thought that it was really important to know what it was like outside of academia. I wanted to provide students with classes that were up to date and provide the information that is really used and needed."

Acknowledgments: Scott would like to acknowledge his family for the time to work on the text and his students and peers at Kettering who offered helpful critiques of the materials that eventually became the book.

 **Joe King** received the B.S. and M.S. degrees from the University of California at Davis. He is a Professor of Computer Engineering at the University of the Pacific, Stockton, CA, where he teaches courses in digital design, computer design, artificial intelligence, and computer networking. Since joining the UOP faculty, Professor King has spent yearlong sabbaticals teaching in Zimbabwe, Singapore, and Finland. A licensed engineer in the state of California, King's industrial experience includes major design projects with Lawrence Livermore National Laboratory, as well as independent consulting projects. Prof. King has had a number of books published with titles including *Matlab*, MathCAD, Exploring Engineering, and Engineering and Society.

 **David C. Kuncicky** is a native Floridian. He earned his Baccalaureate in psychology, Master's in computer science, and Ph.D. in computer science from Florida State University. He has served as a faculty member in the Department of Electrical Engineering at the FAMU–FSU College of Engineering and the Department of Computer Science at Florida State University. He has taught computer science and computer engineering courses for over 15 years. He has published research in the areas of intelligent hybrid systems and neural networks. He is currently the Director of Engineering at Bioreason, Inc. in Sante Fe, New Mexico.

Acknowledgments: Thanks to Steffie and Helen for putting up with my late nights and long weekends at the computer. Finally, thanks to Susan Bassett for having faith in my abilities, and for providing continued tutelage and support.

Ron Larsen is a Professor of Chemical Engineering at Montana State University, and received his Ph.D. from the Pennsylvania State University. He was initially attracted to engineering by the challenges the profession offers, but also appreciates that engineering is a serving profession. Some of the greatest challenges he has faced while teaching have involved non-traditional teaching methods, including evening courses for practicing engineers and teaching through an interpreter at the Mongolian National University. These experiences have provided tremendous opportunities to learn new ways to communicate technical material. Dr. Larsen views modern software as one of the new tools that will radically alter the way engineers work, and his book *Introduction to MathCAD* was written to help young engineers prepare to meet the challenges of an ever-changing workplace.

Acknowledgments: To my students at Montana State University who have endured the rough drafts and typos, and who still allow me to experiment with their classes—my sincere thanks.

Sanford Leestma is a Professor of Mathematics and Computer Science at Calvin College, and received his Ph.D. from New Mexico State University. He has been the long-time co-author of successful textbooks on Fortran, Pascal, and data structures in Pascal. His current research interest are in the areas of algorithms and numerical computation.

Jack Leifer is an Assistant Professor in the Department of Mechanical Engineering at the University of Kentucky Extended Campus Program in Paducah, and was previously with the Department of Mathematical Sciences and Engineering at the University of South Carolina–Aiken. He received his Ph.D. in Mechanical Engineering from the University of Texas at Austin in December 1995. His current research interests include the analysis of ultra-light and inflatable (Gossamer) space structures.

Acknowledgments: I'd like to thank my colleagues at USC–Aiken, especially Professors Mike May and Laurene Fausett, for their encouragement and feedback; and my parents, Felice and Morton Leifer, for being there and providing support (as always) as I completed this book.

Richard M. Lueptow is the Charles Deering McCormick Professor of Teaching Excellence and Associate Professor of Mechanical Engineering at Northwestern University. He is a native of Wisconsin and received his doctorate from the Massachusetts Institute of Technology in 1986. He teaches design, fluid mechanics, an spectral analysis techniques. Rich has an active research program on rotating filtration, Taylor Couette flow, granular flow, fire suppression, and acoustics. He has five patents and over 40 refereed journal and proceedings papers along with many other articles, abstracts, and presentations.

Acknowledgments: Thanks to my talented and hardworking co-authors as well as the many colleagues and students who took the tutorial for a "test drive." Special thanks to Mike Minbiole for his major contributions to Graphics Concepts with SolidWorks. Thanks also to Northwestern University for the time to work on a book. Most of all, thanks to my loving wife, Maiya, and my children, Hannah and Kyle, for supporting me in this endeavor. (Photo courtesy of Evanston Photographic Studios, Inc.)

Larry Nyhoff is a Professor of Mathematics and Computer Science at Calvin College. After doing bachelor's work at Calvin, and Master's work at Michigan, he received a Ph.D. from Michigan State and also did graduate work in computer science at Western Michigan. Dr. Nyhoff has taught at Calvin for the past 34 years—mathematics at first and computer science for the past several years.

 **Paul Riley** is an author, instructor, and designer specializing in graphics and design for multimedia. He is a founding partner of CAD Support Associates, a contract service and professional training organization for computer-aided design. His 15 years of business experience and 20 years of teaching experience are supported by degrees in education and computer science. Paul has taught AutoCAD at the University of Massachusetts at Lowell and is presently teaching AutoCAD at Mt. Ida College in Newton, Massachusetts. He has developed a program,

Computer-aided Design for Professionals that is highly regarded by corporate clients and has been an ongoing success since 1982.

 **Robert Rizza** is an Assistant Professor of Mechanical Engineering at North Dakota State University, where he teaches courses in mechanics and computer-aided design. A native of Chicago, he received the Ph.D. degree from the Illinois Institute of Technology. He is also the author of *Getting Started with Pro/ENGINEER*. Dr. Rizza has worked on a diverse range of engineering projects including projects from the railroad, bioengineering, and aerospace industries. His current research interests include the fracture of composite materials, repair of cracked aircraft components, and loosening of prostheses.

 **Peter Schiavone** is a professor and student advisor in the Department of Mechanical Engineering at the University of Alberta, Canada. He received his Ph.D. from the University of Strathclyde, U.K. in 1988. He has authored several books in the area of student academic success as well as numerous papers in international scientific research journals. Dr. Schiavone has worked in private industry in several different areas of engineering including aerospace and systems engineering. He founded the first Mathematics Resource Center at the University of Alberta, a unit designed specifically to teach new students the necessary *survival skills* in mathematics and the physical sciences required for success in first-year engineering. This led to the Students' Union Gold Key Award for outstanding contributions to the university. Dr. Schiavone lectures regularly to freshman engineering students and to new engineering professors on engineering success, in particular about maximizing students' academic performance.

Acknowledgements: Thanks to Richard Felder for being such an inspiration; to my wife Linda for sharing my dreams and believing in me; and to Francesca and Antonio for putting up with Dad when working on the text.

 **David I. Schneider** holds an A.B. degree from Oberlin College and a Ph.D. degree in Mathematics from MIT. He has taught for 34 years, primarily at the University of Maryland. Dr. Schneider has authored 28 books, with one-half of them computer programming books. He has developed three customized software packages that are supplied as supplements to over 55 mathematics textbooks. His involvement with computers dates back to 1962, when he programmed a special purpose computer at MIT's Lincoln Laboratory to correct errors in a communications system.

 **David I. Schwartz** is an Assistant Professor in the Computer Science Department at Cornell University and earned his B.S., M.S., and Ph.D. degrees in Civil Engineering from State University of New York at Buffalo. Throughout his graduate studies, Schwartz combined principles of computer science to applications of civil engineering. He became interested in helping students learn how to apply software tools for solving a variety of engineering problems. He teaches his students to learn incrementally and practice frequently to gain the maturity to tackle other subjects. In his spare time, Schwartz plays drums in a variety of bands.

Acknowledgments: I dedicate my books to my family, friends, and students who all helped in so many ways.

Many thanks go to the schools of Civil Engineering and Engineering & Applied Science at State University of New York at Buffalo where I originally developed and tested my UNIX and Maple books. I greatly appreciate the opportunity to explore my goals and all the help from everyone at the Computer Science Department at Cornell.

 **John T. Sears** received the Ph.D. degree from Princeton University. Currently, he is a Professor and the head of the Department of Chemical Engineering at Montana State University. After leaving Princeton he worked in research at Brookhaven National Laboratory and Esso Research and Engineering, until he took a position at West Virginia University. He came to MSU in 1982, where he has served as the Director of the College of Engineering Minority Program and Interim Director for BioFilm Engineering. Prof. Sears has written a book on air pollution and economic development, and over 45 articles in engineering and engineering education.

 **Michael T. Snyder** is President of Internet startup company Appointments 123.com. He is a native of Chicago, and he received his Bachelor of Science degree in Mechanical Engineering from the University of Notre Dame. Mike also graduated with honors from Northwestern University's Kellogg Graduate School of Management in 1999 with his Masters of Management degree. Before Appointments123.com, Mike was a mechanical engineer in new product development for Motorola Cellular and Acco Office Products. He has received four patents for his mechanical design work. "Pro/ ENGI-NEER was an invaluable design tool for me, and I am glad to help students learn the basics of Pro/ ENGINEER."

Acknowledgments: Thanks to Rich Lueptow and Jim Steger for inviting me to be a part of this great project. Of course, thanks to my wife Gretchen for her support in my various projects.

 **Jim Steger** is currently Chief Technical Officer and cofounder of an Internet applications company. He graduated with a Bachelor of Science degree in Mechanical Engineering from Northwestern University. His prior work included mechanical engineering assignments at Motorola and Acco Brands. At Motorola, Jim worked on part design for two-way radios and was one of the lead mechanical engineers on a cellular phone product line. At Acco Brands, Jim was the sole engineer on numerous office product designs. His Worx stapler has won design awards in the United States and in Europe. Jim has been a Pro/ENGINEER user for over six years.

Acknowledgments: Many thanks to my co-authors, especially Rich Lueptow for his leadership on this project. I would also like to thank my family for their continuous support.

 **Royce Wilkinson** received his undergraduate degree in chemistry from Rose-Hulman Institute of Technology in 1991 and the Ph.D. degree in chemistry from Montana State University in 1998 with research in natural product isolation from fungi. He currently resides in Bozeman, MT and is involved in HIV drug research. His research interests center on biological molecules and their interactions in the search for pharmaceutical advances.

ESource Reviewers

We would like to thank everyone who helped us with or has reviewed texts in this series.

Christopher Rowe, *Vanderbilt University*
Steve Yurgartis, *Clarkson University*
Heidi A. Diefes-Dux, *Purdue University*
Howard Silver, *Fairleigh Dickenson University*
Jean C. Malzahn Kampe, *Virginia Polytechnic Institute and State University*
Malcolm Heimer, *Florida International University*
Stanley Reeves, *Auburn University*
John Demel, *Ohio State University*
Shahnam Navee, *Georgia Southern University*
Heshem Shaalem, *Georgia Southern University*
Terry L. Kohutek, *Texas A & M University*
Liz Rozell, *Bakersfield College*
Mary C. Lynch, *University of Florida*
Ted Pawlicki, *University of Rochester*
James N. Jensen, *SUNY at Buffalo*
Tom Horton, *University of Virginia*
Eileen Young, *Bristol Community College*
James D. Nelson, *Louisiana Tech University*
Jerry Dunn, *Texas Tech University*
Howard M. Fulmer, *Villanova University*
Naeem Abdurrahman, *University of Texas, Austin*
Stephen Allan, *Utah State University*
Anil Bajaj, *Purdue University*
Grant Baker, *University of Alaska–Anchorage*
William Beckwith, *Clemson University*
Haym Benaroya, *Rutgers University*
John Biddle, *California State Polytechnic University*
Tom Bledsaw, *ITT Technical Institute*
Fred Boadu, *Duk University*
Tom Bryson, *University of Missouri, Rolla*
Ramzi Bualuan, *University of Notre Dame*
Dan Budny, *Purdue University*
Betty Burr, *University of Houston*
Dale Calkins, *University of Washington*
Harish Cherukuri, *University of North Carolina –Charlotte*
Arthur Clausing, *University of Illinois*
Barry Crittendon, *Virginia Polytechnic and State University*
James Devine, *University of South Florida*

Ron Eaglin, *University of Central Florida*
Dale Elifrits, *University of Missouri, Rolla*
Patrick Fitzhorn, *Colorado State University*
Susan Freeman, *Northeastern University*
Frank Gerlitz, *Washtenaw College*
Frank Gerlitz, *Washtenaw Community College*
John Glover, *University of Houston*
John Graham, *University of North Carolina–Charlotte*
Ashish Gupta, *SUNY at Buffalo*
Otto Gygax, *Oregon State University*
Malcom Heimer, *Florida International University*
Donald Herling, *Oregon State University*
Thomas Hill, *SUNY at Buffalo*
A.S. Hodel, *Auburn University*
James N. Jensen, *SUNY at Buffalo*
Vern Johnson, *University of Arizona*
Autar Kaw, *University of South Florida*
Kathleen Kitto, *Western Washington University*
Kenneth Klika, *University of Akron*
Terry L. Kohutek, *Texas A&M University*
Melvin J. Maron, *University of Louisville*
Robert Montgomery, *Purdue University*
Mark Nagurka, *Marquette University*
Romarathnam Narasimhan, *University of Miami*
Soronadi Nnaji, *Florida A&M University*
Sheila O'Connor, *Wichita State University*
Michael Peshkin, *Northwestern University*
Dr. John Ray, *University of Memphis*
Larry Richards, *University of Virginia*
Marc H. Richman, *Brown University*
Randy Shih, *Oregon Institute of Technology*
Avi Singhal, *Arizona State University*
Tim Sykes, *Houston Community College*
Neil R. Thompson, *University of Waterloo*
Raman Menon Unnikrishnan, *Rochester Institute of Technology*
Michael S. Wells, *Tennessee Tech University*
Joseph Wujek, *University of California, Berkeley*
Edward Young, *University of South Carolina*
Garry Young, *Oklahoma State University*
Mandochehr Zoghi, *University of Dayton*

Contents

5   ELECTRICAL CIRCUITS

6   THERMODYNAMICS

7   FLUID MECHANICS

8   DATA ANALYSIS: GRAPHING

9   DATA ANALYSIS: STATISTICS

1

The Role of Analysis in Engineering

1.1   INTRODUCTION

What is **analysis**? A dictionary definition of analysis might read something like this:

> *The separation of a whole into its component parts. An examination of a complex system, its elements, and their relationships.*

Based on this general definition, analysis may refer to everything from the study of a person's mental state (psychoanalysis) to the determination of the amount of certain elements in an unknown metal alloy (elemental analysis). *Engineering analysis*, however, has a specific meaning. A concise working definition of **engineering analysis** is the

> *Analytical solution of an engineering problem, using mathematics and principles of science.*

Engineering analysis relies heavily on **basic mathematics** such as algebra, geometry, trigonometry, calculus, and statistics. **Higher level mathematics** such as linear algebra, differential equations and complex variables may also be used. Principles and laws from the **physical sciences**, particularly physics and chemistry, are key ingredients of engineering analysis.

Engineering analysis involves more than searching for an equation that fits a problem, plugging numbers into the equation, and "turning the crank" to generate an answer. It is not a simple "plug and chug" procedure. Engineering analysis requires logical and systematic thinking about the engineering problem. The engineer must first be able to state the problem clearly, logically, and concisely. The engineer must understand the physical behavior of the system being analyzed and know which scientific principles to apply. He or she must recognize which mathematical tools to use and how to implement them by hand or on a computer. The engineer

OBJECTIVES

After reading this chapter, you will have learned

- What engineering analysis is
- That analysis is a major component of the engineering curriculum
- How analysis is used in engineering design
- How analysis helps engineers prevent and diagnose failures

must be able to generate a solution that is consistent with the stated problem and any simplifying assumptions. The engineer must then ascertain that the solution is reasonable and contains no errors.

Engineering analysis may be regarded as a type of **modeling** or *simulation*. For example, suppose that a civil engineer wants to know the tensile stress in a cable of a suspension bridge that is being designed. The bridge exists only on paper, so a direct stress measurement cannot be made. A scale model of the bridge could be constructed, and a stress measurement taken on the model, but models are expensive and very time-consuming to develop. A better approach is to create an analytical model of the bridge or a portion of the bridge containing the cable. From this model, the tensile stress can be calculated.

Engineering courses that focus on analysis, such as statics, dynamics, strength of materials, thermodynamics, and electrical circuits, are considered *core* courses in the engineering curriculum. Because you will be taking many of these courses, it is vital that you gain a fundamental understanding of what analysis is and, more importantly, how to do analysis properly. As the bridge example illustrates, analysis is an integral part of engineering design. Analysis is also a key part of the study of engineering failures.

Engineers who perform engineering analyses on a regular basis are referred to as *engineering analysts* or *analytic engineers*. These functional titles are used to differentiate analysis from the other engineering functions such as research and development (R&D), design, testing, production, sales, marketing, etc. In some engineering companies, clear distinctions are made between the various engineering functions and the people who work in them. Depending on the organizational structure and the type of products involved, large companies may dedicate a separate department or group of engineers as analysts. Engineers whose work is dedicated to analysis are considered specialists. In this capacity, the engineering analyst usually works in a support role for design engineering. It is not uncommon, however, for design and analysis functions to be combined in a single department because design and analysis are so closely related. In small firms that employ only a few engineers, the engineers often bear the responsibility of many technical functions, including analysis.

PROFESSIONAL SUCCESS: CHOOSING AN ENGINEERING MAJOR

Perhaps the biggest question facing the new engineering student (besides "How much money will I make after I graduate?") is "In which field of engineering should I major?" Engineering is a broad area, so the beginning student has numerous options. The new engineering student should be aware of a few facts. First, all engineering majors have the potential for preparing the student for a satisfying and rewarding engineering career. As a profession, engineering has historically enjoyed a fairly stable and well-paid market. There have been fluctuations in the engineering market in recent decades, but the demand for engineers in all the major disciplines is high, and the future looks bright for engineers. Second, all engineering majors are academically challenging, but some engineering majors are more challenging than others. Study the differences between the various engineering programs. Compare the course requirements of each program by examining the course listings in your college or university catalog. Ask department chairs to discuss the similarities and differences between their engineering programs and the programs in other departments. (Just keep in mind that professors, like everyone else, are biased and will probably tell you that *their* engineering discipline is the best.) Talk with people who are practicing engineers in the various disciplines, and ask them about their educational experiences. Learn all you can from as many sources as you can about the various engineering disciplines. Third, and this is the most important point, try to answer the following question: "What kind of engineering will be the most gratifying for me?" It makes little sense to devote four or more years of intense study of X engineering just because it happens to be the highest paid discipline, just because your uncle Vinny is an X engineer, just because X engineering is the easiest program at your school, or just because someone tells you that they are an X engineer, so you should be one too.

Engineering disciplines may be broadly categorized as either mainstream or narrowly focused. Mainstream disciplines are the broad-based, traditional disciplines that have been in existence for decades (or even centuries) and in which degrees are offered by most of the larger colleges and universities. Many colleges and universities do not offer engineering degrees in some of the narrowly focused disciplines. Chemical, civil, computer, electrical, and mechanical engineering are considered the core mainstream disciplines. These mainstream disciplines are broad in subject content and represent the majority of practicing engineers. Narrowly focused disciplines concentrate on a narrow engineering subject by combining specific components from the mainstream disciplines. For example, biomedical engineering may combine portions of electrical and mechanical engineering plus components from biology.

Construction engineering may combine elements from civil engineering and business or construction trades. Other narrowly focused disciplines include materials, aeronautical and aerospace, environmental, nuclear, ceramic, geological, manufacturing, automotive, metallurgical, corrosion, ocean, and cost and safety engineering.

Should you major in a mainstream area or a narrowly focused area? The safest thing to do, especially if you are uncertain about which discipline to study, is to major in one of the mainstream disciplines. By majoring in a mainstream area, you will graduate with a general engineering education that will make you marketable in a broad engineering industry. On the other hand, majoring in a narrowly focused discipline may lead you into an extremely satisfying career, particularly if your area of expertise, narrow as it may be, is in high demand. Perhaps your decision will be largely governed by geographical issues. The narrowly focused majors may not be offered at the school you wish to attend. These are important issues to consider when selecting an engineering major.

1.2   ANALYSIS AND ENGINEERING DESIGN

Design is the heart of engineering. In ancient times, people recognized a need for protection against the natural elements, for collecting and utilizing water, for finding and growing food, for transportation, and for defending themselves against other people with unfriendly intentions. Today, even though our world is much more advanced and complex than that of our ancestors, our basic needs are essentially the same. Throughout history, engineers have designed various devices and systems that met the changing needs of society. The following is a concise definition of **engineering design**:

> *A process of devising a component, system or operation that meets a specific need.*

The key word in this definition is *process*. The design process is like a road map that guides the designer from need recognition to problem solution. Design engineers make decisions based on a thorough understanding of engineering fundamentals, design constraints, cost, reliability, manufacturability, and human factors. A knowledge of design *principles* can be learned in school from professors and books, but in order to become a good design engineer, you must *practice* design. Design engineers are like artists and architects who harness their creative powers and skills to produce sculptures and buildings. The end products made by design engineers may be more functional than artistic, but their creation still requires knowledge, imagination, and creativity.

Design has always been a part of engineering programs in colleges and universities. Historically, design courses have been taught in the junior or senior years. At some schools, design courses have even been postponed until the senior year when the students would do a "senior design project" or a "capstone design project." In recent years, the traditional practice of placing design courses in the latter half of the curriculum has been scrutinized. Recognizing that design is indeed the heart of engineering and that students need an earlier introduction to the subject, colleges and universities have revised their engineering programs to include design experiences early in the curriculum, perhaps as early as the introductory course. By introducing design courses at the level that introductory mathematics and science courses are taught, students benefit from a more integrated approach to their engineering education. Students gain a better understanding of *how* mathematics and science are used to design engineering systems when they study these courses concurrently. Analysis is becoming embedded in design to teach engineering students more practical, real-world applications of mathematics and science.

What is the relationship between engineering analysis and engineering design? As we defined it earlier, engineering analysis is the *analytical solution of an engineering problem, using mathematics and principles of science*. The false notion that engineering is merely mathematics and applied science is widely held by many beginning engineering students. Because design is the heart of engineering, this notion may lead a student to believe that engineering design is the equivalent of a "story problem" found in high school math books. Hence, engineering design is merely a math problem posed in word form, right? Wrong! Unlike math problems, design problems are "open ended." This means, among other things, that design problems do not have a single "correct" solution. Design problems have many possible solutions, depending on the *decisions* made by the design engineer. The main goal of engineering design is to obtain the *best* or *optimum* solution within the specifications and constraints of the problem.

So, how does analysis fit in? One of the steps in the design process is to obtain a preliminary concept of the *design*. (Note that the word *design* here refers to the actual component, system, or operation that is being created.) At this point, the engineer begins to investigate design alternatives. Alternatives are different approaches, or options, that the design engineer considers to be viable at the conceptual stage of the design. For example, some of these concepts may be used to design a better mousetrap:

- use a mechanical or an electronic sensor
- insert cheese or peanut butter as bait
- construct a wood, plastic, or metal cage
- install an audible or a visible alarm
- kill or catch and release the mouse

Analysis is a *decision-making tool* for evaluating a set of design alternatives. By performing analysis, the design engineer zeroes in on the alternatives that yield the optimum solution, while eliminating alternatives that either violate design constraints or yield inferior solutions. In the mousetrap design, a dynamics analysis may show that a mechanical sensor is too slow, resulting in delaying the closing of a trap door and therefore freeing the mouse. Thus, an electronic sensor is chosen because it yields a superior solution.

The application that follows illustrates how analysis is used to design a machine component.

APPLICATION: DESIGNING A MACHINE COMPONENT

One of the major roles for mechanical engineers is the design of machines. Machines can be very complex systems consisting of numerous moving components. In order for a machine to work properly, each component must be designed so that it performs a specific function in unison with the other components. The components must be designed to withstand specified forces, vibrations, temperatures, corrosion, and other mechanical and environmental factors. An important aspect of machine design is determining the *dimensions* of the mechanical components.

Consider a machine component consisting of a 20-cm-long circular rod, as shown in Figure 1.1. As the machine operates, the rod is subjected to a 100-kN tensile force. One of the design constraints is that the axial deformation (change in length) of the rod cannot exceed 0.5 mm if the rod is to interface properly with a mating component. Taking the rod length and the applied tensile force as given, what is the minimum diameter required for the rod?

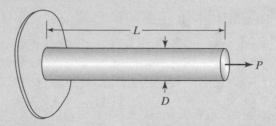

Figure 1.1. A machine component.

To solve this problem, we use a familiar equation from mechanics of materials,

$$\delta = \frac{PL}{AE}$$

where

- δ = axial deformation (m)
- P = axial tensile force (N)
- L = original length of rod (m)
- $A = \pi D^2/4$ = cross-sectional area of rod (m²)
- E = modulus of elasticity (N/m²)

The use of this equation assumes that the material behaves elastically (i.e., it does not undergo permanent deformation when subjected to a force). Upon substituting the formula for the rod's cross-sectional area into the equation and solving for the rod diameter, D, we obtain

$$D = \sqrt{\frac{4\,PL}{\pi\delta E}}$$

We know the tensile force, P, the original rod length, L, and the maximum axial deformation, δ. But to find the diameter, D, we must also know the modulus of elasticity, E. The modulus of elasticity is a material property, a constant defined by the ratio of stress to strain. Suppose we choose 7075-T6 aluminum for the rod. This material has a modulus of elasticity of $E = 72$ GPa. Substituting values into the equation gives the following diameter:

$$D = \sqrt{\frac{4(100 \times 10^3\,\text{N})(0.20\,\text{m})}{\pi(0.0005\,\text{m})(72 \times 10^9\,\text{N/m}^2)}}$$

$$= 0.0266\,\text{m} = 26.6\,\text{mm}$$

As part of the design process, we wish to consider other materials for the rod. Let's find the diameter for a rod made of structural steel ($E = 200$ GPa). For structural steel, the rod diameter is

$$D = \sqrt{\frac{4(100 \times 10^3\,\text{N})(0.20\,\text{m})}{\pi(0.005\,\text{m})(200 \times 10^9\,\text{N/m}^2)}}$$

$$= 0.0160\,\text{m} = 16.0\,\text{mm}.$$

Our analysis shows that the minimum diameter for the rod depends on the material we choose. Either 7075-T6 aluminum or structural steel will work as far as the axial deformation is concerned, but other design issues such as weight, strength, wear, corrosion, and cost should be considered. The important point to be learned here is that analysis is a fundamental step in machine design.

As the application example illustrates, analysis is used to ascertain what design features are required to make the component or system *functional*. Analysis is used to size the cable of a suspension bridge, to select a cooling fan for a computer, to size the heating elements for curing a plastic part in a manufacturing plant, and to design the solar panels that convert solar energy to electrical energy for a spacecraft. Analysis is a crucial part of virtually every design task because it guides the design engineer through a sequence of decisions that ultimately lead to the optimum design. It is important to point out that in design work, it is not enough to simply produce a *drawing* of the component or system. A drawing by itself, while revealing the visual and dimensional characteristics of the design, may say little, or nothing, about the functionality of the design. Analysis must be included in the design process if the engineer is to know whether the design will actually work when it is placed into service. Also, once a working prototype of the design is constructed, testing is performed to validate analysis and to aid in the refinement of the design.

1.3   ANALYSIS AND ENGINEERING FAILURE

With the possible exception of farmers, engineers are probably the most taken for granted people in the world. Virtually all the man-made products and devices that people use in their personal and professional lives were designed by engineers. Think for a moment. What is the first thing you did when you arose from bed this morning? Did you hit the snooze button on your alarm clock? Your alarm clock was designed by engineers. What did you do next, go into the bathroom, perhaps? The bathroom fixtures—the sink, bathtub, shower, and toilet—were designed by engineers. Did you use an electrical appliance to fix breakfast? Your toaster, waffle maker, microwave oven, refrigerator and other kitchen appliances were designed by engineers. Even if you ate cold cereal for breakfast, you still took advantage of engineering because engineers designed the processes by which the cereal and milk were produced, and they even designed the cereal box and milk container! What did you do after breakfast? If you brushed your teeth, you can thank engineers for designing the toothpaste tube and toothbrush and even formulating the toothpaste. Before leaving for school, you got dressed. Because our society does not look favorably upon public nudity, you can thank engineers for designing the textiles from which your clothing is made and the machines that manufactured them. Did you drive a car to school or ride a bicycle? In either case, engineers designed both transportation systems. What did you do when you arrived at school? You sat down in your favorite chair in a classroom, removed a pen or pencil and a note pad from your backpack, and began another day of learning. The chair you sat in, the writing instrument you used to take notes, the note pad you wrote on, and the bulging backpack you use to carry books, binders, paper, pens, and pencils, plus numerous other devices were designed by engineers.

We take engineers for granted, but we expect a lot from them. We expect everything they design, including alarm clocks, plumbing, toasters, automobiles, chairs, and pencils, to work and to work all the time. Unfortunately, they don't. We experience a relatively minor inconvenience when the heating coil in our toaster burns out, but when a bridge collapses, a commercial airliner crashes, or a space shuttle explodes, and people are injured or die, the story makes headline news, and engineers are suddenly thrust into the spotlight of public scrutiny. Are engineers to blame for every failure that occurs? Some failures occur because people misuse the products. For example, if you persist in using a screwdriver to pry lids off cans, to dig weeds from the garden, and to chisel masonry, it may soon stop functioning as a screwdriver. Engineers try to design products that are "people proof." The types of failures that engineers take primary responsibility for are those caused by various types of errors during the design phase. After all, engineering is a human enterprise, and humans make mistakes.

Whether we like it or not, **failure** is part of engineering. It is part of the design process. When engineers design a new product, it seldom works exactly as expected the first time. Mechanical components may not fit properly, electrical components may be connected incorrectly, software glitches may occur, or materials may be incompatible. The list of potential causes of failure is long, and the cause of a specific failure in a design is probably unexpected because otherwise the design engineer would have accounted for it. Failure will always be part of engineering, because engineers cannot anticipate every mechanism by which failures *can* occur. Engineers make every concerted effort to design systems that do not fail. If failures do arise, hopefully they are revealed during the design phase and can be ironed out before the product goes into service. One of the hallmarks of a good design engineer is one who turns failure into success.

The role of analysis in engineering failure is twofold. First, as discussed earlier, analysis is a crucial part of engineering design. It is one of the main decision-making tools the design engineer uses to explore alternatives. Analysis helps establish the functionality of the design. Analysis may therefore be regarded as a *failure prevention* tool. People expect kitchen appliances, automobiles, airplanes, televisions, and other systems to work as they are supposed to work, so engineers make every reasonable attempt to design products that are reliable. As part of the design phase, engineers use analysis to ascertain what the physical characteristics of the system must be in order to prevent system failure within a specified period of time. Do engineers ever design products to fail on purpose? Surprisingly, the answer is yes. Some devices rely on failure for their proper operation. For example, a fuse "fails" when the electrical current flowing through it exceeds a specified amperage. When this amperage is exceeded, a metallic element in the fuse melts, breaking the circuit, thereby protecting personnel or a piece of electrical equipment. Shear pins in transmission systems protect shafts, gears, and other components when the shear force exceeds a certain value. Some utility poles and highway signs are designed to safely break away when struck by an automobile.

The second role of failure analysis in engineering pertains to situations where design flaws escaped detection during the design phase, only to reveal themselves after the product was placed into service. In this role, analysis is utilized to address the questions "Why did the failure occur?" and "How can it be avoided in the future?" This type of detective work in engineering is sometimes referred to as *forensic engineering*. In failure investigations, analysis is used as a diagnostic tool of reevaluation and reconstruction. Following the explosion of the Space Shuttle Challenger in 1986, engineers at Thiokol used analysis (and testing) to reevaluate the joint design of the solid rocket boosters. Their analyses and tests showed that, under the unusually cold conditions on the day of launch, the rubber O-rings responsible for maintaining a seal between the segments of one of the solid rocket boosters lost resiliency and therefore the ability to contain the high-pressure gases inside the booster. Hot gases leaking past the O-rings developed into an impinging jet directed against the external (liquid hydrogen) tank and a lower strut attaching the booster to the external tank. Within seconds, the entire aft dome of the tank fell away, releasing massive amounts of liquid hydrogen. Challenger was immediately enveloped in the explosive burn, destroying the vehicle and killing all seven astronauts. In the aftermath of the Challenger disaster, engineers used analysis extensively to redesign the solid rocket booster joint.

APPLICATION: FAILURE OF THE TACOMA NARROWS BRIDGE

The collapse of the Tacoma Narrows Bridge was one of the most sensational failures in the history of engineering. This suspension bridge was the first of its kind spanning the Puget Sound, connecting Washington State with the Olympic Peninsula. Compared with existing suspension bridges, the Tacoma Narrows Bridge had an unconventional design. It had a narrow two-lane deck, and the stiffened-girder road structure was not very deep. This unusual design gave the bridge a slender, graceful appearance. Although the bridge was visually appealing, it had a problem: It oscillated in the wind. During the four months following its opening to traffic on July 1, 1940, the bridge earned the nickname "Galloping Gertie" from motorists who felt as though they were riding a giant roller coaster as they crossed the 2800-ft center span. (See Figure 1.2.) The design engineers failed to recognize that their bridge might behave more like the wing of an airplane subjected to severe turbulence than an earth-bound structure subjected to a steady load. The engineers' failure to consider the aerodynamic aspects of the design led to the destruction of the bridge on November 7, 1940, during a 42-mile-per-hour wind storm. (See Figure 1.3.) Fortunately, no people were injured or killed. A newspaper editor, who lost control of his car between the towers due to the violent undulations, managed to stumble and crawl his way to safety, only to look back to see the road rip away from the suspension cables and fall, along with his car and presumably his dog, which he could not save, into the Narrows below.

Figure 1.2. The Tacoma Narrows Bridge twisting in the wind.

Even as the bridge was being torn apart by the wind storm, engineers were testing a scale model of the bridge at the University of Washington in an attempt to understand the problem. Within a few days following the bridge's demise, Theodore von Karman, a world renown fluid dynamicist, who worked at the California Institute of Technology, submitted a letter to *Engineering News-Record* outlining an aerodynamic analysis of the bridge. In the analysis, he used a differential equation for an idealized bridge deck twisting like an airplane wing as the lift forces of the wind tend to twist the deck one way, while the steel in the bridge tends to twist it in another way. His analysis showed that the Tacoma Narrows Bridge should indeed have exhibited an aerodynamic instability more pronounced than any existing suspension bridge. Remarkably, von Karman's "back of the envelope" calculations predicted dangerous levels of vibration for a wind speed not 10 miles per hour over the wind speed measured on the morning of November 7, 1940. The dramatic failure of Galloping Gertie forever established the importance of aerodynamic analysis in the design of suspension bridges.

The bridge was eventually redesigned with a deeper and stiffer open-truss structure that allowed the wind to pass through. The new and safer Tacoma Narrows Bridge was reopened on October 14, 1950.

Figure 1.3. The center span of the Tacoma Narrows Bridge plunges into Puget Sound.

PROFESSIONAL SUCCESS: LEARN FROM FAILURE

The Tacoma Narrows Bridge and countless other engineering failures teach engineers a valuable lesson:

Learn from your own failures and the failures of other engineers.

Unfortunately, the designers of the Tacoma Narrows Bridge did not learn from the failures of others. Had they studied the history of suspension bridges dating back to the early 19th century, they would have discovered that 10 suspension bridges suffered severe damage or destruction by winds.

NASA and Thiokol learned that the pressure-seal design in the solid rocket-booster joint of the Space Shuttle Challenger was overly sensitive to a variety of factors such as temperature, physical dimensions, reusability, and joint loading. Not only did they learn some hard-core technical lessons, they also learned some lessons in engineering judgment. They learned that the decision-making process culminating in the launch of Challenger was

flawed. To correct both types of errors, during the two-year period following the Challenger catastrophe, the joint was redesigned, additional safety-related measures were implemented and the decision-making process leading to shuttle launches was improved.

In another catastrophic failure, NASA determined that fragments of insulation that broke away from the external fuel tank during the launch of the Space Shuttle Columbia impacted the left wing of the vehicle, severely damaging the wing's leading edge. The damage caused a breach in the wing's surface, which, upon reentry of Columbia, precipitated a gradual burn-through of the wing, resulting in a loss of vehicle control. Columbia broke apart over the southwestern part of the United States, killing all seven astronauts aboard.

If we are to learn from engineering failures, the *history* of engineering becomes as relevant to our education as design, analysis, science, mathematics, and the liberal arts do. Lessons learned not only from our own experiences, but also from those who have gone before us, contribute enormously to the constant improvement of our technology and the advancement of engineering as a profession. Errors in judgement made by Roman and Egyptian engineers have been repeated in modern times, notwithstanding a greatly improved chest of scientific and mathematical tools. Engineers have and will continue to make mistakes. Learn from them.

KEY TERMS

analysis

basic mathematics

engineering analysis

engineering design

failure

higher level mathematics

modeling

physical sciences

REFERENCES

Adams, J.L., *Flying Buttresses, Entropy and O-rings: The World of an Engineer*: Cambridge, MA: Harvard University Press, 1991.

Ferguson, E.S., "How Engineers Lose Touch," *Invention and Technology*, Vol. 8 (3), Winter 1993, pp. 16–24.

Fogler, H.S. and S.E. LeBlanc, *Strategies for Creative Problem Solving*, Upper Saddle River, NJ: Prentice Hall, 1995.

French, M., *Invention and Evolution: Design in Nature and Engineering*, 2d ed., Cambridge, UK: Cambridge University Press, 1994.

Horenstein, M.N., *Design Concepts for Engineers*, 2d ed., Upper Saddle River, NJ: Prentice Hall, 2002.

Howell, S.K., *Engineering Design and Problem Solving*, 2d ed., Upper Saddle River, NJ: Prentice Hall, 2002.

Hyman, B., *Fundamentals of Engineering Design*, 2d ed., Upper Saddle River, NJ: Prentice Hall, 2002.

Laithwaite, E., *An Inventor in the Garden of Eden*, Cambridge, UK: Cambridge University Press, 1994.

Petroski, H., *Design Paradigms: Case Histories of Error and Judgement in Engineering*, Cambridge, UK: Cambridge University Press, 1994.

Petroski, H., *To Engineer is Human: The Role of Failure in Successful Design*, NY: Vintage Books, 1992.

Wright, P.H., *Introduction to Engineering*, 3d ed., NY: John Wiley and Sons, 2003.

Problems

1. The following basic devices are commonly found in a typical home or office. Discuss how analysis might be used to design these items.

   a. staple remover

   b. scissors

   c. fork

   d. mechanical pencil

   e. door hinge

   f. paper clip

   g. toilet

   h. incandescent light bulb

   i. cereal box

   j. coat hanger

   k. three-ring binder

   l. light switch

   m. doorknob

n. stapler
o. can opener
p. water faucet
q. kitchen sink
r. electrical outlet
s. window
t. door
u. dinner plate
v. chair
w. table
x. plastic CD case
y. drawer slide
z. bookend

2. A 1-m-long beam of rectangular cross section carries a uniform load of $w = 15$ kN/m. The design specification calls for a 5-mm maximum deflection of the end of the beam. The beam is to be constructed of fir ($E = 13$ GPa). By analysis, determine at least five combinations of beam height, h, and beam width, b, that meet the specification. Use the equation

$$y_{max} = \frac{wL^4}{8EI}$$

where

y_{max} = deflection of end of beam (m)
w = uniform loading (N/m)
L = beam length (m)
E = modulus of elasticity of beam (Pa)
$I = bh^3/12$ = area moment of inertia of beam (m^4)

Note: 1 Pa = 1 N/m^2, 1 kN = 10^3 N and 1 GPa = 10^9 Pa.

What design conclusions can you draw about the influence of beam height and width on the maximum deflection? Is the deflection more sensitive to h or b? If the beam were constructed of a different material, how would the deflection change? See Figure P1.2 for an illustration of the beam.

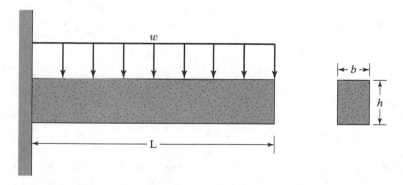

Figure P1.2.

3. Identify a device from your own experience that has failed. Discuss how it failed and how analysis might be used to redesign it.

4. Research the following notable engineering failures. Discuss how analysis was used or could have been used to investigate the failure.

   a. Hyatt Hotel, Kansas City, 1981
   b. Titanic, North Atlantic, 1912
   c. Chernobyl Nuclear Power Plant, Soviet Union, 1986
   d. Three Mile Island Nuclear Power Plant, Pennsylvania, 1979
   e. Hartford Civic Center, Connecticut, 1978
   f. Union Carbide Plant, India, 1984
   g. Hindenburg airship, New Jersey, 1937
   h. Hubble Space Telescope, 1990
   i. Highway I-880, Loma Prieta, California earthquake, 1989
   j. American Airlines DC-10, Chicago, 1979
   k. Skylab, 1979
   l. Apollo I capsule fire, Cape Canaveral, Florida, 1967
   m. Dee Bridge, England, 1847
   n. Green Bank radio telescope, West Virginia, 1989
   o. Boiler explosions, North America, 1870–1910
   p. ValuJet Airlines DC-9, Miami, 1996
   q. Ford Pinto gas tanks, 1970s
   r. Teton dam, Idaho, 1976
   s. Apollo 13, 1970
   t. Mars Climate Orbiter, 1999
   u. Space Shuttle Challenger, 1986
   v. Space Shuttle Columbia, 2003

2

Dimensions and Units

2.1   INTRODUCTION

Suppose for a moment that someone asks you to hurry to the grocery store to buy a few items for tonight's dinner. You get in your car, turn the ignition on, and drive down the road. Immediately you notice something strange. There are no numbers or divisions on your speedometer! As you accelerate and decelerate, the speedometer indicator changes position, but you do not know your speed because there are no markings to read. Bewildered, you notice that the speed limit and other road signs between your house and the store also lack numerical information. Realizing that you were instructed to arrive home with the groceries by 6 P.M., you glance at your digital watch only to discover that the display is blank. Now you are really spooked, but you drive on. Upon arriving at the store, you check your list: 1 pound of lean ground beef, 4 ounces of fresh mushrooms, and a 12-ounce can of tomato paste. You go to the meat counter first. As you scan the meat counter, you can't believe your eyes. The label on each package does not indicate the weight of the product. "This can't be," you mutter under your breath. You hastily grab what appears to be a 1-pound package and scurry to the produce section. Scooping up a bunch of mushrooms, you place them on the scale to weigh them. "Oh no! The scale looks like my speedometer:—It has no markings either!" Once again, you estimate. One item is left: the tomato paste. The canned goods aisle is very large and contains many types of canned items: soup, juice, and fruit. Finally, you locate the tomato paste. "Not again!" you exclaim. The label on the can has no numerical information—no weight, no volume, nothing to let you know the amount of tomato paste in the can. Being mystified and shaken by this whole experience, you still make your purchase, drive home, and deliver the items. Later, you somehow manage to consume a large plate of spaghetti.

OBJECTIVES

After reading this chapter, you will have learned

- How to check equations for dimensional consistency
- The physical standards on which units are based
- Rules for proper usage of SI units
- Rules for proper usage of English units
- The difference between mass and weight
- How to do unit conversions between the SI and English unit systems

The preceding Twilight Zone-like story is, of course, fictitious, but it dramatically illustrates how strange our world would be without measures of physical quantities. Speed is a physical quantity that is measured by the speedometers in our automobiles and the radar gun of a traffic officer. Time is a physical quantity that is measured by the watch on our wrist and the clock on the wall. Weight is a physical quantity that is measured by the scale in the grocery store or at the health spa. The need for measurement was recognized by the ancients, who based standards of length on the breadth of the hand or palm, the length of the foot, or the distance from the elbow to the tip of the middle finger (referred to as a cubit). Such measurement standards were both changeable and perishable because they were based on human dimensions. In modern times, definite and unchanging standards of measurement have been adopted to help us quantify the physical world. These measurement standards are used by engineers and scientists to analyze physical phenomena by applying the laws of nature such as conservation of energy, the second law of thermodynamics, and the law of universal gravitation. As engineers design new products and processes by utilizing these laws, they use dimensions and units to describe the physical quantities involved. For instance, the design of a bridge primarily involves the dimensions of length and force. The units used to express the magnitudes of these quantities are usually either the meter and newton or the foot and pound. The thermal design of a boiler primarily involves the dimensions of pressure, temperature, and heat transfer, which are expressed in units of pascal, degrees Celsius, and watt, respectively. Dimensions and units are as important to engineers as the physical laws they describe. It is vitally important that engineering students learn how to work with dimensions and units. Without dimensions and units, analyses of engineering systems have little meaning.

2.2   DIMENSIONS

To most people, the term dimension denotes a measurement of length. Certainly, length is one type of dimension, but the term dimension has a broader meaning. A **dimension** is a *physical variable that is used to describe or specify the nature of a measurable quantity*. For example, the mass of a gear in a machine is a dimension of the gear. Obviously, the diameter is also a dimension of the gear. The compressive force in a concrete column holding up a bridge is a structural dimension of the column. The pressure and temperature of a liquid in a hydraulic cylinder are thermodynamic dimensions of the liquid. The velocity of a space probe orbiting a distant planet is also a dimension. Many other examples could be given. Any variable that engineers use to specify a physical quantity is, in the general sense, a dimension of the physical quantity. Hence, there are as many dimensions as there are physical quantities. Engineers always use dimensions in their analytical and experimental work. In order to specify a dimension fully, two characteristics must be given. First, the *numerical value* of the dimension is required. Second, the appropriate *unit* must be assigned. A dimension missing either of these two elements is incomplete and therefore cannot be fully used by the engineer. If the diameter of a gear is given as 3.85, we would ask the question, "3.85 what? Inches? Meters?" Similarly, if the compressive force in a concrete column is given as 150,000, we would ask, "150,000 what? Newtons? Pounds?"

Dimensions are categorized as either *base* or *derived*. A **base dimension**, sometimes referred to as a *fundamental* dimension, is a dimension that cannot be broken down or subdivided into other dimensions or a dimension that has been internationally accepted as the most basic dimension of a physical quantity. There are seven base dimensions that have been formally defined for use in science and engineering:

1.   length [L]
2.   mass [M]

3. time [t]
4. temperature [T]
5. electric current [I]
6. amount of substance [N]
7. luminous intensity [i]

A **derived dimension** is obtained by any combination of the base dimensions. For example, volume is length cubed, density is mass divided by length cubed, and velocity is length divided by time. Obviously, there are numerous derived dimensions. Table 2.1 lists some of the most commonly used derived dimensions in engineering, expressed in terms of base dimensions.

TABLE 2.1   Derived Dimensions Expressed in Terms of Base Dimensions.

Quantity	Variable Name	Base Dimensions
Area	A	$[L]^2$
Volume	V	$[L]^3$
Velocity	v	$[L][t]^{-1}$
Acceleration	a	$[L][t]^{-2}$
Density	ρ	$[M][L]^{-3}$
Force	F	$[M][L][t]^{-2}$
Pressure	P	$[M][L]^{-1}[t]^{-2}$
Stress	σ	$[M][L]^{-1}[t]^{-2}$
Energy	E	$[M][L]^2[t]^{-2}$
Work	W	$[M][L]^2[t]^{-2}$
Power	P	$[M][L]^2[t]^{-3}$
Mass flow rate	$\dot{m}$	$[M][t]^{-1}$
Specific heat	c	$[L]^2[t]^{-2}[T]^{-1}$
Dynamic viscosity	μ	$[M][L]^{-1}[t]^{-1}$
Molar mass	M	$[M][N]^{-1}$
Voltage	V	$[M][L]^2[t]^{-3}[I]^{-1}$
Resistance	R	$[M][L]^2[t]^{-3}[I]^{-2}$

The single letters in brackets in Table 2.1 are symbols that designate each base dimension. These symbols are useful for checking the dimensional consistency of equations. Every mathematical relation used in science and engineering must be **dimensionally consistent**, or *dimensionally homogeneous*. This means that the dimension on the left side of the equal sign must be the same as the dimension on the right side of the equal sign. The equality in any equation denotes not only a numerical equivalency but also a dimensional equivalency. To use a simple analogy, you cannot say that five apples equals four apples, nor can you say that five apples equals five oranges. You can only say that five apples equals five apples.

The following examples illustrate the concept of dimensional consistency:

EXAMPLE 2.1

Dynamics is a branch of engineering mechanics that deals with the motion of particles and rigid bodies. The straight-line motion of a particle, under the influence of gravity, may be analyzed by using the equation

$$y = y_0 + v_0 t - \frac{1}{2}gt^2$$

where

$$y = \text{height of particle at time } t$$
$$y_0 = \text{initial height of particle (at } t = 0)$$
$$v_0 = \text{initial velocity of particle (at } t = 0)$$
$$t = \text{time}$$
$$g = \text{gravitational acceleration}$$

Verify that this equation is dimensionally consistent.

SOLUTION

We check the dimensional consistency of the equation by determining the dimensions on both sides of the equal sign. The heights, y_0 and y, are one-dimensional coordinates of the particle; so these quantities have a dimension of length, [L]. The initial velocity, v_0, is a derived dimension consisting of a length, [L], divided by a time, [t]. Gravitational acceleration, g, is also a derived dimension consisting of a length, [L], divided by time squared, $[t]^2$. Of course, time, [t], is a base dimension. Writing the equation in its dimensional form, we have

$$[L] = [L] + [L][t]^{-1}[t] - [L][t]^{-2}[t]^2$$

Note that the factor, $\frac{1}{2}$, in front of the gt^2 term is a pure number, and therefore has no dimension. In the second term on the right side of the equal sign, the dimension [t] cancels, leaving length, [L]. Similarly, in the third term on the right side of the equal sign, the dimension [t] cancels, leaving length, [L]. This equation is dimensionally consistent because all terms have the dimension of length, [L].

EXAMPLE 2.2

Aerodynamics is the study of the performance of bodies moving through air. An aerodynamics analysis could be used to determine the lift force on an airplane wing or the drag force on an automobile. A commonly used equation in aerodynamics relates the total drag force acting on a body to the velocity of the air approaching it. This equation is

$$F_D = \frac{1}{2}C_D A \rho U^2$$

where

$$F_D = \text{drag force}$$
$$C_D = \text{drag coefficient}$$
$$A = \text{frontal area of body}$$
$$\rho = \text{air density}$$
$$U = \text{upstream air velocity}$$

Determine the dimensions of the drag coefficient, C_D.

SOLUTION

The dimension of the drag coefficient, C_D, may be found by writing the equation in dimensional form and simplifying the equation by combining like dimensions. Using the information in Table 2.1, we write the dimensional equation as

$$[M][L][t]^{-2} = C_D[M][L]^{-3}[L]^2[t]^{-2}[L]^2$$
$$= C_D[M][L][t]^{-2}$$

Compare the combination of base dimensions on the left and right sides of the equal sign. They are identical. This can only mean that the drag coefficient, C_D, has no dimension. If it did, the equation would not be dimensionally consistent. Thus, we say that C_D

is *dimensionless*. In other words, the drag coefficient, C_D, has a numerical value, but no dimensional value. This is not as strange as it may sound. In engineering, there are many instances, particularly, in the disciplines of fluid mechanics and heat transfer, where a physical quantity is dimensionless. Dimensionless quantities enable engineers to form special ratios that reveal certain physical insights into properties and processes. In this instance, the drag coefficient is physically interpreted as a "shear stress" at the surface of the body, which means that there is an aerodynamic force acting on the body parallel to its surface that tends to retard the body's motion through the air. If you take a course in fluid mechanics, you will learn more about this important concept.

EXAMPLE 2.3

For the following dimensional equation, find the dimensions of the quantity k:

$$[M][L][t]^{-2} = k[L][t]$$

SOLUTION

To find the dimensions of k, we multiply both sides of the equation by $[L]^{-1}[t]^{-1}$ to eliminate the dimensions on the right side of the equation, leaving k by itself. Thus, we obtain

$$[M][L][t]^{-2}[L]^{-1}[t]^{-1} = k$$

which, after applying a law of exponents, reduces to

$$[M][t]^{-3} = k$$

A closer examination of the given dimensional equation reveals that it is Newton's second law of motion:

$$F = ma$$

Here F is force, m is mass, and a is acceleration. Referring to Table 2.1, force has dimensions of $[M][L][t]^{-2}$, which is a mass $[M]$ multiplied by acceleration, $[L][t]^{-2}$.

PRACTICE!

1. For the following dimensional equation, find the base dimensions of the parameter k:
$$[M][L]^2 = k[L][t][M]^2$$
   Answer: $[L][M]^{-1}[t]^{-1}$

2. For the following dimensional equation, find the base dimensions of the parameter g:
$$[T]^{-1}[t][L] = g[L]^{-2}$$
   Answer: $[L]^3[t][T]^{-1}$

3. For the following dimensional equation, find the base dimensions of the parameter h:
$$[I][t]^{-1}h = [N]$$
   Answer: $[I][N]^{-1}[t]^{-1}$

4. For the following dimensional equation, find the base dimensions of the parameter f:
$$[M][M]^{-3} = \cos(f[L])$$
   Answer: $[L]^{-1}$

5. For the following dimensional equation, find the base dimensions of the parameter p:
$$[T] = [T]\log([T]^{-2}[t]p)$$
   Answer: $[T]^2[t]^{-1}$

2.3   UNITS

A **unit** is an *arbitrarily chosen size subdivision by which the magnitude of a dimension is expressed*. For example, the dimension length, [L], may be expressed in units of meter (m), feet (ft), mile (mi), millimeter (mm), and many others. The dimension temperature, [T], is expressed in units of degrees Celsius (°C), degrees Fahrenheit (°F), degrees rankine (°R), or kelvin (K). (By convention, the degree symbol (°) is not used for the Kelvin temperature scale.) In the United States, there are two unit systems commonly in use. The first unit system, and the one that is internationally accepted as the standard, is the **SI** (System International d'Unites) **unit system**, commonly referred to as the *metric* system. The second unit system is the **English** (or **British**) **unit system**, sometimes referred to as the *United States Customary System (USCS)*. With the exception of the United States, most of the industrialized nations of the world use the SI system exclusively. The SI system is preferred over the English system, because it is an internationally accepted standard and is based on simple powers of 10. To a limited extent, a transition to the SI system has been federally mandated in the United States. Unfortunately, this transition to total SI usage is a slow one, but many American companies are using the SI system to remain internationally competitive. Until the United States makes a complete adaptation to the SI system, U.S. engineering students need to be conversant in both unit systems and know how to make unit conversions.

The seven base dimensions are expressed in terms of SI units that are based on **physical standards**. These standards are defined such that, the corresponding SI units, except the mass unit, can be reproduced in a laboratory anywhere in the world. The reproducibility of these standards is important, because everyone with a suitably equipped laboratory has access to the same standards. Hence, all physical quantities, regardless of where in the world they are measured, are based on identical standards. This universality of physical standards eliminates the ancient problem of basing dimensions on the changing physical attributes of kings, rulers, and magistrates who reigned for a finite time. Modern standards are based on constants of nature and physical attributes of matter and energy.

The seven base dimensions and their associated SI units are summarized in Table 2.2. Note the symbol for each unit. These symbols are the accepted conventions for science and engineering. The discussion that follows outlines the physical standards by which the base units are defined.

TABLE 2.2 Base Dimensions and Their SI Units.

Quantity	Unit	Symbol
Length	meter	m
Mass	kilogram	kg
Time	second	s
Temperature	kelvin	K
Electric current	ampere	A
Amount of substance	mole	mol
Luminous intensity	candela	cd

Length

The unit of length in the SI system is the *meter* (m). As illustrated in Figure 2.1, the meter is defined as the distance traveled by light in a vacuum, during a time interval of 1/299,792,458 s. The definition is based on a physical standard: the speed of light in a vacuum. The speed of light in a vacuum is 299,792,458 m/s. Thus, light travels one

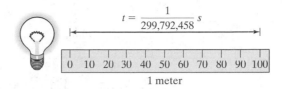

$$t = \frac{1}{299{,}792{,}458} s$$

Figure 2.1. The physical standard for the meter is based on the speed of light in a vacuum.

meter during a time interval of the reciprocal of this number. Of course, the unit of time, the *second* (s), is itself a base unit.

Mass

The unit of mass in the SI system is the *kilogram* (kg). Unlike the other units, the kilogram is not based on a reproducible physical standard. The standard for the kilogram is a cylinder of platinum-iridium alloy, which is maintained by the International Bureau of Weights and Measures in Paris, France. A duplicate of this cylinder is kept in the United States by the National Institute of Standards and Technology (NIST). (See Figure 2.2.)

Mass is the only base dimension that is defined by an artifact. An artifact is a man-made object, not as easily reproduced as the other laboratory-based standards.

Figure 2.2. A duplicate of the kilogram standard is a platinum-iridium cylinder maintained by NIST. (© Copyright Robert Rathe. Courtesy of the National Institute of Standards and Technology, Gaithersburg, MD)

Time

The unit of time in the SI system is the *second* (s). The second is defined as the duration of 9,192,631,770 cycles of radiation corresponding to the transition between the two hyperfine levels of the ground state of the cesium133 atom. An atomic clock incorporating this standard is maintained by NIST. (See Figure 2.3.)

Figure 2.3. The seventh generation atomic clock maintained by NIST keeps time with an accuracy of five parts in 10^{15}, equivalent to about one second in six million years. (Courtesy of the National Institute of Standards and Technology, Boulder, CO.)

Temperature

The unit of temperature in the SI system is the *kelvin* (K). The kelvin is defined as the fraction 1/273.16 of the temperature of the triple point of water. The triple point of water is the combination of pressure and temperature at which water exists as a solid, liquid, and gas at the same time. (See Figure 2.4.) This temperature is 273.16 K, 0.01°C, or 32.002°F. Absolute zero is the temperature at which all molecular activity ceases and has a value of 0 K.

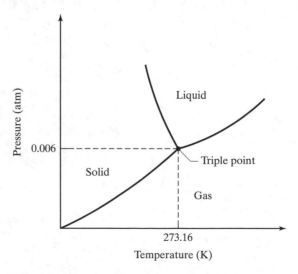

Figure 2.4. A phase diagram for water shows the triple point on which the kelvin temperature standard is based.

Electric Current

The unit of electric current in the SI system is the *ampere* (A). As shown in Figure 2.5, the ampere is defined as the steady current, which, if maintained in two straight parallel wires of infinite length and negligible circular cross section and placed one meter apart in a vacuum, produces a force of 2×10^{-7} newton per meter of wire length. Using Ohm's law, $I = V/R$, one ampere may also be denoted as the current that flows when one volt is applied across a 1-ohm resistor.

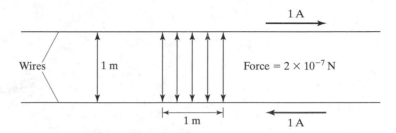

Figure 2.5. The standard for the ampere is based on the electrical force produced between two parallel wires, each carrying 1 A, located 1 m apart.

Amount of Substance

The unit used to denote the amount of substance is the *mole* (mol). One mole contains the same number of elements as there are atoms in 0.012 kg of carbon-12. This number is called Avogadro's number and has a value of approximately 6.022×10^{23}. (See Figure 2.6.)

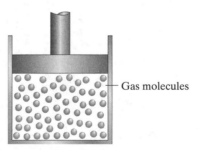

Figure 2.6. A mole of gas molecules in a piston-cylinder device contains 6.022×10^{23} molecules.

Luminous Intensity

The unit for luminous intensity is the *candela* (cd). As illustrated in Figure 2.7, one candela is the luminous intensity of a source emitting light radiation at a frequency of 540×10^{12} Hz that provides a power of 1/683 watt (W) per steradian. A steradian is a solid angle, which, having its vertex in the center of a sphere, subtends (cuts off) an area of the sphere equal to that of a square with sides of length equal to the radius of the sphere.

The unit for luminous intensity, the candela, utilizes the steradian, a dimension that may be unfamiliar to most students. The *radian* and *steradian* are called *supplementary*

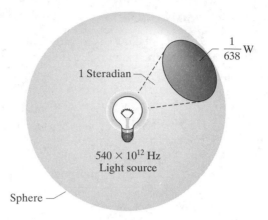

Figure 2.7. The candela standard for luminous intensity.

dimensions. These quantities, summarized in Table 2.3, refer to plane and solid angles, respectively. The radian is frequently used in engineering, and it is defined as the plane angle between two radii of a circle that subtends on the circumference an arc equal in length to the radius. From trigonometry, you may recall that there are 2π radians in a circle (i.e., 2π radians equals 360°). Thus, one radian equals approximately 57.3°. The steradian, defined earlier, is used primarily for expressing radiation quantities such as light intensity and other electromagnetic parameters.

TABLE 2.3 Supplementary Dimensions.

Quantity	Unit	Symbol
Plane angle	radian	rad
Solid angle	steradian	sr

2.4   SI UNITS

Throughout the civilized world there are thousands of engineering companies that design and manufacture products for the benefit of man. The international buying and selling of these products is an integral part of a global network of industrialized countries, and the economic health of these countries, including the United States, depends to a large extent on international trade. Industries such as the automotive and electronics industries are heavily involved in international trade, so these industries have readily embraced the SI unit system in order to be economically competitive. The general adoption of the SI unit system by U.S. companies has been slow, but global economic imperatives are driving them to fall into step with the other industrialized nations of the world. SI units are now commonplace on food and beverage containers, gasoline pumps, and automobile speedometers. The SI unit system is the internationally accepted standard. In the United States, however, the English unit system is still widely used. Hopefully, it is only a matter of time before all U.S. companies use SI units exclusively. Until that time, the burden is upon you, the engineering student, to learn both unit systems. You will gladly discover, however, that most engineering textbooks emphasize SI units, but provide a list of unit conversions between the SI and English systems.

Table 2.2 summarizes the seven base dimensions and their SI units, and Table 2.3 summarizes the supplementary dimensions and their units. Derived dimensions consist of a combination of base and supplementary dimensions. Sometimes, the units of a derived dimension are given a specific name. For example, the derived dimension *force* consists of the SI base units $kg \cdot m \cdot s^{-2}$. This combination of SI base units is called a *newton* and is abbreviated N. Note that the unit name, in honor of Isaac Newton, is not capitalized when spelled out as a unit name. The same rule applies to other units named after people such as hertz (Hz), kelvin (K), pascal (Pa), etc. Another example is the *joule*, the SI unit for energy, work, and heat. The joule unit is abbreviated J and consists of the SI base units $kg \cdot m^2 \cdot s^{-2}$. A summary of the most commonly used SI derived dimensions and the corresponding SI unit names is given in Table 2.4.

Most derived dimensions do not have specific SI unit names, but their units may contain specific SI unit names. For example, the dimension, *moment of force*, usually

TABLE 2.4 Derived Dimensions and SI Units with Specific Names.

Quantity	SI Unit	Unit Name	Base Units
Frequency	Hz	hertz	s^{-1}
Force	N	newton	$kg \cdot m \cdot s^{-2}$
Pressure	Pa	pascal	$kg \cdot m^{-1} \cdot s^{-2}$
Stress	Pa	pascal	$kg \cdot m^{-1} \cdot s^{-2}$
Energy	J	joule	$kg \cdot m^2 \cdot s^{-2}$
Work	J	joule	$kg \cdot m^2 \cdot s^{-2}$
Heat	J	joule	$kg \cdot m^2 \cdot s^{-2}$
Power	W	watt	$kg \cdot m^2 \cdot s^{-3}$
Electric charge	C	coulomb	$A \cdot s$
Electric potential (voltage)	V	volt	$kg \cdot m^2 \cdot s^{-3} \cdot A^{-1}$
Electric resistance	Ω	ohm	$kg \cdot m^2 \cdot s^{-3} \cdot A^{-2}$
Magnetic flux	Wb	weber	$kg^{-1} \cdot m \cdot s^{-2} \cdot A^{-1}$
Luminous flux	lm	lumen	$cd \cdot sr$

referred to simply as *moment*, has the SI units $N \cdot m$, a force multiplied by a distance. There is no special name for this unit—we simply call it "newton–meter." Another example is the dimension *mass flow rate*. Mass flow rate is the mass of a fluid that flows past a point in a given time. The SI units for mass flow rate are $kg \cdot s^{-1}$, which we state as "kilograms per second." Note that units that are located in the denominator, that is, those that have a negative sign on their exponent, may also be written using a divisor line. Thus, the units for mass flow rate may be written as kg/s. Caution must be exercised, however, when utilizing this type of notation for some units. For example, the SI units for thermal conductivity, a quantity used in heat transfer, are $W \cdot m^{-1} \cdot K^{-1}$. How do we write these units with a divisor line? Do we write these units as W/m/K? How about $W/m \cdot K$? The first choice can cause some confusion. Does a "watt per meter per kelvin" mean that the kelvin unit is inverted twice and therefore goes above the divisor line? One glance at the units written as $W \cdot m^{-1} \cdot K^{-1}$ tells us that the temperature unit belongs "downstairs" because K has a negative exponent. If the kelvin unit were placed above the divisor line, and the thermal conductivity were used in an equation, a dimensional inconsistency would result. The second choice is the preferred method of writing units when more than one unit is below the divisor line. Because the meter and kelvin units are located to the right of the divisor line and they are separated by a dot, both units are interpreted as being in the denominator. Sometimes, parentheses are used to group units above or below the divisor line. Units for thermal conductivity would then be written as $W/(m \cdot K)$. In any case, a dot or a dash should always be placed between adjacent units to separate them regardless of whether the units are above or below the divisor line. Some derived dimensions and their SI units are given in Table 2.5.

TABLE 2.5 Derived Dimensions and SI Units.

Quantity	SI Units
Acceleration	$m \cdot s^{-2}$
Angular acceleration	$rad \cdot s^{-2}$
Angular velocity	$rad \cdot s^{-1}$
Area	m^2
Concentration	$mol \cdot m^{-3}$
Density	$kg \cdot m^{-3}$
Electric field strength	$V \cdot m^{-1}$
Energy	$N \cdot m$
Entropy	$J \cdot K^{-1}$
Heat	J
Heat transfer	W
Magnetic field strength	$A \cdot m^{-1}$
Mass flow rate	$kg \cdot s^{-1}$
Moment of force	$N \cdot m$
Radiant intensity	$W \cdot sr^{-1}$
Specific energy	$J \cdot kg^{-1}$
Surface tension	$N \cdot m^{-1}$
Thermal conductivity	$W \cdot m^{-1} \cdot K^{-1}$
Velocity	$m \cdot s^{-1}$
Viscosity, dynamic	$Pa \cdot s$
Viscosity, kinematic	$m^2 \cdot s^{-1}$
Volume	m^3
Volume flow rate	$m^3 \cdot s^{-1}$
Wavelength	m
Weight	N

When a physical quantity has a numerical value that is very large or very small, it is cumbersome to write the number in standard decimal form. The general practice in engineering is to express numerical values between 0.1 and 1000 in standard decimal form. If a value cannot be expressed within this range, a *prefix* should be used. Because the SI unit system is based on powers of 10, it is more convenient to express such numbers by using prefixes. A prefix is a letter in front of a number that denotes multiples of powers of 10. For example, if the internal force in an I-beam is three million seven hundred and fifty thousand newtons, it would be awkward to write this number as 3,750,000 N. It is preferred to write the force as 3.75 MN, which is stated as "3.75 mega-newtons." The prefix "M" denotes a multiple of a million. Hence, 3.75 MN equals 3.75×10^6 N. Electrical current is a good example of a quantity represented by a small number. Suppose the current flowing in a wire is 0.0082 A. This quantity would be expressed as 8.2 mA, which is stated as "8.2 milliamperes." The prefix "m" denotes a multiple of one-thousandth, or 1×10^{-3}. A term we often hear in connection with personal computers is the storage capacity of hard disks. When personal computers first appeared in the early 1980s, most hard disks could hold around 10 or 20 MB (megabytes) of information. Nowadays, hard disks typically can hold around one thousand times that amount. Perhaps, a few years from now, the typical storage capacity of a personal computer's hard disk will be on the order of TB (terabytes). The standard prefixes for SI units are given in Table 2.6.

TABLE 2.6 Standard Prefixes for SI Units

Multiple	Exponential Form	Prefix	Prefix Symbol
1,000,000,000,000	10^{12}	tera	T
1,000,000,000	10^{9}	giga	G
1,000,000	10^{6}	mega	M
1000	10^{3}	kilo	k
0.01	10^{-2}	centi	c
0.001	10^{-3}	milli	m
0.000 001	10^{-6}	micro	μ
0.000 000 001	10^{-9}	nano	n
0.000 000 000 001	10^{-12}	pico	p

As indicated in Table 2.6, the most widely used SI prefixes for science and engineering quantities come in multiples of one thousand. For example, stress and pressure, which are generally large quantities for most structures and pressure vessels, are normally expressed in units of kPa, MPa, or GPa. Frequencies of electromagnetic waves such as radio, television, and telecommunications are also large numbers. Hence, they are generally expressed in units of kHz, MHz, or GHz. Electrical currents, on the other hand, are often small quantities, so they are usually expressed in units of μA or mA. Because frequencies of most electromagnetic waves are large quantities, the wavelengths of these waves are small. For example, the wavelength range of the visible light region of the electromagnetic spectrum is approximately 0.4 μm to 0.75 μm. It should be noted that the SI mass unit, kilogram (kg), is the only base unit that has a prefix.

Here are some rules on how to use SI units properly that every beginning engineering student should know:

1. A unit symbol is never written as a plural with an "s." If a unit is pluralized, the "s" may be confused with the unit second (s).

2. A period is never used after a unit symbol, unless the symbol is at the end of a sentence.

3. Do not use invented unit symbols. For example, the unit symbol for "second" is (s), not (sec), and the unit symbol for "ampere" is (A), not (amp).

4. A unit symbol is always written by using lowercase letters, with two exceptions. The first exception applies to units named after people, such as the newton (N), joule (J), and watt (W). The second exception applies to units with the prefixes M, G, and T. (See Table 2.6.)

5. A quantity consisting of several units must be separated by dots or dashes to avoid confusion with prefixes. For example, if a dot is not used to express the units of "meter-second" (m · s), the units could be interpreted as "millisecond" (ms).

6. An exponential power for a unit with a prefix refers to both the prefix and the unit; for example, $\text{ms}^2 = (\text{ms})^2 = \text{ms} \cdot \text{ms}$.

7. Do not use compound prefixes. For example, a "kilo MegaPascal" (kMPa) should be written as GPa, because the product of "kilo" (10^3) and "mega" (10^6) equals "giga" (10^9).

8. Put a space between the numerical value and the unit symbol.

9. Do not put a space between a prefix and a unit symbol.

10. Do not use prefixes in the denominator of composite units. For example, the units N/mm should be written as kN/m.

Table 2.7 provides some additional examples of these rules.

TABLE 2.7 Correct and Incorrect Ways of Using SI Units.

Correct	Incorrect	Rules
12.6 kg	12.6 kgs	1
450 N	450 Ns	1
36 kPa	36 kPa.	2
1.75 A	1.75 amps	1, 3
10.2 s	10.2 sec	3
20 kg	20 Kg	4
150 W	150 w.	2, 4
4.50 kg/m · s	4.50 kg/ms	5
750 GN	750 MkN	7
6 ms	6 kμs	7
800 Pa · s	800Pa · s	8
1.2 MΩ	1.2 M Ω	9
200 MPa	200 M Pa	9
150 μA	150 μ A	9
6 MN/m	6 N/μm	10

APPLICATION: DERIVING FORMULAS FROM UNIT CONSIDERATIONS

To the beginning engineering student, it can seem as if there is an infinite number of formulas to learn. Formulas contain physical quantities that have numerical values plus units. Because formulas are written as equalities, formulas must be numerically and dimensionally equivalent across the equal sign. Can this feature be used to help us derive formulas that we do not know or have forgotten? Suppose that we want to know the mass of gasoline in an automobile's gas tank. The tank has a volume of 70 L, and a handbook of fluid properties states that the density of gasoline is 736 kg/m³. Thus, we write

$$\rho = 736 \text{ kg/m}^3 \quad V = 70 \text{ L} = 0.070 \text{ m}^3$$

If the tank is completely filled with gasoline, what is the mass of the gasoline? Suppose that we have forgotten that density is defined as mass per volume, $\rho = m/V$. Because our answer will be a mass, the unit of our answer must be kilogram (kg). Looking at the units of the input quantities, we see that if we multiply density, ρ, by volume, V, the volume unit (m³) divides out, leaving mass (kg). Hence, the formula for mass in terms of ρ and V is

$$m = \rho V$$

so the mass of gasoline is

$$m = (736 \text{ kg/m}^3)(0.070 \text{ m}^3) = 51.5 \text{ kg}$$

PROFESSIONAL SUCCESS: USING SI UNITS IN EVERYDAY LIFE

The SI unit system is used commercially to a limited extent in the United States, so the average person does not know the highway speed limit in kilometers per hour, his or her weight in newtons, atmospheric pressure in kilopascals, or the outdoor air temperature in kelvin or degrees Celsius. It is ironic that the leading industrialized nation on earth has yet to embrace this international standard. Admittedly, American beverage containers routinely show the volume of the liquid

product in liters (L) or milliliters (mL), gasoline pumps often show liters of gasoline delivered, speedometers may indicate speed in kilometers per hour (km/h), and automobile tires indicate the proper inflation pressure in kilopascals (kPa) on the sidewall. On each of these products, and many others like them, a corresponding English unit is written along side the SI unit. The beverage container shows pints or quarts, the gasoline pump shows gallons, speedometers show miles per hour, and tires show pounds per square inch. Dual labeling of SI and English units on U.S. products are supposed to help people learn the SI system, "weaning" them from the antiquated English system in anticipation of the time when a full conversion to SI units occurs. This transition is analogous to the process of incrementally quitting smoking. Rather than quitting "cold turkey," we employ nicotine patches, gums, and other substitutes until our habit is broken. So, you may ask, "Why don't we make the total conversion now? Is it as painful as quitting smoking suddenly?" It probably is. As you might guess, the problem is largely an economic one. A complete conversion to SI units may not occur until we are willing to pay the price in actual dollars. People could learn the SI unit system fairly quickly if the conversion were done suddenly, but an enormous financial commitment would have to be made.

As long as dual product labeling of units is employed in the United States, most people will tend to ignore the SI unit and look only at the English unit, the unit with which they are most familiar. In U.S. engineering schools, SI units are emphasized. Therefore, the engineering student is not the average person on the street who does not know, or know how to calculate, his or her weight in newtons. So, what can engineering students in the United States do to accelerate the conversion process? A good place to start is with yourself. Start using SI units in your everyday life. When you make a purchase at the grocery store, look only at the SI unit on the label. Learn by inspection how many milliliters of liquid product are packaged in your favorite sized container. Abandon the use of inches, feet, yards, and miles as much as possible. How many kilometers lie between your home and school? What is 65 miles per hour in kilometers per hour? What is the mass of your automobile in kilograms? Determine your height in meters, your mass in kilograms, and your weight in newtons. How long is your arm in centimeters? What is your waist size in centimeters? What is the current outdoor air temperature in degrees Celsius? Most fast-food restaurants offer a "quarter pounder" on their menu. It turns out that $1 \text{ N} = 0.2248 \text{ lb}$, almost a quarter pound. On the next visit to your favorite fast-food place, order a "newton burger" and fries. (See Figure 2.8.)

Figure 2.8. An engineering student orders lunch (art by Kathryn Hagen).

PRACTICE!

1. A structural engineer states that an I-beam in a truss has a design stress of "five million, six hundred thousand pascals." Write this stress, using the appropriate SI unit prefix.

   Answer: 5.6 MPa

2. The power cord on an electric string trimmer carries a current of 5.2A. How many milliamperes is this? How many microamperes?

   Answer: 5.2×10^3 mA, 5.2×10^6 μA

3. Write the pressure 13.8 GPa in scientific notation.

   Answer: 13.8×10^9 Pa

4. Write the voltage 0.00255 V, using the appropriate SI unit prefix.

   Answer: 2.55 mV

5. In the following list, various quantities are written using SI units incorrectly. Write the quantities, using the correct form of SI units.
   a. 4.5 mw
   b. 8.75 M pa
   c. 200 Joules/sec
   d. 20 W/m^2 K
   e. 3 Amps

   Answer:
   a. 4.5 mW
   b. 8.75 MPa
   c. 200 J/s
   d. 20 W/m$^2 \cdot$ K
   e. 3 A

2.5   ENGLISH UNITS

The English unit system is known by various names. Sometimes it is referred to as the United States Customary System (USCS), the British System or the Foot-Pound-Second (FPS) system. The English unit system is still used extensively in the United States even though the rest of the industrialized world, including Great Britain, has adopted the SI unit system. English units have a long and colorful history. In ancient times, measures of length were based on human dimensions. The foot started out as the actual length of a man's foot. Because not all men were the same size, the foot varied in length by as much as three or four inches. Once the ancients started using feet and arms for measuring distance, it was only a matter of time before they began using hands and fingers. The unit of length that we refer to today as the inch was originally the width of a man's thumb. The inch was also once defined as the distance between the tip to the first joint of the forefinger. Twelve times that distance made one foot. Three times the length of a foot was the distance from the tip of a man's nose to the end of his outstretched arm. This distance closely approximates what we refer to today as the yard. Two yards equaled a fathom, which was defined as the distance across a man's outstretched arms. Half a yard was the 18-inch cubit, which was called a span. Half a span was referred to as a hand.

The pound, which uses the symbol *lb*, is named after the ancient Roman unit of weight called the libra. The British Empire retained this symbol into modern times. Today, there are actually two kinds of pound units, one for mass and one for weight and force. The first unit is called pound-mass (lb$_m$), and the second is called pound-force (lb$_f$). Because mass and weight are not the same quantity, the units lb$_f$ and lb$_m$ are different.

As discussed previously, the seven base dimensions are length, mass, time, temperature, electric current, amount of substance, and luminous intensity. These base dimensions, along with their corresponding English units, are given in Table 2.8. As with SI units, English units are not capitalized. The slug, which has no abbreviated symbol, is the mass unit in the English system, but the pound-mass (lb_m) is frequently used. Electric current is based on SI units of meter and newton, and luminous intensity is based on SI units of watt. Hence, these two base dimensions do not have English units per se, and these quantities are rarely used in combination with other English units.

TABLE 2.8 Base Dimensions and Their English Units.

Quantity	Unit	Symbol
Length	foot	ft
Mass	slug[1]	slug
Time	second	s
Temperature	rankine	°R
Electric current	ampere[2]	A
Amount of substance	mole	mol
Luminous intensity	candela[2]	cd

(1) The unit pound-mass (lb_m) is also used. 1 slug = 32.174 lb_m.
(2) There are no English units for electrical current and luminous intensity. The SI units are given here for completeness only.

Recall that derived dimensions consist of a combination of base and supplementary dimensions. Table 2.9 summarizes some common derived dimensions expressed in English units. Note that Table 2.9 is the English counterpart of the SI version given by Table 2.5. The most notable English unit with a special name is the British thermal unit (Btu), a unit of energy. One Btu is defined as the energy required to change the temperature of 1 lb_m of water at a temperature of 68°F by 1°F. One Btu is approximately the energy released by the complete burning of a single kitchen match. The magnitudes of the kilojoule and Btu are almost equal (1 Btu = 1.055 kJ). Unlike the kelvin (K), the temperature unit in the SI system, the rankine (°R) employs a degree symbol as do the Celsius (°C) and Fahrenheit (°F) units. The same rules for writing SI units apply for English units with one major exception: *Prefixes are generally not used with English units*. Thus, units such as kft (kilo-foot), Mslug (megaslug), and GBtu (gigaBtu) should not be used. Prefixes are reserved for SI units. Two exceptions are the units ksi, which refers to a stress of 1000 psi (pounds per square inch), and kip, which is a special name for a force of 1000 lb_f (pound-force).

There are some non-SI units that are routinely used in the United States and elsewhere. Table 2.10 summarizes some of these units and provides an equivalent value in the SI system. The inch is a common length unit, being found on virtually every student's ruler and carpenter's tape measure in the United States. There are exactly 2.54 centimeters per inch. Inches are still used as the primary length unit in many engineering companies. The yard is commonly used for measuring cloth, carpets, and loads of concrete (cubic yards), as well as ball advancement on the American football field. The ton is used in numerous industries, including shipping, construction, and transportation. Time subdivisions on clocks are measured in hours, minutes, and seconds. Radians and degrees are the most commonly used units for plane angles, whereas minutes and seconds are primarily used in navigational applications when referring to latitude and longitude on the earth's surface. The liter has made a lot of headway into the American culture, being found on beverage and food containers and many gasoline pumps. Virtually every American has seen the liter unit on a product, and many know that there are about four liters in a gallon (actually, 1 gal = 3.7854 L), but fewer people know that 1000 L = 1 m^3.

TABLE 2.9 Derived Dimensions and English Units.

Quantity	English Units
Acceleration	$ft \cdot s^{-2}$
Angular acceleration	$rad \cdot s^{-2}$
Angular velocity	$rad \cdot s^{-1}$
Area	ft^2
Concentration	$mol \cdot ft^{-3}$
Density	$slug \cdot ft^{-3}$
Electric field strength	$V \cdot ft^{-1}$
Energy	Btu
Entropy	$Btu \cdot slug^{-1} \cdot °R^{-1}$
Force	lb_f
Heat	Btu
Heat transfer	$Btu \cdot s^{-1}$
Magnetic field strength	$A \cdot ft^{-1}$
ass flow rate	$slug \cdot s^{-1}$
Moment of force	$lb_f \cdot ft$
Radiant intensity	$Btu \cdot s^{-1} \cdot sr^{-1}$
Specific energy	$Btu \cdot slug^{-1}$
Surface tension	$lb_f \cdot ft^{-1}$
Thermal conductivity	$Btu \cdot s^{-1} \cdot ft^{-1} \cdot °R$
Velocity	$ft \cdot s^{-1}$
Viscosity, dynamic	$slug \cdot ft^{-1} \cdot s^{-1}$
Viscosity, kinematic	$ft^2 \cdot s^{-1}$
Volume	ft^3
Volume flow rate	$ft^3 \cdot s^{-1}$
Wavelength	ft

TABLE 2.10 Non-SI Units Commonly Used in the United States.

Quantity	Unit Name	Symbol	SI Equivalent
Length	inch	in	0.0254 m[1]
	yard	yd	0.9144 m (36 in)
Mass	metric ton	t	1000 kg
	short ton	t	907.18 kg (2000 lb_m)
Time	minute	min	60 s
	hour	h	3600 s
	day	d	86,400 s
Plane angle	degree	°	$\pi/180$ rad
	minute	'	$\pi/10{,}800$ rad
	second	"	$\pi/648{,}000$ rad
Volume	liter	L	10^{-3} m^3
Land area	hectare	ha	10^4 m^2
Energy	electron-volt	eV	1.602177×10^{-19} J

[1]Exact conversion

2.6 MASS AND WEIGHT

The concepts of *mass* and *weight* are fundamental to the proper use of dimensions and units in engineering analysis. Mass is one of the seven base dimensions used in science and engineering. Mass is a base dimension because it cannot be broken down into more fundamental dimensions. **Mass** is defined as a *quantity of matter*. This simple definition of mass may be expanded by exploring its basic properties. All matter possesses mass. The magnitude of a given mass is a measure of its resistance to a change in velocity. This property of matter is called *inertia*. A large mass offers more resistance to a change in velocity than a small mass, so a large mass has a greater inertia than a small mass. Mass may be considered in another way. Because all matter has mass, all matter exerts a gravitational attraction on other matter. Shortly after formulating his three laws of motion, Sir Isaac Newton postulated a law governing the gravitational attraction between two masses. Newton's law of universal gravitation is stated mathematically as

$$F = G\frac{m_1 m_2}{r^2} \tag{2.1}$$

where

F = gravitational force between masses (N)
G = universal gravitational constant = 6.673×10^{-11} m^3/kg·s^2
m_1 = mass of body 1 (kg)
m_2 = mass of body 2 (kg)
r = distance between the centers of the two masses (m)

According to Equation (2.1), between any two masses there exists an attractive gravitational force whose magnitude varies inversely as the square of the distance between the masses. Because Newton's law of universal gravitation applies to *any* two masses, let's apply Equation (2.1) to a body resting on the surface of the earth. Accordingly, we let $m_1 = m_e$, the mass of the earth and $m_2 = m$, the mass of the body. The distance, r, between the body and the earth may be taken as the mean radius of the earth, r_e. The quantities m_e and r_e have the approximate values

$$m_e = 5.979 \times 10^{24}\,\text{kg} \qquad r_e = 6.378 \times 10^6\,\text{m}$$

Thus, we have

$$F = G\frac{m_e m}{r_e^2}$$

$$= \frac{(6.673 \times 10^{-11}\,\text{m}^3/\text{kg}\cdot\text{s}^2)(5.979 \times 10^{24}\,\text{kg})}{(6.378 \times 10^6\,\text{m})^2} m$$

$$= (9.808\,\text{m/s}^2)\,m$$

We can see that upon substituting values, the term Gm_e/r_e^2 yields a number of approximately 9.81 m/s^2, the standard acceleration of gravity on the earth's surface. Redefining this term as g, and letting $F = W$, we express the law of universal gravitation in a special form as

$$W = mg \tag{2.2}$$

where

W = weight of body (N)
m = mass of body (kg)
g = standard gravitational acceleration = 9.81 m/s^2

This derivation clearly shows the difference between mass and weight. We may therefore state the definition of **weight** as *a gravitational force exerted on a body by the earth*. Because mass is defined as a quantity of matter, the mass of a body is independent of its location in the universe. A body has the same mass whether it is located on the earth, the moon, Mars, or in outer space. The weight of the body, however, depends on its location. The mass of an 80 kg astronaut is the same whether or not he is on earth or in orbit about the earth. The astronaut weighs approximately 785 N on the earth, but while in orbit he is "weightless." His weight is zero while he orbits the earth, because he is continually "falling" toward earth. A similar weightless or "zero-g" condition is experienced by a skydiver as he free falls prior to opening the parachute.

The greatest source of confusion about mass and weight to the beginning engineering student is not the physical concept, but the units used to express each quantity. To see how units of mass and weight relate to each other, we employ a well-known scientific principle, **Newton's second law** of motion. Newton's second law of motion states that *a body of mass, m, acted upon by an unbalanced force, F, experiences an acceleration, a, that has the same direction of the force and a magnitude that is directly proportional to the force*. Stated mathematically, this law is

$$F = ma \tag{2.3}$$

where

$$F = \text{force (N)}$$
$$m = \text{mass (kg)}$$
$$a = \text{acceleration (m/s}^2)$$

Note that this relation resembles Equation (2.2). Weight is a particular type of force, and acceleration due to gravity is a particular type of acceleration, so Equation (2.2) is a special case of Newton's second law, given by Equation (2.3). In the SI unit system, the newton (N) is *defined* as the force that will accelerate a 1-kg mass at a rate of 1 m/s^2. Hence, we may write Newton's second law dimensionally as

$$1\,\text{N} = 1\,\text{kg} \cdot \text{m/s}^2$$

In the English unit system, the pound-force (lb$_f$) is *defined* as the force that will accelerate a 1-slug mass at a rate of 1 ft/s^2. Hence, we may write Newton's second law dimensionally as

$$1\,\text{lb}_f = 1\,\text{slug} \cdot \text{ft/s}^2$$

See Figure 2.9 for an illustration of Newton's second law. Confusion arises from the careless interchange of the English mass unit, pound-mass (lb$_m$), with the English force unit, pound-force (lb$_f$). These units are not the same thing! In accordance with our definitions of mass and weight, pound-mass refers to a quantity of matter, whereas pound-force refers to a force or weight. In order to write Newton's second law in terms of pound-mass instead of slug, we rewrite Equation (2.3) as

$$F = \frac{ma}{g_c} \tag{2.4}$$

where g_c is a constant that is required to make Newton's second law dimensionally consistent when mass, m, is expressed in lb$_m$, rather than slug. As stated previously, the English unit for force is lb$_f$, the English unit for acceleration is ft/s^2, and, as indicated in

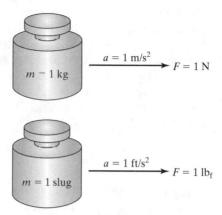

Figure 2.9. Definitions of the force units newton (N) and pound-force (lb_f).

Table 2.8, 1 slug = 32.174 lb_m. Thus, the constant g_c is

$$
\begin{aligned}
g_c &= \frac{ma}{F} \\
&= \frac{(32.174 \ lb_m)(ft/s^2)}{lb_f} \\
&= 32.174 \frac{lb_m \cdot ft}{lb_f \cdot s^2}
\end{aligned}
$$

This value is usually rounded to

$$
g_c = 32.2 \frac{lb_m \cdot ft}{lb_f \cdot s^2}
$$

Note that g_c has the same numerical value as g, the standard acceleration of gravity on the earth's surface. Newton's second law as expressed by Equation (2.4) is dimensionally consistent when the English unit of mass, lb_m, is used.

To verify that Equation (2.4) works, we recall that the pound-force is defined as the force that will accelerate a 1-slug mass at a rate of 1 ft/s². Recognizing that 1 slug = 32.2 lb_m, we have

$$
\begin{aligned}
F &= \frac{ma}{g_c} \\
&= \frac{(32.2 \ lb_m)(1 \ ft/s^2)}{32.2 \dfrac{lb_m \cdot ft}{lb_f \cdot s^2}} = 1 \ lb_f
\end{aligned}
$$

Note that in this expression, all the units, except lb_f, cancel. Hence, the pound-force (lb_f) is *defined* as the force that will accelerate a 32.2-lb_m mass at a rate of 1 ft/s². Therefore, we may write Newton's second law dimensionally as

$$
1 \ lb_f = 32.2 \ lb_m \cdot ft/s^2
$$

To have dimensional consistency when English units are involved, Equation (2.4) *must* be used when mass, m, is expressed in lb_m. When mass is expressed in slug, however, the use of g_c in Newton's second law is not required for dimensional consistency because 1 lb_f is already defined as the force that will accelerate a 1-slug mass at a rate of 1 ft/s². Further-more, because 1 N is already defined as the force that will accelerate a 1-kg mass at a rate of 1 m/s², the use of g_c is not required for dimensional consistency in the SI unit system. *Thus, Equation (2.3) suffices for all calculations, except for those in which mass is expressed in lb$_m$; in that case, Equation (2.4) must be used.* However, Equation (2.4) may be universally used when recognizing that the numerical value and units for g_c can be defined such that any consistent unit system will work. For example, substituting $F = 1$ N, $m = 1$ kg, and $a = 1$ m/s² into Equation (2.4) and solving for g_c, we obtain

$$g_c = \frac{1 \text{ kg} \cdot \text{m}}{\text{N} \cdot \text{s}^2}$$

Since the numerical value of g_c is 1, we can successfully use Equation (2.3) as long as we recognize that 1 N is the force that will accelerate a 1-kg mass at a rate of 1 m/s².

Sometimes, the units pound-mass (lb$_m$) and pound-force (lb$_f$) are casually inter-changed because a body with a mass of 1 lb$_m$ has a weight of 1 lb$_f$ (i.e., the mass and weight are *numerically equivalent*). Let's see how this works: By definition, a body with a mass of 32.2 lb$_m$ (1 slug) when accelerated at a rate of 1 ft/s² has a weight of 1 lb$_f$. Therefore, using Newton's second law in the form, $W = mg$, we can also state that a body with a mass of 1 lb$_m$, when accelerated at a rate of 32.2 ft/s² (the standard value of g), has a weight of 1 lb$_f$. Our rationale for making such a statement is that we maintained the same numerical value on the right side of Newton's second law by assigning the mass, m, a value of 1 lb$_m$ and the gravitational acceleration, g, the standard value of 32.2 ft/s². The numerical values of the mass and weight are equal even though a pound-mass and a pound-force are conceptually different quantities. It must be emphasized, however, that mass in pound-mass and weight in pound-force are numerically equivalent only when the standard value, $g = 32.2$ ft/s², is used. See Figure 2.10 for an illustration. The next example illustrates the use of g_c.

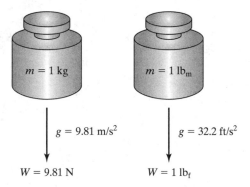

Figure 2.10.　Definitions of weight for the standard value of gravitational acceleration.

EXAMPLE 2.4

Find the weight of some objects with the following masses:

(a)　50 slug

(b)　50 lb$_m$

(c)　75 kg

SOLUTION

To find weight, we use Newton's second law, where the acceleration, a, is the standard acceleration of gravity, $g = 9.81$ m/s^2 = 32.2 ft/s^2.

(a) The mass unit, slug, is the standard unit for mass in the English unit system. The weight is

$$W = mg$$
$$= (50 \text{ slug})(32.2 \text{ ft/s}^2) = 1{,}610 \text{ lb}_f$$

(b) When mass is expressed in terms of lb$_m$, we must use Equation (2.4):

$$W = \frac{mg}{g_c} = \frac{(50 \text{ lb}_m)(32.2 \text{ ft/s}^2)}{32.2 \dfrac{\text{lb}_m \cdot \text{ft}}{\text{lb}_f \cdot \text{s}^2}} = 50 \text{ lb}_f$$

Note that the mass and weight are numerically equivalent. This is true only in cases where the standard value of g is used, which means that an object with a mass of x lb$_m$ will always have a weight of x lb$_f$ on the earth's surface.

(c) The mass unit, kg, is the standard unit for mass in the SI unit system. The weight is

$$W = mg$$

$$= (75 \text{ kg})(9.81 \text{ m/s}^2) = 736 \text{ N}$$

Alternatively, we can find weight by using Equation (2.4):

$$W = \frac{mg}{g_c} = \frac{(75 \text{ kg})(9.81 \text{ m/s}^2)}{1\dfrac{\text{kg} \cdot \text{m}}{\text{N} \cdot \text{s}^2}} = 736 \text{ N}$$ ∎

Now that we understand the difference between mass and weight and know how to use mass and weight units in the SI and English systems, let's revisit the astronaut we discussed earlier. (See Figure 2.11.) The mass of the astronaut is 80 kg, which equals about 5.48 slug. His mass does not change, regardless of where he ventures. Prior to departing

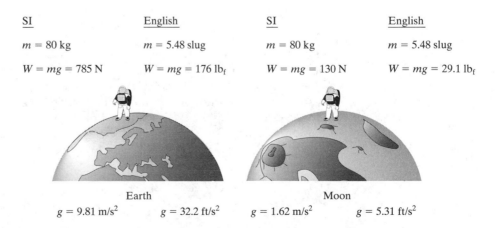

SI	English	SI	English
$m = 80$ kg	$m = 5.48$ slug	$m = 80$ kg	$m = 5.48$ slug
$W = mg = 785$ N	$W = mg = 176$ lb$_f$	$W = mg = 130$ N	$W = mg = 29.1$ lb$_f$

Earth
$g = 9.81$ m/s^2 $g = 32.2$ ft/s^2

Moon
$g = 1.62$ m/s^2 $g = 5.31$ ft/s^2

Figure 2.11. An astronaut's mass and weight on the earth and moon.

on a trip to the moon, he weighs in at 785 N (176 lb$_f$). What is the mass of the astronaut in pound-mass? Three days later, his vehicle lands on the moon, and he begins constructing a permanent base for future planetary missions. The mass of the moon is about one-sixth that of the earth, so the value of the gravitational acceleration on the moon is only 1.62 m/s^2(5.31 ft/s^2). The astronaut's mass is still 80 kg, but his weight is only 130 N (29.1 lb$_f$) due to the smaller value of g. Is the mass and weight of the astronaut in pound-mass and pound-force numerically equivalent? No, because the standard value of g is not used.

EXAMPLE 2.5

Special hoists are used in automotive repair shops to lift engines. As illustrated in Figure 2.12, a 200-kg engine is suspended in a fixed position by a chain attached to the cross member of an engine hoist. Neglecting the weight of the chain itself, what is the tension in portion AD of the chain?

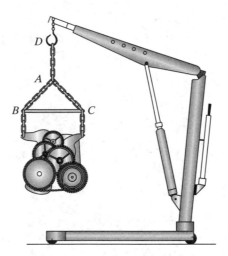

Figure 2.12. Engine hoist for Example 2.5.

SOLUTION

This example is a simple problem in engineering statics. Statics is the branch of engineering mechanics that deals with forces acting on bodies at rest. The engine is held by the chain in a fixed position, so clearly the engine is at rest; that is, it is not in motion. This problem can be solved by recognizing that the entire weight of the engine is supported by portion AD of the chain. (The tension in portions AB and AC could also be calculated, but a thorough equilibrium analysis would be required.) Hence, the tension, which is a force that tends to elongate the chain, is equivalent to the weight of the engine. Using Equation (2.2), we have

$$F = mg$$

$$= (200 \text{ kg})(9.81 \text{ m/s}^2) = 1962 \text{ N}$$

Therefore, the tension in portion AD of the chain is 1962 N, the weight of the engine.

PRACTICE!

1. It has been said that you do not fully understand a basic technical concept, unless you can explain it in terms simple enough that a second grader can understand it. Write an explanation of the difference between mass and weight for a second grader.

2. Which is larger, a slug or a pound-mass?

   Answer: slug

3. Consider a professional linebacker who weighs 310 lb_f. What is his mass in slugs?

   Answer: 9.63 slug

4. A rock ($\rho = 2300$ kg/m^3) is suspended by a single rope. Assuming the rock to be spherical, with a radius of 15 cm, what is the tension in the rope?

   Answer: 319 N

2.7   UNIT CONVERSIONS

Although the SI unit system is the international standard, English units are in widespread use in the United States. Americans as a whole are much more familiar with English units than SI units. Students of science and engineering in U.S. schools primarily use SI units in their course work because most textbooks and the professors who teach out of them, stress SI units. Unfortunately, when students of these disciplines go about their day-to-day activities outside of the academic environment, they tend to slip back into the English unit mode along with everyone else. It seems as if students have a "unit switch" in their brains. When they are in the classroom or laboratory, the switch is turned to the "SI position." When they are at home, in the grocery store, or driving their car, the switch is turned to the "English position." Ideally, there should be no unit switch at all, but as long as science and engineering programs at colleges and universities stress SI units and American culture stresses English units, our cerebral unit switch toggles. In this section, a systematic method for converting units between the SI and English systems is given.

A **unit conversion** enables us to convert from one unit system to the other by using **conversion factors**. A conversion factor is an equivalency ratio that has a *unit value* of 1. Stated another way, a conversion factor simply relates the same physical quantity in two different unit systems. For example, 0.0254 m and 1 in are equivalent length quantities because 0.0254 m = 1 in. The ratio of these two quantities has a unit value of 1 because they are physically the same quantity. Obviously, the numerical value of the ratio is not 1, but depends on the numerical value of each individual quantity. Thus, when we multiply a given quantity by one or more conversion factors, we alter only the numerical value of the result and not its dimension. Table 2.11 summarizes some common conversion factors used in engineering analysis. A more extensive listing of unit conversions is given in Appendix B.

A systematic procedure for converting a quantity from one unit system to the other is as follows:

Unit Conversion Procedure

1. Write the given quantity in terms of its numerical value and units. Use a horizontal line to divide units in the numerator (upstairs) from those in the denominator (downstairs).

2. Determine the units *to* which you want to make the conversion.

TABLE 2.11 Some Common SI-to-English Unit Conversions

Quantity	Unit Conversion
Acceleration	$1 \text{ m/s}^2 = 3.2808 \text{ ft/s}^2$
Area	$1 \text{ m}^2 = 10.7636 \text{ ft}^2 = 1550 \text{ in}^2$
Density	$1 \text{ kg/m}^3 = 0.06243 \text{ lb}_m/\text{ft}^3$
Energy, work, heat	$1055.06 \text{ J} = 1 \text{ Btu} = 252 \text{ cal}$
Force	$1 \text{ N} = 0.22481 \text{ lb}_f$
Length	$1 \text{ m} = 3.2808 \text{ ft} = 39.370 \text{ in}$
	$0.0254 \text{ m} = 1 \text{ in}^{(1)}$
Mass	$1 \text{ kg} = 2.20462 \text{ lb}_m = 0.06852 \text{ slug}$
Power	$1 \text{ W} = 3.4121 \text{ Btu/h}$
	$745.7 \text{ W} = 1 \text{ hp}$
Pressure	$1 \text{ kPa} = 20.8855 \text{ lb}_f/\text{ft}^2 = 0.14504 \text{ lb}_f/\text{in}^2$
Specific heat	$1 \text{ kJ/kg} \cdot {}^\circ\text{C} = 0.2388 \text{ Btu/lb}_m \cdot {}^\circ\text{F}$
Temperature	$T(\text{K}) = T({}^\circ\text{C}) + 273.16 = T({}^\circ\text{R})/1.8 = [T({}^\circ\text{F}) + 459.67]/1.8$
Velocity	$1 \text{ m/s} = 2.2369 \text{ mi/h}$

$^{(1)}$ Exact conversion

3. Multiply the given quantity by one or more conversion factors that, upon cancellation of units, leads to the desired units. Use a horizontal line to divide the units in the numerator and denominator of each conversion factor.

4. Draw a line through all canceled units.

5. Perform the numerical computations on a calculator, retaining infinite decimal place accuracy until the end of the computations.

6. Write the numerical value of the converted quantity by using the desired number of significant figures (three significant figures is standard practice for engineering) with the desired units.

Examples 2.6, 2.7, and 2.8 illustrate the unit conversion procedure.

EXAMPLE 2.6

An engineering student is late for an early morning class, so she runs across campus at a speed of 9 mi/h. Determine her speed in units of m/s.

SOLUTION

The given quantity, expressed in English units, is 9 mi/h, but we want our answer to be in SI units of m/s. Thus, we need a conversion factor between mi and m and a conversion factor between h and s. To better illustrate the unit conversion procedure, we will use two length conversion factors rather than one. Following the procedure outlined, we have

$$9 \frac{\text{mi}}{\text{h}} \times \frac{5280 \text{ ft}}{1 \text{ mi}} \times \frac{1 \text{ m}}{3.2808 \text{ ft}} \times \frac{1 \text{ h}}{3600 \text{ s}} = 4.02 \frac{\text{m}}{\text{s}}$$

given quantity conversion factors answer

The key aspect of the unit conversion process is that the conversion factors must be written such that the appropriate units in the conversion factors cancel those in the given quantity. If we had inverted the conversion factor between ft and mi, writing it instead as 1 mi/5280 ft, the mi unit would not cancel and our unit conversion exercise would not work, because we would end up with units of mi^2 in the numerator. Similarly, the conversion factor between m and ft was written such that the ft unit canceled the ft unit in the first conversion factor. Also, the conversion factor between h and s was written such that the h unit canceled with the h unit in the given quantity. Writing conversion factors with the units in the proper locations, "upstairs" or "downstairs," requires some practice, but after doing several conversion problems, the correct placement of units will become second nature to you. Note that our answer is expressed in three significant figures.

EXAMPLE 2.7

Lead has one of the highest densities of all the pure metals. The density of lead is 11,340 kg/m^3. What is the density of lead in units of lb$_m$/in^3?

SOLUTION

A direct conversion factor from kg/m^3 to lbm/in^3 may be available, but to illustrate an important aspect of converting units with exponents, we will use a series of conversion factors for each length and mass unit. Thus, we write our unit conversion as

$$11{,}340\frac{kg}{m^3} \times \frac{(1\ m)^3}{(3.2808\ ft)^3} \times \frac{(1\ ft)^3}{(12\ in)^3} \times \frac{2.20462\ lb_m}{1\ kg} = 0.410\ lb_m/in^3$$

We used two length conversion factors, one factor between m and ft and the other between ft and in. But the given quantity is a density than has a volume unit. When performing unit conversions involving exponents, *both* the numerical value and the unit must be raised to the exponent. A common error that students make is to raise the unit to the exponent, which properly cancels units, but to forget to raise the numerical value also. Failure to raise the numerical value to the exponent will lead to the wrong numerical answer even though the units in the answer will be correct. Using the direct conversion factor, we obtain the same result:

$$11{,}340\ kg/m^3 \times \frac{(3.6127 \times 10^{-5}\ lb_m/in^3)}{1\ kg/m^3} = 0.410\ lb_m/in^3$$

EXAMPLE 2.8

Specific heat is defined as the energy required to raise the temperature of a unit mass of a substance by one degree. Pure aluminum has a specific heat of approximately 900 J/kg · °C. Convert this value to units of Btu/lb$_m$ · °F.

SOLUTION

By following the unit conversion procedure, we write the given quantity and then multiply it by the appropriate conversion factors, which can be found in Appendix B:

$$\frac{900\ J}{kg \cdot °C} \times \frac{1\ Btu}{1055.06\ J} \times \frac{1\ kg}{2.20462\ lb_m} \times \frac{1°C}{1.8°F} = 0.215\ Btu/lb_m \cdot °F$$

The temperature unit, °C, in the original quantity has a unique interpretation. Because specific heat is the energy required to raise a unit mass of a substance by one degree, the temperature unit in this quantity denotes a temperature *change*, not an absolute temperature value. A temperature change of 1°C is equivalent to a temperature change of 1.8°F. Stated another way, a change of one degree on the Fahrenheit scale is 1.8 times a change of one degree on the Celsius scale, as given by the temperature difference conversion factor, $\Delta T(°C) = \Delta T(°F)/1.8$. Other thermal properties, such as thermal conductivity, involve the same temperature change interpretation. See problem 39, at the end of this chapter, for reference.

This example can also be done by applying a single conversion factor 1 kJ/kg · °C = 0.2388 Btu/lb$_m$ · °F, which yields the same result.

PROFESSIONAL SUCCESS: UNIT CONVERSIONS AND CALCULATORS

Scientific pocket calculators have evolved from simple electronic versions of adding machines to complex portable computers. Today's high-end scientific calculators have numerous capabilities, including programming, graphing, numerical methods, and symbolic mathematics. Most scientific calculators also have an extensive compilation of conversion factors either burned into a chip within the calculator itself or available as a plug-in application module. Why, then, should students learn to do unit conversions by hand when calculators will do the work? This question lies at the root of a more fundamental question: Why should students learn to do *any* computational task by hand when calculators or computers will do the work? Is it because "in the old days" students and practicing engineers did not have the luxury of highly sophisticated computational tools, so professors, who perhaps lived in the "old days," force their students to do things the old fashioned way? Not really.

Students will always need to learn engineering by *thinking* and *reasoning* their way through a problem, regardless of whether that problem is a unit conversion or a stress calculation in a machine component. Computers, and the software that runs on them, do not replace the thinking process. The calculator, like the computer, should never become a "black box" to the student. A black box is a mysterious device whose inner workings are largely unknown, but that, nonetheless, provides output for every input supplied. By the time you graduate with an engineering degree, or certainly by the time you have a few years of professional engineering practice under your belt, you will come to realize that a calculator program or computer software package exists for solving almost any conceivable type of engineering problem. This does not mean that you need to learn every one of these programs and software packages. It means that you should become proficient in the use of those computational tools that pertain to your particular engineering field *after* learning the underlying basis for each. By all means, use a calculator to perform unit conversions, but *first* know how to do them by hand, so you gain confidence in your own computational skills and have a way of verifying the results of your calculator.

PRACTICE!

1. A microswitch is an electrical switch that requires only a small force to operate it. If a microswitch is activated by a 0.25-oz force, what is the force in units of N that will activate it?

    Answer: 0.0695 N

2. At room temperature, water has a density of about 62.4 lb_m/ft^3. Convert this value to units of $slug/in^3$ and kg/m^3.

    Answer: 1.12×10^{-3} $slug/in^3$, 999.5 kg/m^3

3. At launch, the Saturn V rocket that carried astronauts to the moon developed five million pounds of thrust. What is the thrust in units of MN?

    Answer: 22.2 MN

4. Standard incandescent light bulbs produce more heat than light. Assuming that a typical house has twenty 60-W bulbs that are continuously on, how much heat in units of Btu/h is supplied to the house from light bulbs if 90 percent of the energy produced by the bulbs is in the form of heat?

    Answer: 3685 Btu/h

5. Certain properties of animal (including human) tissue can be approximated by using those of water. Using the density of water at room temperature, $\rho = 62.4$ lb_m/ft^3, calculate the weight of a human male by approximating him as a cylinder with a length and diameter of 6 ft and 1 ft, respectively.

    Answer: 294 lb_f

6. The standard frequency for electrical power in the United States is 60 Hz. For an electrical device that operates on this power, how many times does the current alternate during a year?

    Answer: 1.89×10^9

KEY TERMS

base dimension
conversion factors
derived dimension
dimension
dimensionally consistent

English unit system
mass
Newton's second law
physical standards
SI unit system

unit
unit conversion
weight

REFERENCES

Cardarelli, F., *Encyclopaedia of Scientific Units, Weights and Measures: Their SI Equivalences and Origins*, 3d ed., NY: Springer-Verlag, 2003.
Lewis, R., *Engineering Quantities and Systems of Units*, NY: Halsted Press, 1972.
Lide, D.R., Editor, *CRC Handbook of Chemistry and Physics*, 84th ed. Boca Raton, FL: CRC Press, 2003.

Problems

1. For the following dimensional equations, find the base dimensions of the parameter K:

    a. $[M][L][t]^{-2} = k[M][L]^{-1}[t]^{-2}$

    b. $[M][L][t]^{-2}[L]^{-1} = k[L][t]^{-3}$

    c. $[L]^2[t]^{-2} = k[M]^4[T]^2$

    d. $[M][L]^2[t]^{-3} = k[L][T]$

    e. $[N][L][L]^3 k = [T]^2[M]^{-2}[L]$

    f. $[M][I]^2 k = [N][T][M]^{-3}[L]^{-1}$

    g. $[I][L]^2[t] = k^2[M]^4[t]^2$

    h. $k^3[T]^6[M]^3[L]^{-5} = [T]^{-3}[t]^{-6}[L]$

    i. $[T]^{-1/2}[L]^{-1}[I]^2 = k^{-1/2}[t]^4[T]^{-5/2}[L]^{-3}$

    j. $[M][L][t]^{-2} = [M][L][t]^{-2}\sin(k[L]^{-2}[M]^{-1})$

    k. $[T]^2[N] = [T]^2[N]\ln(k[N][T]^{-1})$

2. Is the following dimensional equation dimensionally consistent? Explain.

    $$[M][L] = [M][L]\cos([L][t])$$

3. Is the following dimensional equation dimensionally consistent? Explain.

    $$[t]^2[L][T] = [t][L][T]\log([t][t]^{-1})$$

4. Is the following dimensional equation dimensionally consistent? Explain.

    $$[T][N][T] = [T][N][T]\exp([M][M]^{-1})$$

5. In the following list, various quantities are written using SI units incorrectly: Write the quantities, using the correct form of SI units.

    a. 10.6 secs

    b. 4.75 amp

    c. 120 M hz

    d.   2.5 kw

    e.   0.00846 kg/μs

    f.   90 W/m^2 K

    g.   650 mGPa

    h.   25 MN.

    i.   950 Joules

    j.   1.5 m/s/s

6.   The dimension *moment*, sometimes referred to as *torque*, is defined as a force multiplied by a distance and is expressed in SI units of newton-meter (N $\cdot$ m). In addition to moment, what other physical quantities are expressed in SI units of N $\cdot$ m? What is the special name given to this combination of units?

7.   Consider a 60-W light bulb. A watt (W) is defined as a joule per second (J/s). Write the quantity 60 W in terms of the units newton (N), meter (m), and second (s).

8.   A commonly used formula in electrical circuit analysis is $P = IV$, power (W) equals current (A) multiplied by voltage (V). Using Ohm's law, write a formula for power in terms of current $[I]$ and resistance $[R]$.

9.   A particle undergoes an average acceleration of 5 m/s^2 as it travels between two points during a time interval of 2 s. Using unit considerations, derive a formula for the average velocity of a particle in terms of average acceleration and time interval. Calculate the average velocity of the particle for the numerical values given.

10.   A crane hoists a large pallet of materials from the ground to the top of a building. In hoisting this load, the crane does 100 kJ of work during a time interval of 5 s. Using unit considerations, derive the formula for power in terms of work and time interval. Calculate the power expended by the crane in lifting the load.

11.   A spherical tank with a radius of 0.25 m is filled with water ($\rho = 1000$ kg/m^3). Calculate the mass and the weight of the water in SI units.

12.   A large indoor sports arena is roughly cylindrical in shape. The height and diameter of the cylinder are 120 m and 180 m, respectively. Calculate the mass and weight of air contained in the sports arena in SI units if the density of air is $\rho = 1.20$ kg/m^3.

13.   A 90-kg astronaut biologist searches for microbial life on Mars where the gravitational acceleration is g $= 3.71$ m/s^2. What is the weight of the astronaut in units of N and lb$_f$?

14.   A 90-kg astronaut biologist places a 4-lb$_m$ rock sample on two types of scales on Mars in order to measure the rock's weight. The first scale is a beam balance, which operates by comparing masses. The second scale operates by the compression of a spring. Calculate the weight of the rock sample in units of lb$_f$ using (a) the beam balance and (b) the spring scale.

15.   A stainless steel plate measuring 1.2 m $\times$ 0.8 m $\times$ 3 mm has a density of $\rho = 8000$ kg/m^3. Find the mass and weight of the plate in SI units.

16.   A circular tube of polyethylene plastic ($\rho = 930$ kg/m^3) has an inside radius of 1.2 cm and an outside radius of 4.6 cm. If the cylinder is 40 cm long, what is the mass and weight of the cylinder in SI units?

17.   The density of porcelain is $\rho = 144$ lb$_m$/ft^3. Approximating a porcelain dinner plate as a flat disk with a diameter and thickness of 9 in and 0.2 in, respectively,

find the mass of the plate in units of slug and lb_m. What is the weight of the plate in units of lb_f?

18. In an effort to reduce the mass of an aluminum bulkhead for a spacecraft, a machinist drills an array of holes in the bulkhead. The bulkhead is a triangular-shaped plate with a base and height of 2.5 m and 1.6 m, respectively, and a thickness of 8 mm. How many 5-cm diameter holes must be drilled clear through the bulkhead to reduce its mass by 8 kg? For the density of aluminum, use $\rho = 2800$ kg/m^3.

19. A world-class sprinter can run 100 m in a time of 10 s, an average speed of 10 m/s. Convert this speed to mi/h.

20. A world-class mile runner can run 1 mi in a time of 4 min. What is the runner's average speed in units of mi/h and m/s?

21. The typical home is heated by a forced-air furnace that burns natural gas or fuel oil. If the heat output of the furnace is 120,000 Btu/h, what is the heat output in units of kW?

22. Calculate the temperature at which the Celsius (°C) and Fahrenheit (°F) scales coincide.

23. A large shipping container of ball bearings is suspended by a cable in a manufacturing plant. The combined mass of the container and ball bearings is 2500 lb_m. Find the tension in the cable in units of N.

24. A typical human adult loses about 65 Btu/h · ft^2 of heat while engaged in brisk walking. Approximating the human adult body as a cylinder with a height and diameter of 5.8 ft and 1.1 ft, respectively, find the total amount of heat lost in units of J if the brisk walking is maintained for a period of 2 h. Include the two ends of the cylinder in the surface area calculation.

25. A symmetric I-beam of structural steel ($\rho = 7860$ kg/m^3) has the cross section shown in Figure P2.25. Calculate the weight per unit length of the I-beam in units of N/m and lb_f/ft.

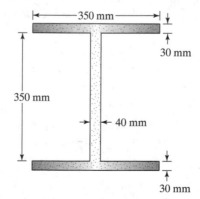

Figure P2.25.

26. A sewer pipe carries waste away from a commercial building at a mass flow rate of 5 kg/s. What is this flow rate in units of lb_m/s and slug/h?

27. The rate at which solar radiation is intercepted by a unit area is called solar heat flux. Just outside the earth's atmosphere, the solar heat flux is approximately 1350 W/m^2. Determine the value of this solar heat flux in units of Btu/h · ft^2.

28.   During a typical summer day in the arid southwest regions of the United States, the outdoor air temperature may range from 115°F during the late afternoon to 50°F several hours after sundown. What is this temperature range in units of °C, K, and °R?

29.   An old saying is "an ounce of prevention is worth a pound of cure." Restate this maxim in terms of the SI unit newton.

30.   How many seconds are there in a leap year?

31.   What is your approximate age in seconds?

32.   A highway sign is supported by two posts as shown in Figure P2.32. The sign is constructed of a high-density pressboard material ($\rho = 900\ \text{kg/m}^3$) and its thickness is 2 cm. Assuming that each post carries half the weight of the sign, calculate the compressive force in the posts in units of N and lb$_f$.

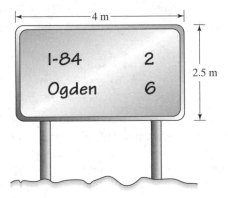

Figure P2.32.

33.   A boiler is a vessel containing water or other fluid at a high temperature and pressure. Consider a boiler containing water at a temperature and pressure of 250°C and 5 MPa, respectively. What is the temperature and pressure in units of K and psi, respectively?

34.   A pressure gauge designed to measure small pressure differences in air ducts has an operating range of 0 to 16 inch H_2O. What is this pressure range in units of Pa and psi?

35.   Resistors are electrical devices that retard the flow of current. These devices are rated by the maximum power they are capable of dissipating as heat to the surrounding area. How much heat does a 25-W resistor dissipate in units of Btu/h if the resistor operates at maximum capacity? Using the formula $P = I^2R$, what is the current flow, I, in the resistor if is has a resistance, R, of 100 Ω?

36.   Chemical reactions can generate heat. This type of heat generation is often referred to as volume heat generation because the heat is produced internally by every small parcel of chemical. Consider a chemical reaction that generates heat at the rate of 50 MW/m^3. Convert this volume heat generation to units of Btu/h · ft^3.

37.   A sport-utility vehicle has an engine that delivers 250 hp. How much power does the engine produce in units of kW and Btu/h?

38.   An underground pipe carries culinary water to a home at a volume flow rate of 5 gal/min. Determine the flow rate in units of m^3/s and ft^3/h.

39.   Thermal conductivity is a property that denotes the ability of a material to conduct heat. A material with a high thermal conductivity readily transports heat, whereas a material with a low thermal conductivity tends to retard heat flow. Fiberglass insulation and silver have thermal conductivities of 0.046 W/m · °C and 429 W/m · °C, respectively. Convert these values to units of Btu/h · ft · °F.

40.   A standard incandescent 75-W light bulb has an average life of 1000 h. What is the total amount of energy that this light bulb produces during its lifetime? Express the answer in units of J, Btu, and cal.

41.   A steam power plant produces 400 MW of power. How much energy does the power plant produce in a year? Express your answer in units of J and Btu.

42.   It is estimated that about 50 million Americans go on a new diet each year. If each of these people cuts 300 cal from their diets each day, how many 100-W light bulbs could be powered by this energy?

43.   The standard acceleration of gravity at the earth's surface is $g = 9.81$ m/s^2. Convert this acceleration to units of ft/h^2 and mi/s^2.

44.   At room temperature, air has a specific heat of 1.007 kJ/kg · °C. Convert this value to units of J/kg · K and Btu/lb$_m$ · °F.

45.   The yield stress for structural steel is approximately 250 MPa. Convert this value to units of psi.

3

Analysis Methodology

3.1   INTRODUCTION

One of the most important things an engineering student learns during his or her program of study is how to approach an engineering problem in a systematic and logical fashion. In this respect, the study of engineering is somewhat similar to the study of science in that a science student learns how to think like a scientist by employing the scientific method. The scientific method is a process by which hypotheses about the physical world are stated, theories formulated, data collected and evaluated, and mathematical models constructed. The **engineering method** may be thought of as a problem-solving process by which the needs of society are met through design and manufacturing of devices and systems. Engineering analysis is a major part of this problem-solving process. Admittedly, engineering and science are not the same, because they each play a different role in our technical society. Science seeks to explain how nature works through fundamental investigations of matter and energy. The objective of engineering is more pragmatic. Engineering, using science and mathematics as tools, seeks to design and build products and processes that enhance our standard of living. Generally, the scientific principles underlying the function of any engineering device were derived and established *before* the device was designed. For example, Newton's laws of motion and Kepler's orbital laws were well established scientific principles long before spacecraft orbited the earth or the other planets. Despite their contrasting objectives, both engineering and science employ tried-and-true methodologies that enable people working in each field to solve a variety of problems. To do science, the scientist must know how to employ the scientific method. To do engineering, the engineer must know how to employ the "engineering method."

OBJECTIVES

After reading this chapter, you will have learned

- How to make order-of-magnitude calculations
- The proper use of significant figures
- How to perform an analysis systematically
- The proper method of analysis presentation
- Advantages and disadvantages of using computers for analysis

Engineering analysis is the solution of an engineering problem by using mathematics and principles of science. Because of the close association between analysis and design, analysis is one of the key steps in the design process. Analysis also plays a major role in the study of engineering failures. The engineering method for conducting an analysis is a logical, systematic procedure characterized by a well-defined format. This procedure, when consistently and correctly applied, leads to the successful solution of any analytical engineering problem. Practicing engineers have been using this analysis procedure successfully for decades, and engineering graduates are expected to know how to apply it upon entering the technical workforce. Therefore, it behooves the engineering student to learn the analysis methodology as thoroughly as possible. The best way to do so is to practice solving analytical problems. As you advance in your engineering course work, you will have ample opportunities to apply the analysis methodology outlined in this chapter. Courses such as statics, dynamics, strength of materials, thermodynamics, fluid mechanics, heat and mass transfer, electrical circuits, and engineering economics are analysis intensive. These courses, and others like them, focus almost exclusively on solving engineering problems that are analytical in nature. That is the character of these engineering subjects. The analysis methodology presented here is a *general* procedure that can be used to solve problems in any analytical subject. Clearly, engineering analysis heavily involves the use of numerical calculations.

3.2   NUMERICAL CALCULATIONS

As a college student, you are well aware of the rich diversity of academic programs and courses offered at institutions of higher learning. Because you are an engineering major, you are perhaps more familiar with the genre of engineering, science, and mathematics courses than liberal arts courses such as sociology, philosophy, psychology, music, and languages. The tenor of liberal arts is vastly different than that of engineering. Suppose for a moment that you are enrolled in a literature class, studying Herman Melville's great book, *Moby Dick*. While discussing the relationship between the whale and Captain Ahab, your literature professor asks the class, "What is your impression of Captain Ahab's attitude toward the whale?" As an engineering major, you are struck by the apparent looseness of this question. You are accustomed to answering questions that require a quantitative answer, not an "impression." What would engineering be like if our answers were "impressions"? Imagine an engineering professor asking a thermodynamics class, "What is your impression of the superheated steam temperature at the inlet of the turbine?" A more appropriate question would be, "What *is* the superheated steam temperature at the inlet of the turbine?" Obviously, literature and the other liberal arts disciplines operate in a completely different mode than engineering. By its very nature, engineering is based on specific, quantitative information. An answer of "hot" to the second thermodynamics question would be quantitative, but not specific, and therefore insufficient. The temperature of the superheated steam at the inlet of the turbine could be calculated by conducting a thermodynamic analysis of the turbine, thereby providing a *specific* value for the temperature; 400°C, for example. The analysis by which the temperature was obtained may consist of several numerical calculations involving different thermodynamic quantities. Numerical calculations are mathematical operations on numbers that represent physical quantities such as temperature, stress, voltage, mass, flow rate, etc. In this section, you will learn the proper numerical calculation techniques for engineering analysis.

3.2.1   Approximations

It is often useful, particularly during the early stages of design, to calculate an approximate answer to a given problem when the given information is uncertain or when little information is available. An approximation can be used to establish the cursory aspects of a design and to determine whether a more precise calculation is required. Approximations are usually based on assumptions, which must be modified or eliminated during the latter stages of the design. Engineering approximations are sometimes referred to as "guesstimates," "ballpark calculations," or "back-of-the-envelope calculations." A more appropriate name for them is **order-of-magnitude** calculations. The term order of magnitude means a *power of* 10. Thus, an order-of-magnitude calculation refers to a calculation involving quantities whose numerical values are estimated to within a factor of 10. For example, if the estimate of a stress in a structure changes from about 1 kPa to about 1 MPa, we say that the stress has changed by three orders of magnitude, because 1 MPa is one thousand (10^3) times 1 kPa. Engineers frequently conduct order-of-magnitude calculations to ascertain whether their initial design concepts are feasible. Order-of-magnitude calculations are therefore a useful decision-making tool in the design process. Order-of-magnitude calculations do not require the use of a calculator because all the quantities have simple power-of-10 values, so the arithmetic operations can be done by hand with pencil and paper or even in your head. The example that follows illustrates an order-of-magnitude calculation.

EXAMPLE 3.1

A warehouse with the approximate dimensions 200 ft $\times$ 150 ft $\times$ 20 ft is ventilated with 12 large industrial blowers. In order to maintain acceptable air quality in the warehouse, the blowers must provide two air changes per hour, meaning that the entire volume of air within the warehouse must be replenished with fresh outdoor air two times per hour. Using an order-of-magnitude analysis, find the required volume flow rate that each blower must deliver, assuming the blowers equally share the total flow rate.

SOLUTION

To begin, we estimate the volume of the warehouse. The length, width, and height of the warehouse is 200 ft, 150 ft, and 20 ft, respectively. These lengths have order-of-magnitude values of 10^2, 10^2, and 10^1, respectively. Two air changes per hour are required. Thus, the total volume flow rate of air for the warehouse, including the factor of two air changes per hour, is

$$Q_t \approx (10^2 \text{ ft})(10^2 \text{ ft})(10^1 \text{ ft})(2 \text{ air changes/h}) = 2 \times 10^5 \text{ ft}^3/\text{h}$$

(Note that "air changes" is not a unit, so it does not appear in the answer.) The number of blowers (12) has an order-of-magnitude value of 10^1. Based on the assumption that each blower delivers the same flow rate, the flow rate per blower is the total volume flow rate divided by the number of blowers:

$$Q = Q_t/N = (2 \times 10^5 \text{ ft}^3/\text{h})(10^1 \text{ blowers}) = 2 \times 10^4 \text{ ft}^3/\text{h} \cdot \text{blower}$$

Our order-of-magnitude calculation shows that each blower must supply 2×10^4 ft^3/h of outdoor air to the warehouse.

How does our order-of-magnitude answer compare with the exact answer? The exact answer is

$$Q = (200 \text{ ft})(150 \text{ ft})(20 \text{ ft})(2 \text{ air changes/h})/(12 \text{ blowers}) = 1 \times 10^5 \text{ ft}^3/\text{h} \cdot \text{blower}$$

By dividing the exact answer by the approximate answer, we see that the approximate answer differs from the exact answer by a factor of five, which is within an order of magnitude.

3.2.2   Significant Figures

After order-of-magnitude calculations have been made, engineers conduct more precise calculations to refine their design or to more fully characterize a particular failure mode. Accurate calculations demand more of the engineer than simply keeping track of powers of 10. Final design parameters must be determined with as much accuracy as possible to achieve the optimum design. Engineers must determine how many digits in their calculations are significant. A **significant figure** or *significant digit* in a number is defined as *a digit that is considered reliable as a result of a measurement or calculation*. The number of significant figures in the answer of a calculation indicates the number of digits that can be used with confidence, thereby providing a way of telling the engineer how accurate the answer is. No physical quantity can be specified with infinite precision because no physical quantity is *known* with infinite precision. Even the constants of nature such as the speed of light in a vacuum, c, and the gravitational constant, G, are known only to the precision with which they can be measured in a laboratory. Similarly, engineering material properties such as density, modulus of elasticity, and specific heat are known only to the precision with which these properties can be measured. A common mistake is to use more significant figures in an answer than are justified, giving the impression that the answer is more accurate than it really is. No answer can be more accurate than the numbers used to generate that answer.

How do we determine how many significant figures (colloquially referred to as "sig figs") a number has? A set of rules has been established for counting the number of significant figures in a number. (All significant figures are underlined in the examples given for each rule.)

Rules for Significant Figures

1. All digits *other than zero* are significant. Examples: 8.936, 456, 0.257.
2. All zeroes *between* significant figures are significant. Examples: 14.06, 5.0072.
3. For nondecimal numbers greater than one, all zeroes placed *after* the significant figures are *not* significant. Examples: 2500, 8,640,000. These numbers can be written in scientific notation as 2.5×10^3 and 8.64×10^6, respectively.
4. If a decimal point is used *after* a nondecimal number larger than one, the zeroes are significant. The decimal point establishes the precision of the number. Examples: 3200., 550,000.
5. Zeroes placed *after* a decimal point that are *not necessary* to set the decimal point are significant. The additional zeroes establish the precision of the number. Examples: 359.00, 1000.00.
6. For numbers smaller than one, all zeroes placed *before* the significant figures are *not* significant. These zeroes only serve to establish the location of the decimal point. Examples: 0.0254, 0.000609

Do not confuse the number of significant figures with the number of decimal places in a number. The number of significant figures in a quantity is established by the precision with which a measurement of that quantity can be made. The primary exception to this are numbers such as π and the Naperian base, e, that are derived from mathematical relations. These numbers are accurate to an infinite number of significant figures.

Let's see how the rules for significant figures are used in calculations.

EXAMPLE 3.2

We wish to calculate the weight of a 25-kg object. Using Newton's second law, $W = mg$, find the weight of the object in units of N. Express the answer, by using the appropriate number of significant figures.

SOLUTION

We have $m = 25$ kg and $g = 9.81$ m/s^2. Suppose that our calculator is set to display six places to the right of the decimal point. We then multiply the numbers 25 and 9.81. In the display of the calculator, we see the number 245.250000. How many digits in this answer are we justified in writing? The number in the calculator's display implies that the answer is accurate to six decimal places (i.e., to within one-millionth of a newton). Obviously, this kind of accuracy is not justified. The rule for significant figures for *multiplication* and *division* is that *the product or quotient should contain the number of significant figures that are contained in the number with the fewest significant figures.*

Another way to state this rule is to say that the quantity with the fewest number of significant figures *governs* the number of significant figures in the answer. The mass, m, contains two significant figures, and the acceleration of gravity, g, contains three. Therefore, we are only justified in writing the weight by using two significant figures, which is the fewest number of significant figures in our given values. Our answer can be written in two ways. First, we can write the weight as 250 N. According to rule 3, the zero is not significant, so our answer contains two significant figures, the "2" and the "5." Second, we can write the weight by using scientific notation as 2.5×10^2 N. In this form, we can immediately see that two significant figures are used without referring to the rules. Note that in both cases we *rounded* the answer *up* to the nearest tens place, because the value of the first digit dropped is 5 or greater. If our answer had been lower than 245 N, we would have rounded *down* to 240 N. If our answer had been precisely 250 N, the rules of rounding suggest rounding up, so our answer would again be 250 N.

The preceding example shows how significant figures are used for multiplication or division, but how are significant figures used for *addition* and *subtraction*?

EXAMPLE 3.3

Two collinear forces (forces that act in the same direction) of 875.4 N and 9.386 N act on a body. Add these two forces, expressing the result in the appropriate number of significant figures.

SOLUTION

The best way to show how significant figures are used in addition or subtraction is to do the problem by hand. We have

$$
\begin{array}{r}
875.4\ \ \text{N} \\
+\ \ \ 9.356\ \text{N} \\
\hline
884.786\ \text{N}
\end{array}
$$

Both forces have four significant figures, but the first force reports one place past the decimal point, whereas the second force reports three places past the decimal point. The answer is written with six significant figures. Are six significant figures justified? Because addition and subtraction are arithmetic operations that require decimal point alignment, the rule for significant figures for *addition* and *subtraction* is different than for multiplication and division. For addition and subtraction, the answer should show *significant figures only as far to the right as is seen in the least precise number in the calculation.* The least precise number in the calculation is the 875.4-N force, because it reports accuracy to the first decimal place, whereas the second force, 9.386 N, reports accuracy to the third decimal place. We are not justified in writing the answer as 884.786 N. We may only write the answer by using the same number of places past the decimal point as seen in the least precise force. Hence, our answer, reported to the appropriate number of significant figures, is 884.8 N. Once again, we rounded the answer up because the value of the first digit dropped is 5 or greater.

In *combined* operations where multiplication and division are performed in the same operation as addition and subtraction, the multiplications and divisions should be performed first, establishing the proper number of significant figures in the intermediate answers, perform the additions and subtractions, and then round the answer to the proper number of significant figures. This procedure, while applicable to operations performed by hand, should not be used in calculator or computer applications, because intermediate rounding is cumbersome and may lead to a serious error in the answer. Perform the entire calculation, letting the calculator or computer software manage the numerical precision, and then express the final answer in the desired number of significant figures:

> *It is standard engineering practice to express final answers in three (or sometimes four) significant figures, because the given input values for geometry, loads, material properties, and other quantities are typically reported with this precision.*

Calculators and computer software such as spreadsheets and equation solvers keep track of and can display a large number of digits. How many digits will your calculator display? The number of digits displayed by a scientific calculator can be set by fixing the decimal point or specifying the numerical format. For example, by fixing the number of decimal places to one, the number 28.739 is displayed as 28.7. Similarly, the number 1.164 is displayed as 1.2. Because the first digit dropped is greater than 5, the calculator automatically rounds the answer up. Small and large numbers should be expressed in scientific notation. For example, the number 68,400 should be expressed as 6.84×10^4, and the number 0.0000359 should be expressed as 3.59×10^{-5}. Scientific calculators also have an *engineering notation* display setting because SI unit prefixes are primarily defined by multiples of one thousand (10^3). In engineering notation, the number 68,400 may be displayed as 68.4×10^3, and the number 0.0000359 may be displayed as 35.9×10^{-6}. Regardless of how numbers are displayed by calculators or computers, the engineering student who uses these computational tools must understand that significant figures have a physical meaning based on our ability to measure engineering and scientific quantities. The casual or sloppy handling of significant figures in engineering analysis may lead to solutions that are inaccurate at best and completely wrong at worst.

APPLICATION: CALCULATING VISCOSITY BY USING THE FALLING-SPHERE METHOD

You know by experience that some fluids are thicker or more "gooey" than others. For example, pancake syrup and motor oil are thicker than water and alcohol. The technical term we use to describe the magnitude of a fluid's thickness is *viscosity*. Viscosity is a fluid property that characterizes the fluid's resistance to flow. Water and alcohol flow more readily than pancake syrup and motor oil under the same conditions. Hence, pancake syrup and motor oil are more viscous than water and

alcohol. Gases have viscosities, too, but their viscosities are much smaller than those of liquids.

One of the classical techniques for measuring viscosities of liquids is called the *falling-sphere method*. In the falling-sphere method, the viscosity of a liquid is calculated by measuring the time it takes for a small sphere to fall a prescribed distance in a large container of the liquid, as illustrated in Figure 3.1. As the sphere falls in the liquid under the influence of gravity, it accelerates

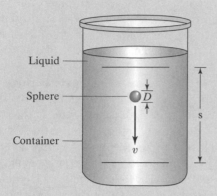

Figure 3.1. Experimental setup of the falling-sphere method for measuring viscosity.

until the downward force (the sphere's weight) is exactly balanced by the buoyancy force and drag force that act upward. From this time forward, the sphere falls with a constant velocity, referred to as terminal velocity. The buoyancy force, which is equal to the weight of the liquid displaced by the sphere, is usually small compared with the drag force, which is caused directly by viscosity. The terminal velocity of the sphere is inversely proportional to viscosity, since the sphere takes longer to fall a given distance in a very viscous liquid, such as motor oil, than in a less viscous liquid, such as water. By employing a force balance on the sphere and invoking some simple relations from fluid mechanics, we obtain the formula

$$\mu = \frac{(\gamma_s - \gamma_f)D^2}{18v}$$

where

μ = dynamic viscosity of liquid (Pa·s)
γ_s = specific weight of sphere (N/m^3)
γ_f = specific weight of liquid (N/m^3)
D = sphere diameter (m)
v = terminal velocity of sphere (m/s)

Note that the quantity *specific weight* is similar to *density*, except that it is a weight per volume, rather than a mass per volume. The word *dynamic* is used to avoid confusion with another type of viscosity known as *kinematic* viscosity.

Using the falling-sphere method, let's calculate the viscosity of glycerine, a very viscous liquid used to make a variety of chemicals. We set up a large glass cylinder

and place two marks, spaced 200 mm apart, on the outside surface. The marks are placed low enough on the cylinder to assure that the sphere will achieve terminal velocity before reaching the top mark. For the sphere, we use a steel (γ_s = 76,800 N/m^3) ball bearing with a diameter of 2.381 mm (measured with a micrometer). From a previous measurement, the specific weight of the glycerin is γ_f = 12,400 N/m^3. Now, we hold the steel sphere above the surface of the glycerin at the center of the cylinder and release the sphere. As accurately as we can determine with our eye, we start a handheld stopwatch when the sphere reaches the top mark. Similarly, we stop the watch when the sphere reaches the bottom mark. Our stopwatch is capable of displaying hundredths of a second, and it reads 11.32 s. Even though the stopwatch is capable of measuring time to the second decimal place, our crude visual timing method does not justify using a time interval with this precision. Sources of uncertainty such as human reaction time and thumb response do not justify the second decimal place. Thus, our time interval is reported as 11.3 s, which has three significant figures. We know that terminal velocity is distance divided by time:

$$v = \frac{s}{t} = \frac{0.200 \text{ m}}{11.3 \text{ s}} = 0.0177 \text{ m/s}$$

The distance was measured to the nearest millimeter, so the quantity, s, has three significant figures. Thus, terminal velocity may be written in three significant figures. (Remember that the zero, according to rule 6, is not significant.) Values of the given quantities for our calculation are summarized as follows:

$$\gamma_s = 76,800 \text{ N/m}^3 = 7.68 \times 10^4 \text{ N/m}^3$$
$$\gamma_f = 12,400 \text{ N/m}^3 = 1.24 \times 10^4 \text{ N/m}^3$$
$$v = 0.0177 \text{ m/s} = 1.77 \times 10^{-2} \text{ m/s}$$
$$D = 2.381 \text{ mm} = 2.381 \times 10^{-3} \text{ m}$$

Each quantity, with the exception of D, which has four significant figures, has three significant figures. Upon substituting values into the equation for dynamic viscosity, we obtain

$$\mu = \frac{(\gamma_s - \gamma_f)D^2}{18v}$$
$$= \frac{(76,800 - 12,400)\text{N/m}^3 (2.381 \times 10^{-3} \text{ m})}{18(0.0177 \text{ m/s})}$$
$$= 1.1459 \text{ Pa·s}$$

(Where did the pressure unit, Pa, come from?) According to the rules of significant figures for multiplication and division, our answer should contain the same number of significant figures as the number with the fewest significant figures. Our answer should therefore have three significant figures, so the dynamic viscosity of glycerin, expressed in the proper number of significant figures, is reported as

$$\mu = 1.15 \text{ Pa} \cdot \text{s}$$

Note that, because the value of the first digit dropped is 5, we rounded our answer up.

PROFESSIONAL SUCCESS: LEARN HOW TO USE YOUR CALCULATOR

As an engineering student, your best friend is your scientific calculator. If you do not yet own a quality scientific calculator, purchase one as soon as you can and begin learning how to use it. You cannot succeed in school without one. Do not scrimp on cost. You will probably only need one calculator for your entire academic career, so purchase one that offers the greatest number of functions and features. Professors and fellow students may offer advice on which calculator to buy. Your particular engineering department or college may even require that you use a particular calculator because they have heavily integrated calculator usage into the curriculum, and it would be too cumbersome to accommodate several types of calculators. Your college bookstore or local office supply store carry two or three name brands that have served engineering students and professionals for many years. Today's scientific calculators are remarkable engineering tools. A high-end scientific calculator has hundreds of built-in functions a large storage capacity graphics capabilities, and communication links to other calculators or personal computers.

Regardless of which scientific calculator you own or plan to purchase, *learn how to use it*. Begin with the basic arithmetic operations and the standard mathematical and statistical functions. Learn how to set the number of decimal places in the display and how to display numbers in scientific and engineering notation. After you are confident with performing unit conversions by hand, learn how to do them with your calculator. Learn how to write simple programs on your calculator. This skill will come in handy numerous times throughout your course work. Learn how to use the equation solving functions, matrix operations and calculus routines. By the time you learn most of the calculator's operations, you will probably have devoted a few hundred hours. The time spent mastering your calculator is perhaps as valuable as the time spent attending lectures, conducting experiments in a laboratory, doing homework problems, or studying for exams. Knowing your calculator thoroughly will help you succeed in your engineering program. Your engineering courses will be challenging enough. Do not make them an even bigger challenge by failing to adequately learn how to use your principal computational asset, your calculator.

PRACTICE!

1. Using an order-of-magnitude analysis, estimate the surface area of your body in units of m^2.

2. Using an order-of-magnitude analysis, estimate the number of hairs on your head.

3. Use an order-of-magnitude analysis to estimate the number of telephones in use in the United States.

4. Use an order-of-magnitude analysis to estimate the electrical energy in kWh used by your city in one month.

5. Underline the significant figures in the following numbers (the first number is done for you):
   a. 0.00<u>254</u>
   b. 29.8
   c. 2001
   d. 407.2

    e. 0.0303

    f. 2.006

    Answer: b. 29.8 c. 2001 d. 407.2 e. 0.0303 f. 2.006

6.   Perform the following calculations, reporting the answers with the correct number of significant figures:

    a. 5.64/1.9

    b. 500./0.0025

    c. $(45.8 - 8.1)/1.922$

    d. $2\pi/2.50$

    e. $(5.25 \times 10^4)/(100 + 10.5)$

    f. $0.0008/(1.2 \times 10^{-5})$

    Answer: a. 3.0 b. 2.0×10^5 c. 19.6 d. 3 e. 473 f. 70

7.   A ball bearing is reported to have a radius of 3.256 mm. Using the correct number of significant figures, what is the weight of this bearing in units of N if its density is $\rho = 1675$ kg/m^3?

    Answer: 2.38×10^{-3} N

8.   The cylinder of an internal combustion engine is reported to have a diameter of 4.000 in. If the stroke (length) of the cylinder is 6.25 in, what is the volume of the cylinder in units of in^3? Write the answer by using the correct number of significant figures.

    Answer: 25.0 in^3

3.3   GENERAL ANALYSIS PROCEDURE

Engineers are problem solvers. In order to solve an engineering analysis problem thoroughly and accurately, engineers employ a solution method that is systematic, logical, and orderly. This method, when consistently and correctly applied, leads the engineer to a successful solution of the analytical problem at hand. The problem-solving method is an integral part of a good engineer's thought process. To the engineer, the procedure is second nature. When challenged by a new analysis, a good engineer knows precisely how to approach the problem. The problem may be fairly short and simple or extremely long and complex. Regardless of the size or complexity of the problem, the same solution method applies. Because of the *general* nature of the procedure, it applies to analytical problems associated with *any* engineering discipline: chemical, civil, electrical, mechanical, etc. Practicing engineers in all disciplines have been using the **general analysis procedure** in one form or another for a long time, and the history of engineering achievements is a testament to its success. While you are a student, it is vitally important that you learn the steps of the general analysis procedure. After you have learned the steps in the procedure and feel confident that you can use the procedure to solve problems, apply it in your analytical course work. Apply it religiously. Practice the procedure over and over again until it becomes a habit. Establishing good habits while still in school will make it that much easier for you to make a successful transition into professional engineering practice.

General Analysis Procedure

The general analysis procedure consists of the following seven steps:

1. PROBLEM
   STATEMENT

The problem statement is a written description of the analytical problem to be solved. It should be written clearly, concisely, and logically. The problem statement summarizes the given information, providing all necessary input data to solve the problem. The problem statement also states what is to be determined by performing the analysis.

2. DIAGRAM

The diagram is a sketch, drawing or schematic of the system being analyzed. Typically, it is a simplified pictorial representation of the actual system, showing only those aspects of the system that are necessary to perform the analysis. The diagram should show all given information contained in the problem statement such as geometry, applied forces, energy flows, mass flows, electrical currents, temperatures, or other physical quantities as required.

3. ASSUMPTIONS

Engineering analysis almost always involves some assumptions. Assumptions are special assertions about the physical characteristics of the problem that simplify or refine the analysis. A very complex analytical problem would be difficult or even impossible to solve without making some assumptions.

4. GOVERNING EQUATIONS

All physical systems may be described by mathematical relations. Governing equations are those mathematical relations that specifically pertain to the physical system being analyzed. These equations may represent physical laws, such as Newton's laws of motion, conservation of mass, conservation of energy, and Ohm's law; or they may represent fundamental engineering definitions such as velocity, stress, moment of force, and heat flux. The equations may also be basic mathematical or geometrical formulas involving angles, lines, areas, and volumes.

5. CALCULATIONS

In this step, the solution is generated. First, the solution is developed algebraically as far as possible. Then numerical values of known physical quantities are substituted for the corresponding algebraic variables. All necessary calculations are performed, using a calculator or computer, to produce a numerical result with the correct units and the proper number of significant figures.

6. SOLUTION CHECK

This step is crucial. Immediately after obtaining the result, examine it carefully. Using established knowledge of similar analytical solutions and common sense, try to ascertain whether the result is reasonable. However, whether the result seems reasonable or not, double-check every step of the analysis. Flush out defective diagrams, bad assumptions, erroneously applied equations, incorrect numerical manipulations, and improper use of units.

7. DISCUSSION

After the solution has been thoroughly checked and corrected, discuss the result. The discussion may include an assessment of the assumptions, a summary of the main conclusions, a proposal on how the result may be verified experimentally in a laboratory, or a parametric study, demonstrating the sensitivity of the result to a range of input parameters.

Now that the seven-step procedure has been summarized, further discussion of each step is warranted:

1. Problem statement   In your engineering textbook, the problem statement will generally be supplied to you in the form of a problem or question at the end of each chapter. These problem statements are written by the textbook authors, professors or practicing engineers, who have expertise in the subject area. The great majority of end-of-chapter problems in engineering texts are well organized and well written, so you do not have to fret too much about the problem statement. Alternatively, your engineering professor may give you problem statements from sources outside your textbook or from his or her own engineering experience. In either case, the problem statement should be well posed, contain all the necessary input information, and clearly state what is to be determined by the analysis. What is known and what is unknown in the problem should be clearly identified. If the problem statement is flawed in any way, a meaningful analysis is impossible.

2. Diagram   The old saying, "One picture is worth a thousand words," is certainly applicable to engineering analysis. A complete diagram of the system being analyzed is critical. A good diagram helps the engineer visualize the physical processes or characteristics of the system. It also helps the engineer identify reasonable assumptions and the appropriate governing equations. A diagram might even reveal flaws in the problem statement or alternative methods of solution. Engineers use a variety of diagrams in their analytical work. One of the most widely used diagrams in engineering is the *free-body diagram*. Free-body diagrams are used to solve engineering mechanics (statics, dynamics, strength of materials) problems. These diagrams are called "free-body" diagrams because they represent a specific body, isolated from all other bodies that are in physical contact with, or that may be in the vicinity of, the body in question. The influences of nearby bodies are represented as external forces acting on the body being analyzed. Hence, a free-body diagram is a sketch of the body in question, showing all external forces applied to the body. A free-body diagram is a pictorial representation of a "force balance" on the body. Diagrams are also used in the analysis of thermal systems. Unlike a free-body diagram, which shows forces applied to the body, a diagram of a thermal system shows all the various forms of energy entering and leaving the system. This type of diagram is a pictorial representation of an "energy balance" on the system. Another type of diagram represents a system that transports mass at known rates. Common examples include pipe and duct systems, conveyors and storage systems. A diagram for these systems shows all the mass entering and leaving the system. This type of diagram is a pictorial representation of a "mass balance" on the system. Still another type of diagram is an electrical circuit schematic. Electrical schematics show how components are connected and the currents, voltages and other electrical quantities in the circuit. Some examples of diagrams used in analysis are given in Figure 3.2.

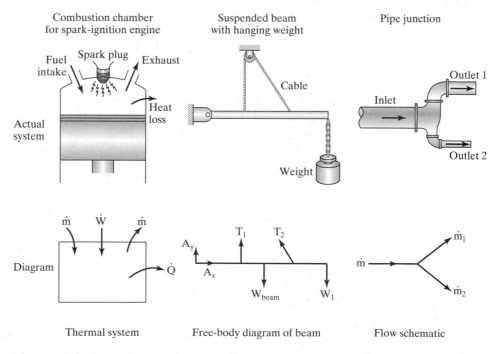

Figure 3.2. Examples of typical diagrams used in engineering analysis.

3. Assumptions I attended a lecture given by a physicist who referred to himself as an "atmospheric scientist" who studied various processes that occur in the upper atmosphere. He recounted an accomplishment that seemed truly remarkable. After convincing the audience that atmospheric processes are some of the most complex phenomena in physics, he boasted that he had developed, over the space of a few months, an analytical model of the upper atmosphere that contained *no* assumptions. There was only one problem: His model had no solution either. By including every physical mechanism to the minutest detail in the model, his analysis was so mathematically convoluted that it could not generate a solution. Had he made some simplifying assumptions, his atmospheric model could have worked even though the results would have been approximate.

Engineers and scientists routinely employ assumptions to simplify a problem. As my story illustrates, an approximate answer is better than no answer at all. Failure to invoke one or more simplifying assumptions in the analysis, particularly a complex one, can increase the complexity of the problem by an order of magnitude, leading the engineer down a very long road, only to reach a dead end. How do we determine which assumptions to use and whether our assumptions are good or bad? To a large extent, the application of good assumptions is an acquired skill, a skill that comes with engineering experience. However, you can begin to learn this skill in school through repeated application of the general analysis procedure in your engineering courses. As you apply the procedure to a variety of engineering problems, you will gain a basic understanding of how assumptions are used in engineering analysis. Then, after you graduate and accept a position with an engineering firm, you can refine this skill as you apply the analysis procedure to solve problems that are specific to the company. Sometimes, a problem can be overly constrained by assumptions such that the problem is simplified to the point where it becomes grossly inaccurate or even meaningless. The engineer must therefore be able to apply the proper *number* as well as the proper *type* of assumptions in a given analysis. A common assumption made in the stress analysis of a column is shown in Figure 3.3.

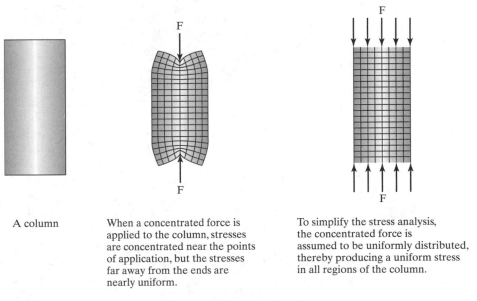

A column

When a concentrated force is applied to the column, stresses are concentrated near the points of application, but the stresses far away from the ends are nearly uniform.

To simplify the stress analysis, the concentrated force is assumed to be uniformly distributed, thereby producing a uniform stress in all regions of the column.

Figure 3.3.   A common assumption made in the stress analysis of a column.

4. Governing equations   The governing equations are the "workhorses" of the analysis. To a very limited extent, we may be able to afford some sloppiness in the other steps in the analysis procedure, but not in the governing equations. The governing equations are either right or they are wrong—there is no middle ground. They either describe the physical problem at hand, or they do not. If the wrong governing equations are used, the analysis will most certainly lead to a result that does not reflect the true physical nature of the problem, or the analysis will not be possible at all because the governing equations are not in harmony with the problem statement or assumptions. When using a governing equation to a solve a problem, the engineer must ascertain that the equation being used *actually* applies to the specific problem at hand. As an extreme (and probably absurd) example, imagine an engineer attempting to use Newton's second law, $F = ma$, to calculate the heat loss from a boiler? How about trying to apply Ohm's law, $V = IR$, to find the stress in a concrete column that supports a bridge deck? The problem of matching governing equations to the problem at hand is usually more subtle. In thermodynamics, for example, the engineer must determine whether the thermal system is "closed" or "open" (i.e., whether the system allows mass to cross the system boundary). After the type of thermal system has been identified, the thermodynamic equations which apply to that type of system are chosen, and the analysis proceeds. Governing equations must also be consistent with the assumptions. It is counterproductive to invoke simplifying assumptions if the governing equations do not make allowances for them. Some governing equations, particularly those that are experimentally derived, have built-in restrictions that limit the use of the equations to specific numerical values of key variables. A common mistake made in the application of a governing equation in this situation is failing to recognize the restrictions by forcing the equation to accept numerical values that lie outside the equation's range of applicability.

5. Calculations   A common practice, particularly among beginning students, is to substitute numerical values of quantities into equations *too early* in the calculations. It seems that students are more comfortable working with *numbers* than *algebraic variables*, so their first impulse is to substitute numerical values for all parameters at the beginning of the calculation. Avoid this impulse. To the extent that it is practical, develop the solution *analytically* prior to assigning physical quantities their numerical values. Before rushing to "plug" numbers into equations, carefully examine the equations to see if they can be mathematically manipulated to yield simpler expressions. A variable from one equation can often be substituted into another equation to reduce the total number of variables. Perhaps an expression can be simplified by factoring. By developing the solution analytically first, you might uncover certain physical characteristics about the system or even make the problem easier to solve. The analytical skills you learned in your algebra, trigonometry and calculus courses are meant to be used for performing mathematical operations on *symbolic* quantities, not numbers. When doing engineering analysis, do not put your mathematics skills on a shelf to collect dust—*use* them.

The calculations step demands more of an engineer than the ability to simply "crunch numbers" on a calculator or computer. The numbers have to be meaningful, and the equations containing the numbers must be fully understood and properly used. All mathematical relations must be dimensionally consistent, and all physical quantities must have a numerical value plus the correct units. Here is a tip concerning units that will save you time and help you avoid mistakes: *If the quantities given in the problem statement are not expressed in terms of a consistent set of units, convert all quantities to a consistent set of units before performing any calculations.* If some of the input parameters are expressed as a mixture of SI units and English units, convert all parameters to either SI units or English units, and then perform the calculations. Students tend to make more mistakes when they attempt to perform unit conversions *within* the governing equations. If all unit conversions are done prior to substituting numerical values into the equations, unit consistency is assured throughout the remainder of the calculations, because a consistent set of units is established at the onset. Dimensional consistency should still be verified, however, by substituting all quantities along with their units into the governing equations.

6. *Solution check* This step is perhaps the easiest one to overlook. Even good engineers sometimes neglect to thoroughly check their solution. The solution may "look" good at first glance, but a mere glance is not good enough. Much effort has gone into formulating the problem statement, constructing diagrams of the system, determining the appropriate number and type of assumptions, invoking governing equations and performing a sequence of calculations. All this work may be for naught if the solution is not carefully checked. Checking the solution of an engineering analysis is analogous to checking the operation of an automobile immediately following a major repair. It's always a good idea if the mechanic checks the overhauled transmission to verify that it works before returning the vehicle to its owner.

There are two main aspects of the solution check. First, the result itself should be checked. Ask the question, "Is this result reasonable?" There are several ways to answer this question. The result must be consistent with the information given in the problem statement. For example, suppose you wish to calculate the temperature of a microprocessor chip in a computer. In the problem statement, the ambient air temperature is given as 25°C, but your analysis indicates that the chip temperature is only 20°C. This result is not consistent with the given information because it is physically impossible for a heat-producing component, a microprocessor chip in this case, to have a lower temperature than the surrounding environment. If the answer had been 60°C, it is at least consistent with the problem statement, but it may still be incorrect. Another way to check the result is to compare it with that of similar analyses performed by you or other engineers. If the result of a similar analysis is not available, an alternative analysis that utilizes a different solution approach may have to be conducted. In some cases, a laboratory test may be needed to verify the solution experimentally. Testing is a normal part of engineering design anyway, so a test to verify an analytical result may be customary.

The second aspect of the solution check is a thorough inspection and review of each step of the analysis. Returning to our microprocessor example, if no mathematical or numerical errors are committed, the answer of 60°C may be considered correct insofar as the calculations are concerned, but the answer could still be in error. How? By applying bad assumptions. For example, suppose that the microprocessor chip is air cooled by a small fan, so we assert that forced convection is the dominant mechanism by which heat is transferred from the chip. Accordingly, we assume that conduction and radiation heat transfer are negligible, so we do not include these mechanisms in the analysis. A temperature of 60°C seems a little high, so we revise our assumptions. A second analysis that includes conduction and radiation reveals that the microprocessor chip is much cooler, about 42°C. Knowing whether assumptions are good or bad comes through increased knowledge of physical processes and practical engineering experience.

7. *Discussion* This step is valuable from the standpoint of communicating to others what the results of the analysis mean. By discussing the analysis, you are in effect writing a "minitechnical report." This report summarizes the major conclusions of the analysis. In the microprocessor example given earlier, the main conclusion may be that 42°C is below the recommended operating temperature for the chip, and therefore, the chip will operate reliably in the computer for a minimum of 10,000 hours before failing. If the chip temperature was actually measured at 45°C shortly after performing the analysis, the discussion might include an examination of why the predicted and measured temperatures differ and particularly why the predicted temperature is lower than the measured temperature. A brief parametric study may be included that shows how the chip temperature varies as a function of ambient air temperature. The discussion may even include an entirely separate analysis that predicts the chip temperature in the event of a fan failure. In the discussion step, the engineer is given one last opportunity to gain additional insights into the problem.

PROFESSIONAL SUCCESS: REAL-WORLD PROBLEM STATEMENTS

Engineering programs strive to give students a sense of what is it like to actually practice engineering in the "real world." But *studying* engineering in school and *practicing* engineering in the real world are not the same thing. One difference is amply illustrated by considering the origins of problem statements for analysis. In school, problem statements are typically found at the end of each chapter of your engineering texts. (The answers to many of these problems are even provided at the back of the book.) Sometimes your professors obtain problem statements from other texts or invent new ones (especially for exams). In any case, problem statements are supplied to you in a nice, neat little package all ready for you to tackle the problem. If textbooks and professors supply problem statements to students in school, who or what supplies problem statements to practicing engineers in industry? Real-world engineering problems are not typically found in textbooks (answers are never found in the back of the book, either), and your engineering professors are not going to follow you around after you graduate. So, where do the real-world problem statements come from? They are *formulated* by the engineer who is going to perform the analysis. As stated before, analysis is an integral part of engineering design. As a design matures, quantitative parameters that characterize the design begin to emerge. When an analysis is called for, these parameters are woven into a problem statement from which an analysis may be conducted. The engineer must be able to formulate a coherent, logical problem statement from the design information available. Because engineering design is an iterative process, the values of some or all of the input parameters may be uncertain. The engineer must therefore be able to write the problem statement in such a way as to allow for these uncertainties. The analysis will have to be repeated several times until the parameters are no longer in a state of flux, at which time the design is complete.

The seven-step procedure for performing an engineering analysis is a time-tested method. In order to effectively communicate an analysis to others, the analysis must be presented in a format that can be readily understood and followed. Engineers are known for their ability to present analyses and other technical information with clarity in a thorough, neat, and careful manner. As an engineering student, you can begin to develop this ability by consistently applying the analysis procedure outlined in this section. Your engineering professors will insist that you follow the procedure, or a procedure similar to it, in your engineering courses. You will probably be graded not only on how well you perform the analysis itself, but how well you *present* the analysis on paper. This grading practice is meant to convince students of the importance of presentation standards in engineering and to assist them in developing good presentation skills. An engineering analysis is of little value to anyone unless it can be read and understood. A good analysis is one that can be easily read by others. If your analysis resembles "hen scratchings" or "alien hieroglyphics" that require an interpreter, the analysis is useless. Apply the presentation guidelines given in this section to the point where they become second nature. Then, after you graduate and begin practicing engineering, you can hone your presentation skills as you gain industrial experience.

The 10 guidelines that follow will help you present an engineering analysis in a clear and complete manner. These guidelines are applicable to analysis work in school as well as industrial engineering practice. It should be noted that the guidelines apply specifically to analyses performed by hand with the use of pencil and paper, as opposed to computer-generated analyses.

Analysis Presentation Guidelines

1. A standard practice of engineers who do analysis is to use a special type of *paper*. This paper is usually referred to as "engineer calculation pad" or "engineer's computation paper." The paper is light green in color, and should be available in your college or university bookstore. The back side of the paper is ruled horizontally and vertically with five squares per inch, with only heading and margin rulings on the front side. The rulings on the back side are faintly visible through the paper to help the engineer maintain the proper position and orientation for lettering, diagrams and graphs. (See Figure 3.4.) All work is to be done on the *front* side of the paper. The back side is not used. The paper usually comes prepunched with a standard three-hole pattern at the left edge for placement in a three-ring binder.

Front side   Back side

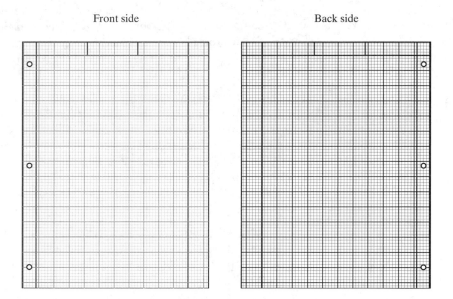

Figure 3.4. Engineer's computation paper is standard issue for analysis work.

2. No more than *one* problem should be placed on a page. This practice helps maintain clarity by keeping different problems separate. Even if a problem occupies a small fraction of a page, the next problem should be started on a separate page.

3. The *heading* area at the top of the page should indicate your name, date, course number, and assignment number. The upper right corner of the heading area is usually reserved for page numbers. To alert the reader to the total number of pages present, page numbers are often reported, for example, as "1/3", which is read as "page 1 of 3." Page 1 is the current page, and there are a total of three pages. When multiple pages are used, they should be stapled in the upper left corner. Each page should nonetheless be identified with your name, in the unlikely event the pages become separated.

4. The *problem statement* should be written out completely, not summarized or condensed. All figures that accompany the problem statement should be shown. If the problem statement originates from a textbook, it should be

written *verbatim* so the reader does not have to refer back to the textbook for the full version. One way to do this is to photocopy the problem statement, along with any figures given, and then cut and paste it by using rubber cement directly beneath the heading area on the engineer's computation paper. The problem statement could also be electronically scanned and printed directly onto the paper.

5. Work should be done in *pencil*, not ink. Everyone makes mistakes. If the analysis is written in pencil, mistakes can be easily erased and corrected. If the analysis is written in ink, mistakes will have to be crossed out, and the presentation will not have a neat appearance. To avoid smudges, use a pencil lead with the appropriate hardness. All markings should be dark enough to reproduce a legible copy if photocopies are needed. If you still use a standard wooden pencil, throw it out. Mechanical pencils are superior. They do not require sharpening, contain several months worth of lead, have replaceable erasers, produce no waste, and come in a range of lead diameters to suit your own writing needs. Mechanical pencils are also durable. (I have been using the same mechanical pencil since 1977!)

6. Lettering should be *printed*. The lettering style should be consistent throughout.

7. Correct *spelling* and *grammar* must be used. Even if the technical aspects of the presentation are flawless, the engineer will lose some credibility if the writing is poor.

8. There are seven steps in the general analysis procedure. These steps should be sufficiently *spaced* so that the reader can easily follow the analysis from problem statement to discussion. A horizontal line drawn across the page is one way of providing this separation.

9. Good *diagrams* are a must. A straight edge, drawing templates and other manual drafting tools should be used. All pertinent quantitative information such as geometry, forces, energy flows, mass flows, electrical currents, pressures, etc., should be shown on the diagrams.

10. Answers should be *double underlined* or *boxed* for ready identification. To enhance the effect, colored pencils may be used.

These 10 guidelines for analysis presentation are recommended to the engineering student. You may find that your particular engineering department or professors may advocate guidelines that are slightly different. By all means, follow the guidelines given to you. Your professors may have special reasons for teaching their students certain methods of analysis presentation. Methods may vary somewhat from professor to professor, but should still reflect the major points contained in the guidelines given in this section.

The next four examples illustrate the general analysis procedure and the recommended guidelines for analysis presentation. Each example represents a basic analysis taken from the subject areas of statics, electrical circuits, thermodynamics, and fluid mechanics. You probably have not yet taken courses in these subjects, so do not be overly concerned if you do not understand all the technical aspects of the examples. Therefore, do not focus on the theoretical and mathematical details. Focus instead on how the general analysis procedure is used to solve problems from different engineering areas and the systematic manner in which the analyses are presented.

EXAMPLE 3.4

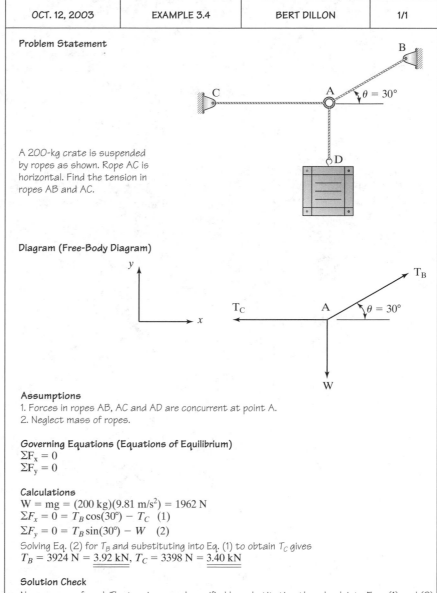

| OCT. 12, 2003 | EXAMPLE 3.4 | BERT DILLON | 1/1 |

Problem Statement

A 200-kg crate is suspended by ropes as shown. Rope AC is horizontal. Find the tension in ropes AB and AC.

Diagram (Free-Body Diagram)

Assumptions
1. Forces in ropes AB, AC and AD are concurrent at point A.
2. Neglect mass of ropes.

Governing Equations (Equations of Equilibrium)
$\Sigma F_x = 0$
$\Sigma F_y = 0$

Calculations
$W = mg = (200 \text{ kg})(9.81 \text{ m/s}^2) = 1962 \text{ N}$
$\Sigma F_x = 0 = T_B \cos(30°) - T_C$ (1)
$\Sigma F_y = 0 = T_B \sin(30°) - W$ (2)
Solving Eq. (2) for T_B and substituting into Eq. (1) to obtain T_C gives
$T_B = 3924 \text{ N} = \underline{3.92 \text{ kN}}$, $T_C = 3398 \text{ N} = \underline{3.40 \text{ kN}}$

Solution Check
No errors are found. The tensions can be verified by substituting them back into Eqs. (1) and (2):

$3924 \cos(30°) - 3398 = 0.3 \approx 0$
$3924 \sin(30°) - 1962 = 0$

The negligible nonzero result in Eq. (1) is due to roundoff.

Discussion
As θ increases, T_B increases and T_C decreases. When $\theta = 90°$,
$T_C = 0$ (rope AC is slack) and $T_B = W = 1962 \text{ N}$.

EXAMPLE 3.5

JAN. 03, 2004	EXAMPLE 3.5	MARIE NORTON	1/2

Problem Statement

Two resistors with resistances of $5\,\Omega$ and $50\,\Omega$ are connected in parallel across a $10\,V$ battery. Find the current in each resistor.

Diagram (Electrical Schematic)

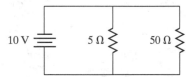

Assumptions

1. Neglect resistance of wires.
2. Battery voltage is a constant $10\,V$.

Governing Equations (Ohm's law)

$$V = IR \qquad \begin{aligned} V &= \text{Voltage (V)} \\ I &= \text{Current (A)} \\ R &= \text{Resistance } (\Omega) \end{aligned}$$

Calculations

Rearranging Ohm's law: $I = \dfrac{V}{R}$.

Define: $R_1 = 5\ \Omega, R_2 = 50\ \Omega$
Because resistors are connected in parallel with battery,
$V = V_1 = V_2 = 10\ V$.

$$\therefore I_1 = \frac{V_1}{R_1} = \frac{10\ V}{5\ \Omega} = \underline{\underline{2\ A}}, I_2 = \frac{V_2}{R_2} = \frac{10\ V}{50\ \Omega} = \underline{\underline{0.2\ A}}$$

Solution Check (no errors found)

JAN. 03, 2004	EXAMPLE 3.5	MARIE NORTON	2/2

Discussion

Current flow in a resistor is inversely proportional to the resistance.
Total current is split according to the ratio of resistances:

$$\frac{I_1}{I_2} = \frac{R_2}{R_1} = \frac{2\ A}{0.2\ A} = \frac{50\ \Omega}{5\ \Omega} = 10$$

Total current:

$$\begin{aligned} I_T &= I_1 + I_2 \\ &= 2\ A + 0.2\ A = 2.2\ A \end{aligned}$$

Total current may also be found by finding total resistance and then using Ohm's law.

Resistors in parallel as follows:

$$R_T = \frac{1}{\dfrac{1}{R_1} + \dfrac{1}{R_2}} = \frac{1}{\dfrac{1}{5} + \dfrac{1}{50}}$$

$$R_T = 4.5455\ \Omega$$

$$I_T = \frac{V}{R_T} = \frac{10\ V}{4.5455\ \Omega} = 2.2\ A$$

EXAMPLE 3.6

MAR. 24, 2004	EXAMPLE 3.6	CY BRAYTON	1/2

Problem Statement

A classroom occupied by 50 students is to be air conditioned with window-mounted air conditioning units with a 4-kW rating. There are 20 florescent lights in the room, each rated at 60 W. While sitting at their desks, each student dissipates 100 W. If the heat transfer to the classroom through the roof, walls and windows is 5 kW, how many air conditioning units are required to maintain the classroom at a constant temperature of 22°C?

Diagram (Thermodynamic System)

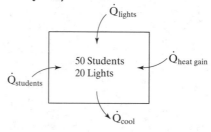

Assumptions

1. Classroom is a closed system, i.e., no mass flows.
2. All heat flows are steady.
3. No other heat sources in classroom such as computers, TVs, etc.

Governing Equations (Conservation of Energy)

$$\dot{E}_{in} - \dot{E}_{out} = \Delta E_{system}$$

Calculations

$$\dot{E}_{in} = \dot{Q}_{students} + \dot{Q}_{lights} + \dot{Q}_{heat\ gain}$$
$$= (50)(100\text{ W}) + (20)(60\text{ W}) + 5000\text{ W} = 11{,}200\text{ W} = 11.2\text{ kW}$$

$\Delta E_{system} = 0$ (*Classroom is maintained at constant temperature*)

$$\dot{E}_{in} - \dot{E}_{out} = \dot{Q}_{cool}$$

$$\text{Number of A.C. units required} = \frac{\dot{Q}_{cool}}{4\text{ kW}} = \frac{11.2\text{ kW}}{4\text{ kW}} = 2.8$$

Fractions of A.C. units are impossible, so round up answer to next integer.

MAR. 24, 2004	EXAMPLE 3.6	CY BRAYTON	2/2

Number of A.C. units required = <u>3.</u>

Solution check (no errors found)

Discussion

The classroom temperature of 22°C was not used in the calculation because this temperature, as well as the outdoor air temperature, are inferred in the given heat gain by a prior heat transfer analysis.

Suppose that the classroom was a computer lab containing 30 computers each dissipating 250 W. We eliminate assumption 3 by including heat input by the computers.

$$\dot{Q}_{cool} = \dot{Q}_{students} + \dot{Q}_{lights} + \dot{Q}_{heat\ gain} + \dot{Q}_{computers}$$
$$= 11{,}200\text{ W} + 30(250\text{ W}) = 18{,}700\text{ W} = 18.7\text{ kW}$$

$$\text{Number of A.C. units required} = \frac{\dot{Q}_{cool}}{4\text{ kW}} = \frac{18.7\text{ kW}}{4\text{ kW}} = 4.7$$

Number of A.C. units required = <u>5.</u>

This example illustrates the effect computers have on air-conditioning requirements.

EXAMPLE 3.7

MAY 17, 2004	EXAMPLE 3.7	EDDIE POWERS	1/2

Problem Statement

Water enters a pipe junction at a mass flow rate of 3.6 kg/s. If the mass flow rate in the small branch is 1.4 kg/s, what is the mass flow rate in the large pipe branch? If the inside diameter of the large pipe branch is 5 cm, what is the velocity in the large pipe branch?

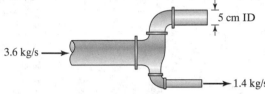

3.6 kg/s ⟶

5 cm ID

⟶ 1.4 kg/s

Diagram (Flow Schematic)

$\dot{m}_2$

$\dot{m} = 3.6$ kg/s ⟶

$\dot{m}_1 = 1.4$ kg/s

Assumptions
1. Steady, incompressible flow
2. Density of water: $\rho = 1000$ kg/m^3

Governing Equations

Conservation of mass: $\dot{m}_{in} = \dot{m}_{out}$

mass flow rate: $\dot{m} = \rho A v$

$\dot{m}$ = mass flow rate (kg/s)
ρ = fluid density (kg/m^3)
A = flow cross-sectional area (m^2)
v = velocity (m/s)

Calculations

$\dot{m} = \dot{m}_1 + \dot{m}_2$
$\dot{m}_2 = \dot{m} - \dot{m}_1 = 3.6$ kg/s $- 1.4$ kg/s
$\qquad\qquad = 2.2$ kg/s

MAY 17, 2004	EXAMPLE 3.7	EDDIE POWERS	2/2

$$\dot{m}_2 = \rho A_2 v_2 = \rho \frac{\pi D_2^2}{4} v_2$$

$$v_2 = \frac{4 \dot{m}_2}{\pi \rho D_2^2} = \frac{4 (2.2 \text{ kg/s})}{\pi (1000 \text{ kg/m}^3)(0.05 \text{ m})^2}$$

$$= \underline{1.12 \text{ m/s}}$$

Solution check (no errors found)

Discussion
The velocity calculated is an average value because there is a velocity profile across the pipe. The velocity profile is caused by viscosity. If the flow condition is laminar, the velocity profile is parabolic, as shown in the following sketch.

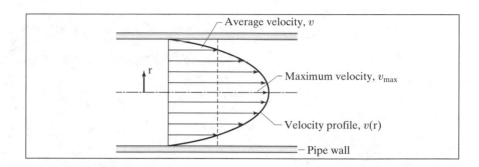

A good engineer is a person who solves an engineering analysis problem by reasoning through it, rather than simply following a prepared "recipe" consisting of step-by-step instructions written by someone else. Similarly, a good engineering student is a person who learns engineering analysis by thinking conceptually about each problem, rather than simply memorizing a collection of disjointed solution sequences and mathematical formulas. This "cookbook" learning approach is a detour on the road of engineering education. Furthermore, the cookbook learning style promotes fragmented rather than integrative learning. A student who embraces this type of learning method will soon discover that it will be difficult and take a long time to solve new engineering problems, unless identical or very similar problems have been previously solved by using an established recipe. An analogy may be drawn from the familiar maxim "Give a man a fish, and you have fed him for a day. Teach a man to fish, and you have fed him for a lifetime." A recipe enables a student to solve only one specific type of problem, whereas a more general conceptual-based learning approach enables a student to solve many engineering problems.

PRACTICE!

Use the general analysis procedure to solve the following problems (present the analysis by using the guidelines for analysis presentation covered in this section):

1. Radioactive waste is to be permanently encased in concrete and buried in the ground. The vessel containing the waste measures 30 cm $\times$ 30 cm $\times$ 80 cm. Federal regulations dictate that there must be a minimum concrete thickness of 50 cm surrounding the vessel on all sides. What is the minimum volume of concrete required to safely encase the radioactive waste?
    Answer: 2.97 m^3

2. An elevator in an office building has an operating capacity of 15 passengers with a maximum weight of 180 lb$_f$ each. The elevator is suspended by a special pulley system with four cables, two of which support 20 percent of the total load and two of which support 80 percent of the total load. Find the maximum tension in each elevator cable.
    Answer: 270 lb$_f$, 1080 lb$_f$

3. A technician measures a voltage drop of 25 V across a 100-Ω resistor by using a digital voltmeter. Ohm's law states that $V = IR$. What is the current flow through the resistor? How much power is consumed by the resistor? (*Hint*: $P = I^2R$.)
    Answer: 250 mA, 6.25 W

4. Air flows through a main duct at a mass flow rate of 4 kg/s. The main duct enters a junction that splits into two branch ducts, one with a cross section of

20 cm $\times$ 30 cm and one with a cross section of 40 cm $\times$ 60 cm. If the mass flow rate in the large branch is 2.8 kg/s, what is the mass flow rate in the small branch? If the density of air is $\rho = 1.16$ kg/m^3, what is the velocity in each branch?

Answer: 1.2 kg/s, 10.1 m/s, 17.2 m/s

3.4 THE COMPUTER AS AN ANALYSIS TOOL

Computers are an integral part of the civilized world. They affect virtually every aspect of our everyday lives, including communications, transportation, financial transactions, information processing, food production, and health care, among others. The world is a much different place today than it was prior to the advent of computers. People use computers for accessing and processing information, word processing, electronic mail, entertainment, and on-line shopping. Like everyone else, engineers use computers in their personal lives in the same ways just mentioned, but they also depend heavily on computers in their professional work. To the engineer, the computer is an indispensable tool. Why do engineers need computers? Without the computer, engineers would not be able to do their work accurately or efficiently. The primary advantage of the computer to engineers is its ability to perform various functions extremely rapidly. For example, a complex sequence of calculations that would take days with a slide rule can be carried out in a few seconds by a computer. Furthermore, the numerical precision of the computer enables engineers to make calculations that are much more accurate. Engineers use computers for computer-aided design (CAD), word processing, communications, information access, graphing, process control, simulation, data acquisition, and, of course, analysis.

The computer is one of the most powerful analysis tools available to the engineer, but the computer does not replace the engineer's thinking. When faced with a new analysis, the engineer must reason through the problem by using sound scientific principles, applied mathematics, and engineering judgement. A computer is only a machine, and, as yet, no machine has been developed that can outthink a human being (except at playing chess, perhaps). A computer can only carry out the instructions supplied to it, but it does so with remarkable speed and efficiency. A computer yields wrong answers just as quickly as it yields right ones. The burden is upon the engineer to supply the computer with correct input. An often-used engineering acronym is *GIGO* (*Garbage In, Garbage Out*), which refers to a situation in which erroneous input data is supplied to a computer, thereby producing erroneous output. When GIGO is at work, the calculations are numerically correct, but the results of those calculations are meaningless, because the engineer supplied the computer with bad input, or the computer program that the engineer wrote is flawed. The computer is capable of accurately performing enormous numbers of computations in a very short time, but it is incapable of composing a problem statement, constructing a diagram of the engineering system, formulating assumptions, selecting the appropriate governing equations, checking the reasonableness of the solution, or discussing and evaluating the results of the analysis. Thus, the only step in the analysis procedure for which a computer is perfectly suited is step 5: calculations. This is not to say that a computer cannot be used to write problem statements, assumptions and equations, as well as draw diagrams. These steps may also be performed by using the computer, but they must be developed by the engineer, whereas calculations are performed automatically once the equations and numerical inputs are supplied.

Engineers use analysis primarily as a design tool and as a means of predicting or investigating failures. Specifically, how does an engineer use the computer to perform an analysis? Steps 1 through 4 and steps 6 and 7 of the analysis procedure are largely unchanged, whether a computer is employed or not. So, exactly how are the calculations in step 5 carried out on a computer? There are basically five categories of computer tools for doing engineering analysis work:

1. Spreadsheets
2. Equation solvers and mathematics software
3. Programming languages
4. Specialty software
5. Finite element software

3.4.1 Spreadsheets

The term **spreadsheet** originally referred to a special type of paper, divided into rows and columns, for doing financial calculations. The computer-based spreadsheet is a modern electronic version of the paper spreadsheet and was initially used for business and accounting applications. By virtue of their general structure, spreadsheets are useful not only for doing financial calculations, but can also be used for performing a variety of scientific and engineering calculations. Like the original paper version, the computer-based spreadsheet consists of any array of rows and columns. The intersection of a row with a column is called a *cell*. Cells serve as locations for input and output data such as text, numbers, or formulas. For example, a cell may contain an equation representing Newton's second law of motion, $F = ma$. A nearby cell would contain a number for the mass, m, while another cell would contain a number for the acceleration, a. Immediately after entering these two input values in their respective cells, the spreadsheet automatically evaluates the formula, inserting the numerical value of the force, F, in the cell containing the formula for Newton's second law. If the values of the mass or acceleration are changed, the spreadsheet automatically updates the value of the force. This example is very simple, but spreadsheets are capable of doing calculations that involve hundreds or even thousands of variables. Suppose that our analysis involves 100 variables and that we want to know how changing only *one* of those variables affects the solution. We simply change the variable of interest and the entire spreadsheet automatically updates all calculations to reflect the change. The spreadsheet is an excellent analysis tool for rapidly answering "what if" questions. Numerous design alternatives can be efficiently investigated by performing the analysis on a spreadsheet. In addition to numerical functions, spreadsheets also have graphics capabilities. Excel,[1] Quattro Pro,[2] and Lotus 1-2-3[3] are popular spreadsheet products. Figure 3.5 shows a simple example of calculating force using Newton's second law by using Excel.

3.4.2 Equation Solvers and Mathematics Software

Equation solvers and **mathematics software** packages are general-purpose scientific and engineering tools for solving equations and performing symbolic mathematical operations. Equation solvers are primarily designed for solving problems that involve *numerical* inputs and outputs, whereas mathematics packages are primarily suited for performing *symbolic* mathematical operations much like you would do in a mathematics course. Equation solvers accept a set of equations that represent the mathematical model of the analytical problem. The equations can be linear or nonlinear. The equations may be written in their original form without prior mathematical manipulation to isolate the unknown quantities on one side of the equals sign. For example, Newton's second law would be written in its original form as $F = ma$ even if the unknown quantity was the acceleration a. Solving this problem by hand, however, we would have to write the equation as $a = F/m$ because we are solving for the acceleration. This is not necessary when we use equation solvers.

---

[1]Excel is a registered trademark of Microsoft® Corporation.
[2]Quattro® Pro is a registered trademark of Corel® Corporation.
[3]Lotus 1-2-3 is a registered trademark of Lotus® Development Corporation part of IBM®.

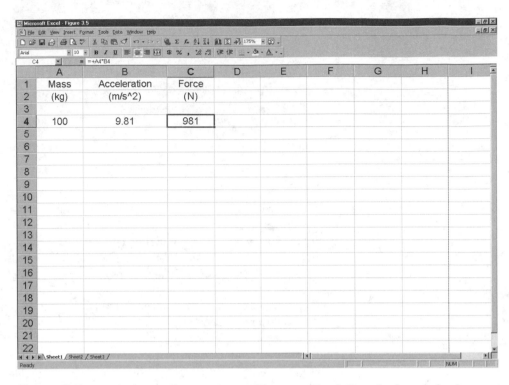

Figure 3.5. A calculation of Newton's second law using Excel. Note the formula for the force, + A4*B4, entered in cell C4.

After we supply the numerical values for the known quantities, equation solvers solve for the remaining unknown values. Equation solvers have a large built-in library of functions for use in trigonometry, linear algebra, statistics, and calculus. Equation solvers can perform a variety of mathematical operations, including differentiation, integration, and matrix operations. In addition to these mathematical features, equation solvers also do unit conversions. Equation solvers also have the capability of displaying results in graphical form. Programming can also be done within equation solvers. Although all equation solvers have some symbolic capabilities, some have the capacity for data acquisition, image analysis, and signal processing. Popular equation solvers are TK Solver,[4] Mathcad,[5] and Matlab.[6]

The strength of mathematics packages is their ability to perform symbolic mathematical operations. A symbolic mathematical operation is one that involves the manipulation of symbols (variables), using mathematical operators such as the vector product, differentiation, integration, and transforms. These packages are capable of performing very complex and sophisticated mathematical procedures. They also have extensive graphical capabilities. Even though mathematics packages are primarily designed for symbolic operations, they can also perform numerical computations. Mathematica[7] and Maple[8] are popular mathematics software products.

3.4.3 Programming Languages

Spreadsheets, equation solvers, and mathematics software packages may not always meet the computational demands of every engineering analysis. In such cases, engineers may

---

[4] TK Solver is a registered trademark of Universal Technical Systems, Incorporated.
[5] Mathcad® is a registered trademark of Mathsoft™, Incorporated.
[6] MATLAB® is a registered trademark of The MathWorks, Incorporated.
[7] MATHEMATICA® is a registered trademark of Wolfram Research, Incorporated.
[8] Maple™ is a registered trademark of Maplesoft™, a division of Waterloo Maple, Incorporated.

choose to write their own computer programs with the use of a programming language. **Programming languages** refer to sequential instructions supplied to a computer for carrying out specific calculations. Computer languages are generally categorized according to their level. *Machine language* is a low-level language, based on a binary system of "zeroes" and "ones." Machine language is the most primitive language, because computers are digital devices whose rudimentary logic functions are carried out by using solid state switches in the "on" or "off" positions. *Assembly language* is also a low-level language, but its instructions are written in English-like statements rather than binary. Assembly language does not have many commands, and it must be written specifically for the computer hardware. Computer programs written in low-level languages run very fast because these languages are tied closely to the hardware, but writing the programs is very tedious.

Due to the tediousness of writing programs in low-level languages, engineers usually write programs in high-level languages that consist of straightforward, English-like commands. The most commonly used high-level languages by engineers are Fortran, C, C++, Pascal, Ada, and BASIC. Fortran is the patriarch of all scientific programming languages. The first version of Fortran (FORmula TRANslation) was developed by IBM between 1954 and 1957. Since its inception, Fortran has been the workhorse of scientific and engineering programming languages. It has undergone several updates and improvements and is still in widespread use today. The C language evolved from two languages, BCPL and B, which were developed during the late 1960s. In 1972, the first C program was compiled. The C++ language grew out of C and was developed during the early 1980s. Both C and C++ are popular programming languages for engineering applications because they use powerful commands and data structures. Pascal was developed during the early 1970s and is a popular programming language for beginning computer science students who are learning programming for the first time. The U.S. Department of Defense prompted the development of Ada during the 1970s in order to have a high-level language suitable for embedded computer systems. BASIC (Beginner's All-purpose Symbolic Instruction Code) was developed during the mid-1960s as a simple learning tool for secondary school students as well as college students. BASIC is often included as part of the operating software for personal computers.

Writing programs in high-level languages is easier than writing programs in low-level languages, but the high-level languages utilize a larger number of commands. Furthermore, high-level languages must be written with specific grammatical rules, referred to as *syntax*. Rules of syntax govern how punctuation, arithmetic operators, parentheses, and other characters are used in writing commands. To illustrate the syntactical differences between programming languages, equation solvers and mathematics packages, Table 3.1 shows how a simple equation is written. Note the similarities and differences in the equals sign, the constant π, and the operator for exponentiation.

TABLE 3.1 Comparison of Computer Statements for the Equation, $V = 4/3\pi R^3$, the Volume of a Sphere

Computer Tool	Statement
Mathcad	$V := 4/3 {}^{*}\pi {}^{*}R^{\wedge}3$
TK Solver	$V = 4/3 {}^{*}pi() {}^{*}R^{\wedge}3$
MATLAB	$V = 4/3 {}^{*}pi {}^{*}R^{\wedge}3;$
MATHEMATICA	$V = 4/3 {}^{*}Pi {}^{*}R^{\wedge}3$
maple	$V := 4/3 {}^{*}pi {}^{*}R^{\wedge}3;$
Fortran	$V = 4/3 {}^{*}3.141593 {}^{*}R^{**}3$
C, C++	$V = 4/3 {}^{*}3.141593 {}^{*}pow(R,3);$
Pascal	$V := 4/3 {}^{*}3.141593 {}^{*}R {}^{*}R {}^{*}R;$
Ada	$V := 4/3 {}^{*}3.141593 {}^{*}R^{**}3;$
BASIC	$V = 4/3 {}^{*}3.141593 {}^{*}R^{\wedge}3$

3.4.4 Specialty Software

Considered as a whole, engineering is a broad field that covers a variety of disciplines and careers. Some of the main engineering disciplines are chemical engineering, civil engineering, electrical and computer engineering, environmental engineering, and mechanical engineering. Primary engineering career fields include research, development, design, analysis, manufacturing, and testing. Given the variety of specific problems that engineers who work in these fields encounter, it comes as no surprise that numerous *specialty software* packages are available to help the engineer analyze specific problems relating to a particular engineering system. For example, specialty software packages are available to electrical engineers for analyzing and simulating electrical circuits. Mechanical and chemical engineers can take advantage of software packages designed specifically for calculating flow parameters in pipe networks. Special software is available to civil and structural engineers for calculating forces and stresses in trusses and other structures. Other specialty software packages are available for performing analyses of heat exchangers, machinery, pressure vessels, propulsion systems, turbines, pneumatic and hydraulic systems, manufacturing processes, mechanical fasteners, and many others too numerous to list. After you graduate and begin working for a company that produces a specific product or process, you will probably become familiar with one or more of these specialty software packages.

3.4.5 Finite Element Software

Some engineering analysis problems are far too complex to solve using any of the aforementioned computer tools. *Finite element* software packages enable the engineer to analyze systems that have irregular configurations, variable material properties, complex conditions at the boundaries, and nonlinear behavior. The finite element method originated in the aerospace industry during the early 1950s when it was used for stress analysis of aircraft. Later, as the method matured, it found application in other analysis areas such as fluid flow, heat transfer, vibrations, impacts, acoustics, and electromagnetics. The basic concept behind the finite-element method is to subdivide a continuous region (i.e., the system to be analyzed is divided into a set of simple geometric shapes called "finite elements"). The elements are interconnected at common points called "nodes." Material properties, conditions at the system boundaries, and other pertinent inputs are supplied. With the use of an advanced mathematical procedure, the finite element software calculates the value of parameters such as stress, temperature, flow rate, or vibration frequency at each node in the region. Hence, the engineer is provided with a set of output parameters at discrete points that approximates a continuous distribution of those parameters for the entire region. The finite element method is an advanced analysis method and is normally introduced in colleges and universities at the senior level or the first-year graduate level.

PROFESSIONAL SUCCESS: PITFALLS OF USING COMPUTERS

The vital role that computers play in engineering analysis cannot be overstated. Given the tremendous advantages of using computers for engineering analysis, however, it may be difficult to accept the fact that there are also pitfalls. A common hazard that entangles some engineers is the tendency to treat the computer as a "black box," a wondrous electronic device whose inner workings are largely unknown, but that nonetheless provides output for every input supplied. Engineers who treat the computer as a black box are not effectively employing the general analysis procedure and in so doing are in danger of losing their ability to systematically reason their way through a problem. The computer is a remarkable computational machine, but it does not replace the engineer's thinking, reasoning, and judgement. Computers, and the software that runs on them, produce output that *precisely* reflects the input supplied to them. If the input is good, the output will be good. If the input is bad, the output will be bad. Computers are not smart enough to compensate for an engineer's inability to make good assumptions or employ the correct governing equations. Engineers must have a thorough understanding of the physical aspects of the problem at hand and the underlying mathematical principles *before* implementing the solution on the computer. A good engineer understands *what* the computer does when it "crunches the numbers" in the analysis. A good engineer is confident that the input data will result in reasonable output because a lot of sound thinking and reasoning has gone into the formulation of that input.

Can the computer be used too much? In a sense, it can. The tendency of some engineers is to use the computer to analyze problems that may not require a computer at all. Upon beginning a new problem, their first impulse is to set up the problem on the computer without even checking to see whether the problem can be solved by hand. For example, a problem in engineering statics may be represented by the quadratic equation, $x^2 + 4x - 12 = 0$.

This problem can be solved analytically by factoring, $(x + 6)(x - 2) = 0$, which yields the two roots, $x = -6$ and $x = 2$. To use the computer in a situation like this is to rely on the computer as a "crutch" to compensate for weak analytical skills. Continued reliance on the computer to solve problems that do not require a computer will gradually dull your ability to solve problems with pencil and paper. Do not permit this to happen. Examine the equations carefully to see whether a computer solution is justified. If it is, use one of the computer tools discussed earlier. If not, solve the problem by hand. Then, if you have time and wish to check your solution with the use of the computer, by all means do so.

APPLICATION: COMPUTERS FOR NUMERICAL ANALYSIS

Most of the equations that you will encounter in school can be solved analytically; that is, they can be solved by employing standard algebraic operations to isolate the desired variable on one side of the equation. Some equations, however, cannot be solved analytically with standard algebraic operations. These equations are referred to as *transcendental* equations because they contain one or more transcendental functions such as a logarithm or trigonometric function. Transcendental equations occur often in engineering analysis work, and techniques for solving them are known as *numerical methods*. For example, consider the transcendental equation

$$e^x - 3x = 0$$

This equation looks straightforward enough, but try solving it by hand. If we add $3x$ to both sides and take the natural logarithm of both sides to undo the exponential function, we obtain

$$x = \ln(3x) \tag{a}$$

which, unfortunately, does not isolate the variable, x, because we still have the term $\ln(3x)$ on the right side of the equation. If we add $3x$ to both sides and then divide both sides by 3, we obtain

$$\frac{e^x}{3} = x \tag{b}$$

The variable, x, is still not isolated without leaving a transcendental function in the equation. Clearly, this equation cannot be solved analytically, so it must be solved numerically. To solve it numerically, we utilize a method called *iteration*, a process by which we repeat the calculation until an answer is obtained.

Before solving this problem by using the computer, we will work it manually to illustrate how iteration works. To begin, we rewrite Equation (a) in the iterative form

$$x_{i+1} = \ln(3x_i)$$

The "i" and the "$i + 1$" subscripts refer to "old" and "new" values of x, respectively. The iteration process requires that we begin the calculation by immediately substituting a number into the iteration formula. This first number constitutes an estimate for the root (or roots) of the variable, x, that satisfy the formula. To keep track of the iterations, we use an iteration table, illustrated in Table 3.2. To start the iterations, we estimate a value of x by letting $x_i = 1$. We now substitute this number into the right side of the formula, yielding a new value of $x_{i+1} = 1.098612$. We then assign this new value of x to the old variable, x_i, and substitute it into the right side of the formula, yielding the second new value, $x = 1.192660$. Substituting this number into the right side of the formula, we obtain the third new value,

TABLE 3.2 Iteration Table for Finding One Root of the Equation $e^x - 3x = 0$

Iteration	x_i	x_{i+1}
1	1	1.098612
2	1.098612	1.192660
3	1.192660	1.274798
4	1.274798	1.341400
5	1.341400	1.392326
.		
.		
.		1.512134
41	1.512134	1.512135

$x_{i+1} = 1.274798$. This process is repeated until the value of x stops changing by the desired amount. At this point, we say that the calculation has *converged* to an answer. Table 3.2 shows the first five iterations and indicates that 41 iterations are required for the calculation to converge to an answer that is accurate to the sixth decimal place. Upon substituting $x = 1.512135$ into the original equation, we see that the equation is satisfied. As this example illustrates, numerous iterations may be required to obtain an accurate solution. The accuracy of the answer depends on how many iterations are taken. Some equations converge to a precise answer in a few iterations, but others, like this one, require several iterations. It is important to note that 1.512135 is not the only root of this equation. The equation has a second root at $x = 0.619061$. If we attempt to find this root by using Equation (a), we discover that our calculation either converges again to 1.512135 or does not converge at all by leading us to an illegal operation; that is, taking the logarithm of a negative number. To find the second root, we iterate on Equation (b), writing it in the iterative form

$$x_{i+1} = \frac{e^{x_i}}{3}$$

With numerical methods, there are often no guarantees that a certain iteration formula will converge rapidly or even converge at all. The success of the iteration formula may also depend on the initial estimate chosen to start the iterations. If our initial estimate for Equation (a) is less than $\frac{1}{3}$, the new value of x immediately goes negative, leading to an illegal operation. If our initial estimate for Equation (b) is too large, the new value of x grows large very rapidly, leading to an exponential overflow. These and other kinds of numerical difficulties can occur whether the iterations are performed by hand or by using a computer.

As Table 3.2 suggests, performing iterations by hand can be a long and tedious task. The computer is tailor-made for performing repetitive calculations. The roots of our transcendental equation can readily be found by using one of the computer tools discussed earlier. Figure 3.6 shows a computer program, written in the BASIC language, for finding the first root, $x = 1.512135$. In the first line the user inputs an initial estimate, which is assigned the variable name XOLD. The program then executes what is referred to as a DO loop that performs the iterations. Each time through the loop, a new value of x is calculated from the old value and an absolute value of the difference between the old and new values is calculated. This value is called DIFF. While DIFF is larger than a preselected convergence tolerance of 0.0000001, the new value of x, XNEW, is reset to the old value, XOLD, and looping continues. When DIFF is less than, or equal to, the convergence tolerance, convergence has been achieved, and looping is halted. The root is then printed. The same program, with the third line replaced with XNEW = EXP(XOLD)/3, could be used to find the second root. There are more sophisticated numerical methods for finding roots than the simple iteration technique illustrated here, and you will study them in your engineering or mathematics courses.

```
INPUT "ESTIMATE = ", XOLD
DO
    XNEW = LOG (3*XOLD)
    DIFF = ABS (XNEW - XOLD)
    XOLD = XNEW
LOOP WHILE DIFF > 0.0000001
PRINT XNEW
END
```

Figure 3.6. BASIC computer program for finding one root of the equation $e^x - 3x = 0$.

PRACTICE!

Using one of the computer tools discussed in this section, work the following problems:
(Note: These problems are identical to those in Section 3.3.)

1. Radioactive waste is to be permanently encased in concrete and buried in the ground. The vessel containing the waste measures 30 cm × 30 cm × 80 cm.

Federal regulations dictate that there must be a minimum concrete thickness of 50 cm surrounding the vessel on all sides. What is the minimum volume of concrete required to safely encase the radioactive waste?

Answer: 2.97 m^3

2. An elevator in an office building has an operating capacity of 15 passengers with a maximum weight of 180 lb$_f$ each. The elevator is suspended by a special pulley system with four cables, two of which support 20 percent of the total load and two of which support 80 percent of the total load. Find the maximum tension in each elevator cable.

Answer: 270 lb$_f$, 1080 lb$_f$

3. A technician measures a voltage drop of 25 V across a 100-Ω resistor by using a digital voltmeter. Using Ohm's law, we find that $V = IR$. What is the current flow through the resistor? How much power is consumed by the resistor? (*Hint*: $P = I^2R$.)

Answer: 250 mA, 6.25 W

4. Air flows through a main duct at a mass flow rate of 4 kg/s. The main duct enters a junction that splits into two branch ducts, one with a cross section of 20 cm × 30 cm and one with a cross section of 40 cm × 60 cm. If the mass flow rate in the large branch is 2.8 kg/s, what is the mass flow rate in the small branch? If the density of air is $\rho = 1.16$ kg/m^3, what is the velocity in each branch?

Answer: 1.2 kg/s, 10.1 m/s, 17.2 m/s

KEY TERMS

engineering method
equation solver
general analysis procedure
mathematics software

order of magnitude
programming language
significant figure
spreadsheet

REFERENCES

Bahder, T.B., *Mathematica for Scientists and Engineers*, NY: Addison-Wesley, 1995.

Dubin, D., *Numerical and Analytical Methods for Scientists and Engineers Using Mathematica*, NY: John Wiley & Sons, 2003.

Etter, D.M. *Introduction to C++*, Upper Saddle River, NJ: Prentice Hall, 1999.

Etter, D.M. and Kuncicky, D.C. *Introduction to Matlab 6*, Upper Saddle River, NJ: Prentice Hall, 2004.

Ferguson, R.J., *TK Solver for Engineers*, NY: Addison-Wesley, 1996.

Kuncicky, D.C., *Introduction to Excel 2002*, Upper Saddle River, NJ: Prentice Hall, 2003.

Larsen, R.W., *Introduction to Mathcad 11*, Upper Saddle River, NJ: Prentice Hall, 2004.

Nyhoff, L. and S. Leestma, *Introduction to FORTRAN 90*, 2d ed., Upper Saddle River, NJ: Prentice Hall, 1999.

Schwartz, D.I., *Introduction to Maple 8*, Upper Saddle River, NJ: Prentice Hall, 2003.

Problems

1. Using an order-of-magnitude analysis, estimate the number of gallons of gasoline used by all automobiles in the United States each year.

2. Using an order-of-magnitude analysis, estimate the number of 4 ft × 8 ft plywood sheets required for the floor, roof, and exterior sheathing of a 3000-ft^2 house.

3. Using an order-of-magnitude analysis, estimate the number of basketballs (fully inflated) that would fit in your engineering classroom.

4. Using an order-of-magnitude analysis, estimate the number of spam e-mail messages received by residents of the United States each year.

5. Using an order-of-magnitude analysis, estimate the number of breaths you will take during your lifetime.

6. Use an order-of-magnitude analysis to estimate the number of short tons of human waste produced worldwide each year.

7. The earth has a mean radius of about 6.37×10^6 m. Assuming the earth is made of granite ($\rho = 2770$ kg/m^3), estimate the mass of the earth, using an order-of-magnitude analysis.

8. The solar radiation flux just outside the earth's atmosphere is about 1350 W/m^2. Using an order-of-magnitude analysis, estimate the amount of solar energy that is intercepted by the United States each year.

9. Using an order-of-magnitude analysis, estimate the total textbook expenditure incurred by all engineering majors at your school per year.

10. Underline the significant figures in the following numbers (the first number is done for you):
    a. <u>345</u>0
    b. 9.807
    c. 0.00216
    d. 5000
    e. 7000.
    f. 12.00
    g. 2066
    h. 106.07
    i. 0.02880
    j. 523.91
    k. 1.207×10^{-3}

11. Perform the following calculations, reporting the answers with the correct number of significant figures:
    a. (8.14)(260)
    b. 456/4.9
    c. (6.74)(41.07)/8.72
    d. (10.78 − 4.5)/300
    e. (10.78 − 4.50)/300.0
    f. (65.2 − 13.9)/240.0
    g. $(1.2 \times 10^6)/(4.52 \times 10^3 + 988)$
    h. $(1.764 - 0.0391)/(8.455 \times 10^4)$
    i. $1000/(1.003 \times 10^9)$

j. $(8.4 \times 10^{-3})/5000$

k. $(8.40 \times 10^{3})/5000.0$

l. 8π

m. $(2\pi - 5)/10$

12. A 250-kg mass hangs by a cable from the ceiling. Using the standard value of gravitational acceleration, $g = 9.81$ m/s^2, what is the tension in the cable? Express your answer with the correct number of significant figures.

13. A 9-slug mass hangs by a rope from the ceiling. Using the standard value of gravitational acceleration, $g = 32.2$ ft/s^2, what is the tension in the rope? Express your answer with the correct number of significant figures. Redo the problem, using a mass of 9.00 slug. Is the answer different? Why?

14. A 175 mA current flows through a 62-Ω resistor. Using Ohm's law, $V = IR$, what is the voltage across the resistor? Express your answer with the correct number of significant figures.

15. A rectangular building lot is reported to have the dimensions 200 ft $\times$ 300 ft. Using the correct number of significant figures, what is the area of this lot in units of acre?

    For problems 16 through 31, use the general analysis procedure of (1) problem statement, (2) diagram, (3) assumptions, (4) governing equations, (5) calculations, (6) solution check, and (7) discussion.

16. An excavation crew digs a hole in the ground measuring 20 yd $\times$ 30 yd $\times$ 6 yd to facilitate a basement for a small office building. Five dump trucks, each with a capacity of 20 yd^3, are used to haul the material away. How many trips must each truck make to remove all the material?

17. Find the current in each resistor and the total current for the circuit shown in Figure P3.17.

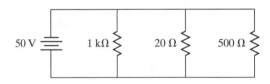

50 V   1 kΩ 20 Ω 500 Ω

Figure P3.17.

18. For easy handling, long sheets of steel for manufacturing automobile body panels are tightly rolled up into a cylinder-shaped package. Consider a roll of steel with an inside and outside diameter of 45 cm and 1.6 m, respectively, that is suspended by a single cable. If the length of the roll is 2.25 m and the density of steel is $\rho = 7850$ kg/m^3, what is the tension in the cable?

19. In a chemical processing plant, glycerin flows toward a pipe junction at a mass flow rate of 30 kg/s as shown in Figure P3.19. If the mass flow rate in the small pipe branch is 8 kg/s, find the velocity in both branches. The density of glycerin is $\rho = 1260$ kg/m^3.

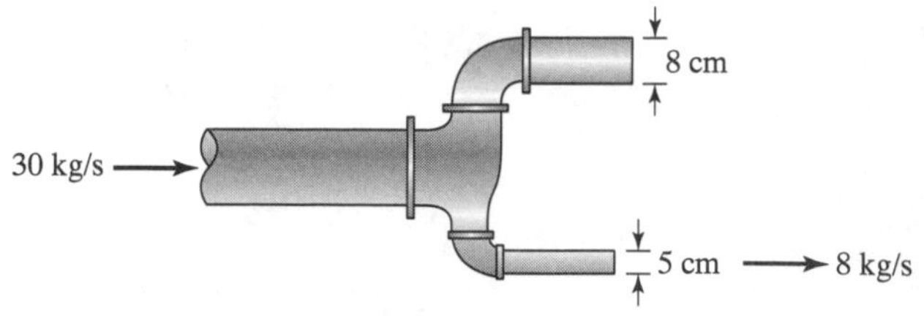

Figure P3.19.

20. A portable classroom is heated with small propane heating units with a capacity of 3 kW each. The portable classroom is occupied by 24 students, each dissipating 120 W, and is lighted by 10 light fixtures that dissipate 100 W each. If the heat loss from the portable classroom is 15 kW, how many heating units are required to maintain the classroom at a temperature of 20°C?

21. A man pushes on a barrel with a force of $P = 80 \, \text{lb}_f$ as shown. Assuming that the barrel does not move, what is the friction force between the barrel and the floor? (*Hint:* The friction force acts parallel to the floor toward the man. See Figure P3.21.)

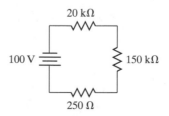

Figure P3.21.

22. The total resistance for resistors connected in series is the arithmetic sum of the resistances. Find the total resistance for the series circuit shown in Figure P3.22. Because the resistors are connected in series, the current is the same in each resistor. Using Ohm's law, find this current. Also, find the voltage drop across each resistor.

20 kΩ

100 V   150 kΩ

250 Ω

Figure P3.22.

23.   The pressure exerted by a static liquid on a vertical submerged surface is calculated from the relation

$$P = \rho g h$$

where

P = pressure (Pa)

ρ = density of the liquid (kg/m^3)

g = gravitational acceleration = 9.81 m/s^2

h = height of vertical surface that is submerged (m)

Consider the dam shown in Figure P3.23. What is the pressure exerted on the dam's surface at depths of 2 m, 6 m, and 20 m? For the density of water, use ρ = 1000 kg/m^3.

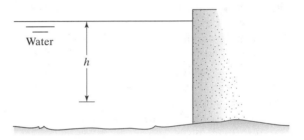

Figure P3.23.

24.   Work Problem 3.16, using one of the computer tools discussed in this chapter.

25.   Work Problem 3.17, using one of the computer tools discussed in this chapter.

26.   Work Problem 3.18, using one of the computer tools discussed in this chapter.

27.   Work Problem 3.19, using one of the computer tools discussed in this chapter.

28.   Work Problem 3.20, using one of the computer tools discussed in this chapter.

29.   Work Problem 3.21, using one of the computer tools discussed in this chapter.

30.   Work Problem 3.22, using one of the computer tools discussed in this chapter.

31.   Work Problem 3.23, using one of the computer tools discussed in this chapter.

4

Mechanics

4.1   INTRODUCTION

Mechanics is one of the most important fields of study in engineering. Mechanics was the first analytical science, and its historical roots can be traced to such great mathematicians and scientists as Archimedes (287–212 B.C.), Galileo Galilei (1564–1642), and Isaac Newton (1642–1727). **Mechanics** is the *study of the state of rest or motion of bodies that are subjected to forces*. As a discipline, mechanics is divided into three general areas: *rigid-body mechanics*, *deformable-body mechanics*, and *fluid mechanics*. As the term implies, rigid-body mechanics deals with the mechanical characteristics of bodies that are rigid (i.e., bodies that do not deform under the influence of forces). Rigid-body mechanics is subdivided into two main areas: *statics* and *dynamics*. Statics deals with rigid bodies in equilibrium. Equilibrium is a state in which a body is at rest with respect to its surroundings. When a body is in equilibrium, the forces that act on it are balanced, resulting in no motion. A state of equilibrium also exists when a body moves with a constant velocity, but this type of equilibrium is a dynamic equilibrium, not a static equilibrium. Dynamics deals with rigid bodies that are in motion with respect to its surroundings or to other rigid bodies. The body may have a constant velocity, in which case the acceleration is zero; but, generally, the body undergoes an acceleration due to the application of an unbalanced force. Deformable-body mechanics, often referred to as *mechanics of materials* or *strength of materials*, deals with solid bodies that deform under the application of external forces. In this branch of mechanics, the relationships between externally applied forces and the resulting internal forces and deformations are studied. Deformable-body mechanics is often subdivided into two specific areas: *elasticity* and *plasticity*. Elasticity deals with the behavior of solid materials that return to their original size and shape after a force is removed, whereas plasticity deals

OBJECTIVES

After reading this chapter, you will have learned

- The importance of mechanics in engineering
- The difference between a scalar and a vector
- How to perform basic vector operations
- How to add forces vectorially
- How to construct free-body diagrams
- How to use equilibrium principles to find unknown forces on a particle
- How to calculate normal stress, strain and deformation
- How to apply a factor of safety to stress

with the behavior of solid materials that experience a permanent deformation after a force is removed. Fluid mechanics deals with the behavior of liquids and gases at rest and in motion. The study of fluids at rest is called *fluid statics*; the study of fluids in motion is called *fluid dynamics*. Even though fluids are, strictly speaking, deformable materials, deformable-body mechanics is set apart from fluid mechanics because deformable-body mechanics deals exclusively with *solid* materials that have the ability, unlike fluids, to sustain shear forces. The topical structure of engineering mechanics is shown schematically in Figure 4.1.

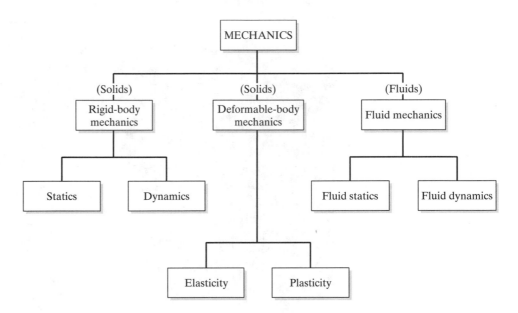

Figure 4.1. Topical structure of engineering mechanics.

In most colleges and universities, the branches of mechanics just outlined are generally taught as separate and distinct engineering courses. Hence, a typical engineering program consists of individual courses in statics, dynamics, mechanics of materials, and fluid mechanics. Other analytically oriented courses such as electrical circuits and thermodynamics are also offered. Mechanics is so essential to engineering education that students majoring in "nonmechanical" fields such as electrical engineering, environmental engineering, and chemical engineering gain a deeper understanding of energy, power, potential, equilibrium, and stability by first studying these principles in their mechanical contexts. However, depending on the specific curricular policies of your school or department, students in all engineering majors may or may not be required to take all of the aforementioned mechanics courses. In any case, the main purpose of this chapter is to introduce the beginning engineering student to the most fundamental principles of mechanics and to show how the general analysis procedure is applied to mechanics problems. In order to focus on the basics and to assist the student in the transition to more advanced material, our treatment of mechanics in this chapter is limited to a few fundamental principles of statics and mechanics of materials. However, dynamics is not covered in this book.

Engineers use principles of mechanics to analyze and design a wide variety of devices and systems. Look around you. Are you reading this book in a building? The structural members in the floor, roof, and walls were designed by structural or civil engineers to withstand forces exerted on them by the contents of the building, winds, earthquakes, snow, and other structural members. Bridges, dams, canals, underground pipelines, and other large, earthbound structures are designed with the use of mechanics. Do you see any mechanical devices nearby? The design of simple mechanisms such as staple removers, paper punches, door locks, and pencil sharpeners involves principles of mechanics. The automobile is an excellent example of a single engineering system that embodies virtually every branch of engineering mechanics, as well as other engineering disciplines. The chassis, bumpers, suspension system, power train, brakes, steering system, engine, air bag, doors, trunk, and even the windshield wipers were designed with the use of mechanics. Principles of mechanics are used to analyze and design virtually every type of engineering system that can be devised. Figures 4.2, 4.3, and 4.4 show some familiar engineering systems that involved the use of engineering mechanics in their design.

Figure 4.2. Engineers use principles of engineering mechanics to design spacecraft, such as the Hubble Space Telescope. (Photo courtesy of NASA).

4.2 SCALARS AND VECTORS

Every physical quantity used in mechanics, and in all of engineering and science, is classified as either a **scalar** or a **vector**. A scalar is a *quantity having magnitude, but no direction*. Having magnitude only, a scalar may be positive or negative, but has no directional characteristics. Common scalar quantities are length, mass, temperature, energy, volume, and density. A vector is a *quantity having both magnitude and direction*. A vector may be positive or negative and has a specified direction in space. Common vector quantities are displacement, force, velocity, acceleration, stress, and momentum. A scalar quantity can be fully defined by a single parameter, its magnitude, whereas a vector requires that both its magnitude and direction be specified. For example, speed is a scalar, but velocity is a vector. A typical speedometer of an automobile indicates how fast the

Figure 4.3.   Principles of engineering mechanics are used to design heavy construction equipment.

Figure 4.4.   Engineers used principles of engineering mechanics to design the Normandie Bridge in LeHavre, France. Completed in 1995, this bridge has one of the longest spans (856 m) of any cable-stayed bridge in the world.

vehicle is traveling, but does not reveal the direction of travel. The temperature of water boiling in an open container at sea level can be completely defined by a single number, 100°C. The force exerted on a beam used as a floor joist, however, must be defined by specifying a magnitude, 2 kN, for example, and a direction, down. The effect of the force on the beam (i.e., the stress and deformation) cannot be determined unless the direction of the force is specified. A force directed along the axis of the beam, for example, would produce a completely different stress and deformation than a normal downward force. Table 4.1 is a summary of some scalar and vector quantities.

TABLE 4.1 Scalar and Vector Quantities.

Scalar	Vector
Length	Force
Mass	Pressure
Time	Stress
Temperature	Moment of force
Speed	Velocity
Density	Acceleration
Volume	Momentum
Energy	Impulse
Work	Electric field
Resistance	Magnetic field

When writing scalars and vectors, standard nomenclature should be followed. Scalars are often written as a standard letter in italic font such as m for mass, T for temperature, ρ for density, etc. To differentiate vectors from scalars, vectors are written in a special way. For handwritten work, a vector in usually written as a letter with a bar ($^{-}$), arrow ($\vec{\ }$), or caret ($^{\wedge}$) over it such as $\overline{A}$, $\vec{A}$, and $\hat{A}$. In books and other printed matter, vectors are typically written in boldface type. For example, **A** is used to denote a vector "A." The magnitude of a vector, which is always a positive quantity, is normally written by hand, using "absolute value" notation. Thus, the magnitude of **A** is written as $|A|$. In books and other printed matter, the magnitude of **A** is usually written in italic type as A.

As shown in Figure 4.5, a vector is represented graphically by a straight arrow with a specified *magnitude, direction*, and *sense*. The magnitude is the length of the arrow, the

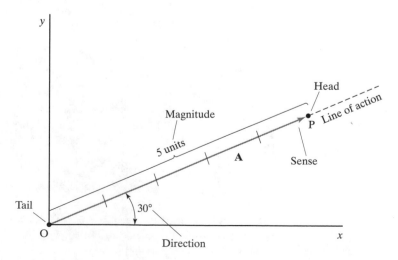

Figure 4.5. A vector has magnitude, direction and sense.

direction is defined by the angle between the arrow and a reference axis, and the sense is defined by the orientation of the arrowhead. The *line of action* of the vector is a line that is collinear with the vector. The vector **A** in Figure 4.5 has a magnitude of 5 units, a direction of 30° with respect to the *x*-axis, and a sense that is upward and to the right. Point O is called the *tail* of the vector, and point P is called the *head* of the vector. The units of the vector depend on what physical quantity the vector represents. For example, if the vector is a force, the units would be N or lb$_f$. The 30° angle indicates that the force acts in a direction defined by a line of action coincident with the arrow. The orientation of the arrowhead indicates that the force acts upward and to the right, rather than downward and to the left.

4.2.1 Vector Operations

In order to utilize principles of mechanics to carry out an analysis, engineers must be able to mathematically manipulate vector quantities. Due to the directional character of vectors, the rules for performing algebraic operations with vectors are different from those of scalars. The product of a scalar *k* and a vector **A**, denoted by *k***A**, has the effect of changing the length of the vector **A**, but its line of action is unaffected. For example, the product 3**A** increases the magnitude of the vector **A** by a factor of three, but the direction and sense of **A** is the same. The product −2**A** increases the magnitude of **A** by a factor of two, but reverses the direction of **A**, because the scalar is negative. Graphical examples of the product of scalars and a vector are illustrated in Figure 4.6. Two vectors, **A** and **B**, are *equal* if they have the same magnitude, direction, and sense regardless of the location of their tails and heads. As shown in Figure 4.6, **A** = **B**.

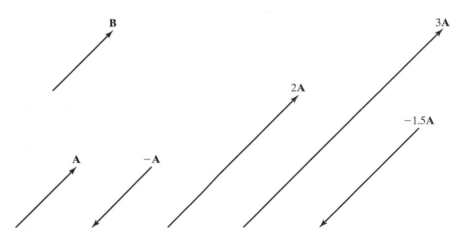

Figure 4.6. Scalar multiplication and vector equality, **A** = **B**.

The addition of two scalars results in a simple algebraic sum, such as c = a + b. The addition of two vectors, however, cannot be obtained by simply adding the magnitudes of each vector. Vectors must be added such that their directions as well as their magnitudes are accounted for. Consider the vectors **A** and **B** in Figure 4.7(a). Vectors **A** and **B** may be added by using the *parallelogram law*. To form this sum, **A** and **B** are joined at their tails. Parallel lines are drawn from the head of each vector, intersecting at a common point, forming adjacent sides of a parallelogram. The vector sum of **A** and **B**, referred to as the *resultant vector* or simply **resultant**, is the diagonal of the parallelogram that extends from the vector tails to the intersection point, as illustrated in Figure 4.7(b). Hence, we may write the vector sum as **R** = **A** + **B**, where **R** is the resultant. The vector sum may also be obtained by constructing a triangle, which is actually half of a parallelogram. In this technique, the tail of **B** is connected to the head of **A**. The resultant **R** = **A** + **B** extends from the tail of **A** to the head of **B**, as shown in Figure 4.7(c).

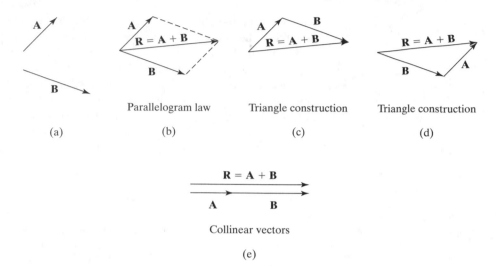

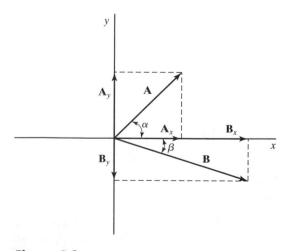

Figure 4.7. Vector addition.

Alternatively, the triangle may be constructed such that the tail of **A** is connected to the head of **B**, in which case we have **R** = **B** + **A**, as shown in Figure 4.7(d). In both triangles, the same resultant is obtained, so we conclude that vector addition is *commutative* (i.e., the vectors can be added in either order). Hence, **R** = **A** + **B** = **B** + **A**. A special case of the parallelogram law is when the two vectors are collinear (i.e., they have the same line of action). In that case, the parallelogram is degenerate, and the vector sum reduces to a scalar sum $R = A + B$, as indicated in Figure 4.7(e).

4.2.2   Vector Components

A powerful method for finding the resultant of two vectors is to first find the *rectangular components* of each vector and then add the corresponding components to obtain the resultant. To see how this method works, we draw the two vectors, **A** and **B**, in Figure 4.7 on a set of (x, y) coordinate axes, shown in Figure 4.8. For convenience, both vectors are drawn with their tails at the origin, and the directions of **A** and **B** with respect to the positive x-axis are defined by the angles α and β, respectively. For the moment, let's consider each vector separately. Using a modified form of the parallelogram law, we draw lines

Figure 4.8. Vector components.

parallel to the x- and y-axes such that the vector **A** becomes the diagonal of a rectangle, which is a special type of parallelogram. The sides of the rectangle that lie along the x and y-axes are called the rectangular components of vector **A**, and are denoted $\mathbf{A}_x$ and $\mathbf{A}_y$ respectively. Because vector **A** is the diagonal of the rectangle, **A** becomes the resultant of vectors $\mathbf{A}_x$ and $\mathbf{A}_y$. Thus, we may write the vector as $\mathbf{A} = \mathbf{A}_x + \mathbf{A}_y$. Similarly, lines parallel to the x- and y-axes are drawn such that the vector **B** becomes the diagonal of a rectangle. The sides of the rectangle that lie along the x- and y-axes are the rectangular components of vector **B** and are denoted $\mathbf{B}_x$ and $\mathbf{B}_y$, respectively. Hence, we may write the vector as $\mathbf{B} = \mathbf{B}_x + \mathbf{B}_y$. The resultant of **A** and **B** may now be written as

$$\mathbf{R} = \mathbf{A} + \mathbf{B} = (\mathbf{A}_x + \mathbf{B}_x) + (\mathbf{A}_y + \mathbf{B}_y) \tag{4.1}$$

The magnitude of the components of **A** and **B** may be written in terms of the angles that define the vectors' directions. From the definitions of the trigonometric functions for cosine and sine, the x and y components of **A** are

$$A_x = A \cos \alpha \tag{4.2}$$

and

$$A_y = A \sin \alpha \tag{4.3}$$

where A is the magnitude of **A**. Similarly, the x and y components of **B** are

$$B_x = B \cos \beta \tag{4.4}$$

and

$$B_y = B \sin \beta \tag{4.5}$$

where B is the magnitude of **B**. Alternatively, we can also see from trigonometry that

$$A_y = A_x \tan \alpha \tag{4.6}$$

and

$$B_y = B_x \tan \beta \tag{4.7}$$

The magnitudes of **A** and **B** form the hypotenuse of their respective right triangles, so from the theorem of Pythagoras, we may write

$$A = \sqrt{A_x^2 + A_y^2} \tag{4.8}$$

and

$$B = \sqrt{B_x^2 + B_y^2} \tag{4.9}$$

4.2.3 Unit Vectors

The justification for grouping the x components of each vector and the y components of each vector in Equation (4.1) is based on the concept of *unit vectors*. A unit vector is a *dimensionless vector of unit length used to specify a given direction*. Unit vectors have no other physical meaning. The most common unit vectors are the *rectangular* or **Cartesian unit vectors**, denoted **i**, **j**, and **k**. The unit vectors **i**, **j**, and **k** coincide with the positive x-, y-, and z-axes, respectively, as shown in Figure 4.9. The rectangular unit vectors form a set of mutually perpendicular vectors and are used to specify the direction of a vector in three-dimensional space.

If the quantity of interest can be described by a two-dimensional vector, only the **i** and **j** unit vectors are required. The vectors **A** and **B**, shown in Figure 4.8, lie in the x–y plane, so they can be represented by the **i** and **j** unit vectors. The x component of **A** has a magnitude of A_x, and the y component of **A** has a magnitude of A_y. Note that the quantities A_x and A_y are not vectors, but scalars, because they represent magnitudes only.

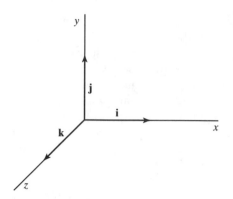

Figure 4.9. Rectangular unit vectors.

The vector components, $\mathbf{A}_x$ and $\mathbf{A}_y$, can be written as products of a scalar and a unit vector as $\mathbf{A}_x = A_x\,\mathbf{i}$ and $\mathbf{A}_y = A_y\,\mathbf{j}$. Thus, vector $\mathbf{A}$ is expressed as

$$\mathbf{A} = A_x\mathbf{i} + A_y\mathbf{j} \tag{4.10}$$

and vector $\mathbf{B}$ is expressed as

$$\mathbf{B} = B_x\mathbf{i} + B_y\mathbf{j} \tag{4.11}$$

Rewriting Equation (4.1) in terms of the x and y component groups, the resultant of $\mathbf{A}$ and $\mathbf{B}$ is

$$\mathbf{R} = \mathbf{A} + \mathbf{B} = (A_x + B_x)\mathbf{i} + (A_y + B_y)\mathbf{j} \tag{4.12}$$

The rectangular components of the resultant vector $\mathbf{R}$ are given by

$$R_x = A_x + B_x \tag{4.13}$$

and

$$R_y = A_y + B_y \tag{4.14}$$

Hence, Equation (4.12) can be written

$$\mathbf{R} = R_x\mathbf{i} + R_y\mathbf{j} \tag{4.15}$$

where R_x and R_y are the x and y components of $\mathbf{R}$, as shown in Figure 4.10. By trigonometry, we may write

$$R_x = R \cos\theta \tag{4.16}$$

$$R_y = R \sin\theta \tag{4.17}$$

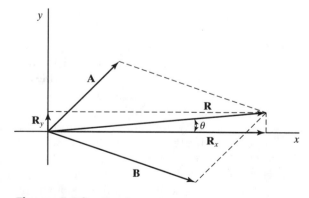

Figure 4.10. Resultant vector.

and

$$R_y = R_x \tan \theta \tag{4.18}$$

The magnitude of **R** forms the hypotenuse of a right triangle, hence, from the theorem of Pythagoras, we have

$$R = \sqrt{R_x^2 + R_y^2} \tag{4.19}$$

EXAMPLE 4.1

Two vectors have magnitudes of $A = 8$ and $B = 6$ and directions as shown in Figure 4.11(a). Find the resultant vector, using (a) the parallelogram law and (b) by resolving the vectors into their x and y components.

SOLUTION

(a) Parallelogram law

The parallelogram for vectors **A** and **B** is shown in Figure 4.11(b). In order to find the magnitude and direction of the resultant vector **R**, some angles must be determined. By subtraction, the acute angle between the vectors is 45°. The sum of the interior angles of a quadrilateral is 360°, so the adjacent angle is found to be 135°. The magnitude of **R** may be found by using the law of cosines:

$$R = \sqrt{6^2 + 8^2 - 2(6)(8)\cos 135°}$$
$$R = \sqrt{36 + 64 - 96(-0.7071)}$$
$$= 13.0$$

The direction of **R** is found by calculating the angle θ. Using the law of sines, we have

$$\frac{\sin \theta}{6} = \frac{\sin 135°}{12.96}$$
$$\sin \theta = 0.3274$$
$$\theta = \sin^{-1}(0.3274) = 19.1°$$

Thus, the angle of **R** with respect to the positive x-axis is

$$\phi = 19.1° + 15° = 34.1°$$

The resultant vector **R** has now been completely defined, because both its direction and magnitude have been determined.

(b) Vector components

In Figure 4.11(c), vectors **A** and **B** are resolved into their x and y components. The magnitudes of these components are

$$A_x = A \cos 15° = 8 \cos 15° = 7.7274$$
$$A_y = A \sin 15° = 8 \sin 15° = 2.0706$$
$$B_x = B \cos 60° = 6 \cos 60° = 3$$
$$B_y = B \sin 60° = 6 \sin 60° = 5.1962$$

Vectors **A** and **B** may now be written in terms of the unit vectors **i** and **j**:

$$\mathbf{A} = A_x\mathbf{i} + A_y\mathbf{j} = 7.7274\mathbf{i} + 2.0706\mathbf{j}$$
$$\mathbf{B} = B_x\mathbf{i} + B_y\mathbf{j} = 3\mathbf{i} + 5.1962\mathbf{j}$$

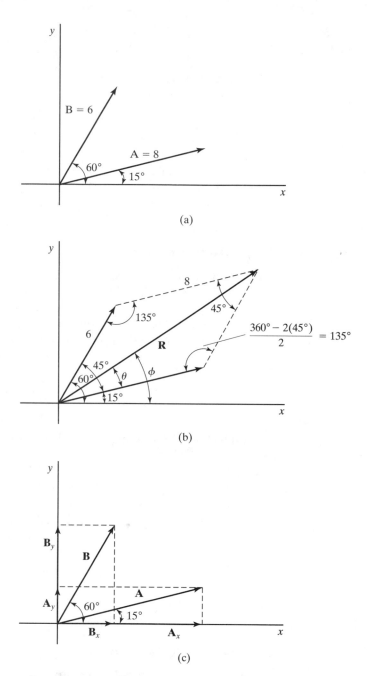

Figure 4.11. Example 4.1.

The resultant vector **R** is

$$\mathbf{R} = \mathbf{A} + \mathbf{B} = R_x\mathbf{i} + R_y\mathbf{j} = (7.7274 + 3)\mathbf{i} + (2.0706 + 5.1962)\mathbf{j}$$
$$= 10.7274\mathbf{i} + 7.2668\mathbf{j}$$

This is the answer, but to compare it with the answer obtained by the parallelogram law, we must find the magnitude of **R** and its direction with respect to the positive x-axis. Using the theorem of Pythagoras, we find that the magnitude of **R** is

$$R = \sqrt{10.7274^2 + 7.2668^2} = 13.0$$

The direction is given by

$$R_y = R_x \tan \phi$$

Solving for the angle ϕ, we get

$$\phi = \tan^{-1}(R_y/R_x) = \tan^{-1}(7.2668/10.7274) = 34.1°$$

We have obtained the same result with two different methods of vector addition. The second method may appear to involve more work. However, many mechanics problems involve more than two vectors, and the problem may be three dimensional. In these cases, resolving the vectors into their rectangular components is the preferred approach, since the parallelogram law is too cumbersome. In the example, to ensure that both methods yielded the same answers to three significant figures, four decimal places were used.

EXAMPLE 4.2

For the vectors $\mathbf{A} = 3\mathbf{i} - 6\mathbf{j} + \mathbf{k}$, $\mathbf{B} = 5\mathbf{i} + \mathbf{j} - 2\mathbf{k}$, and $\mathbf{C} = -2\mathbf{i} + 4\mathbf{j} + 3\mathbf{k}$, find the resultant vector and its magnitude.

SOLUTION

These vectors, unlike those in the previous example, are three-dimensional. They are already expressed in terms of the Cartesian unit vectors $\mathbf{i}$, $\mathbf{j}$, and $\mathbf{k}$, so it is a straightforward matter to add them vectorially. Recall that the $\mathbf{i}$, $\mathbf{j}$, and $\mathbf{k}$ unit vectors correspond to the positive x, y, and z directions, respectively. To find the resultant, we simply add the x components, the y components, and the z components of each vector. To help us avoid errors as we perform the addition, it is useful to write the vectors with their components aligned in columns:

$$\mathbf{A} = 3\mathbf{i} - 6\mathbf{j} + 1\mathbf{k}$$
$$\mathbf{B} = 5\mathbf{i} + 1\mathbf{j} - 2\mathbf{k}$$
$$\mathbf{C} = -2\mathbf{i} + 4\mathbf{j} + 3\mathbf{k}$$

Performing the additions, the resultant vector is

$$\mathbf{R} = (3 + 5 - 2)\mathbf{i} + (-6 + 1 + 4)\mathbf{j} + (1 - 2 + 3)\mathbf{k}$$
$$= 6\mathbf{i} - \mathbf{j} + 2\mathbf{k}$$

The magnitude of the resultant vector is found by extending the theorem of Pythagoras to three dimensions:

$$R = \sqrt{R_x^2 + R_y^2 + R_z^2}$$
$$= \sqrt{6^2 + (-1)^2 + 2^2} = 6.40$$

4.3   FORCES

From our early childhood experiences, we all have a basic understanding of the concept of force. We commonly use terms such as *push, pull,* and *lift* to describe forces that we encounter in our daily lives. *Mechanics* is the study of the state of rest or motion of bodies that are subjected to forces. To the engineer, **force** is defined as *an influence that causes a body to deform or accelerate.* For example, when you push or pull on a lump of clay, the clay deforms into a different shape. When you pull on a rubber band, the rubber band increases in length. The forces required to deform clay and rubber bands are much smaller than those required to deform engineering structures such as buildings, bridges, dams, and machines, but these objects deform nevertheless. What happens when you push on the wall with your hand? Unless you are exceptionally strong, the wall does not move and neither does your hand. In accordance with Newton's third law, as you push on the wall, the wall pushes back on your hand with the same force. When you push on a book in an attempt to slide it across the table, the book will not move unless the frictional force between the table and book is exceeded by the horizontal pushing force. These types of situations are encountered in virtually all engineering systems that are in static equilibrium. Forces are present, but motion does not occur because the forces cause the body to be in a state of balance. When the forces acting on a body are unbalanced, the body undergoes an acceleration. For example, the propulsive force delivered to the wheels of an automobile can exceed the frictional forces that tend to retard the automobile's motion, so the automobile accelerates. Similarly, the thrust and lift forces acting on an aircraft can exceed the weight and drag forces, thereby allowing the aircraft to accelerate vertically and horizontally.

Forces commonly encountered in the majority of engineering systems may be generally categorized as a *contact force, gravitational force, cable force, pressure force,* or *fluid dynamic force*. These five types of forces are depicted in Figure 4.12. A contact force is a force produced by two or more bodies in direct contact. The force produced by pushing on a wall is a contact force because the hand is in direct contact with the wall. When two billiard balls collide, a contact force is produced at the region where the balls touch each other. Friction is a type of contact force. A gravitational force, referred to as *weight*, is exerted on an object on or near the earth's surface. Gravitational forces are directed downward, toward the center of the earth, and act through a point in the body called the *center of gravity*. For a body that is uniform in density, the center of gravity lies at the geometric center of the body. This point is referred to as the *centroid*. The force in a cable is actually a special type of contact force, since the cable is in contact with a body, but it occurs so frequently that it deserves a separate definition. Cables, ropes, and cords are used in pulley systems, suspension bridges, and other engineering structures. A cable, due to its limp and flexible nature,

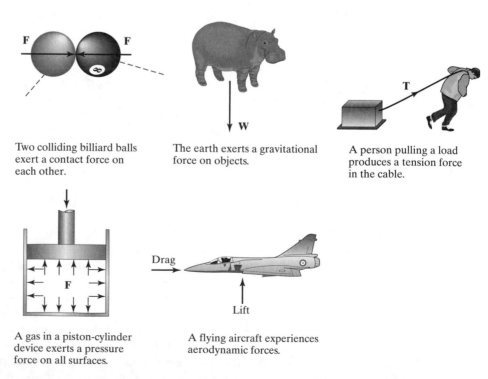

Two colliding billiard balls exert a contact force on each other.

The earth exerts a gravitational force on objects.

A person pulling a load produces a tension force in the cable.

A gas in a piston-cylinder device exerts a pressure force on all surfaces.

A flying aircraft experiences aerodynamic forces.

Figure 4.12. Types of forces commonly encountered in engineering applications.

can support tension forces only. Forces in cables are always directed along the axis of the cable, regardless of whether the cable is straight or not. Pressure forces are normally associated with static fluids. A gas in a cylinder exerts a pressure force on all surfaces of the cylinder. A static liquid, such as the water behind a dam, exerts a pressure force on the dam. Pressure forces always act in a direction normal to the surface. A fluid dynamic force is produced when a fluid flows around a body or through a pipe or conduit. When a fluid flows around a body (or when a body moves through a fluid) aerodynamic forces act on the body. There are basically two types of aerodynamic forces: pressure forces and viscous forces. Pressure forces are caused by pressure distributions around the body and are produced by certain fluid-related mechanisms and body geometry. Viscous forces, sometimes called friction or shear forces, are caused by fluid viscosity. Any object (for example, airplane, missile, ship, submarine, automobile, baseball, etc.) that moves through a fluid experiences aerodynamic forces. When a fluid flows through a pipe, a friction force is produced between the fluid and the inside surface of the pipe. This friction force, which is caused by fluid viscosity, has the effect of retarding the flow. The five types of forces just mentioned are the most common, but there are other kinds of forces that engineers sometimes encounter. These include electric, magnetic, nuclear, and surface tension forces.

Forces are vectors, so all the mathematical operations and expressions that apply to vectors apply to forces. Because a force is a vector, a force has magnitude, direction, and sense. For example, the weight of a 170-pound person is a vector with a magnitude of 170 lb$_f$, a direction of 90° with respect to the horizontal, and a downward sense. A situation in which more than one force acts on a body is referred to as a **force system**. A system of forces is *coplanar* or *two dimensional* if the lines of action of the forces lie in the same plane. Otherwise, the system of forces is *three dimensional*. Forces are *concurrent* if their lines of action pass through the same point and *parallel* if their lines of action are parallel. *Collinear* forces have their lines of action along the same line. These force concepts are illustrated in Figure 4.13.

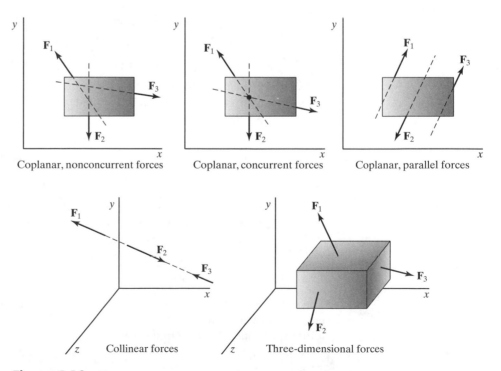

Figure 4.13. Force systems.

EXAMPLE 4.3

Three coplanar forces act as shown in Figure 4.14. Find the resultant force, its magnitude, and its direction with respect to the positive x-axis.

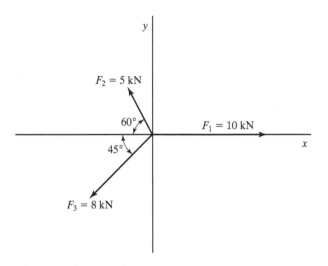

Figure 4.14. Concurrent forces for Example 4.3.

SOLUTION

We have three coplanar forces that act concurrently at the origin. Note that force, $\mathbf{F}_1$, lies along the x-axis. First, we resolve the forces into their x and y components:

$$F_{1x} = F_1 \cos 0° = 10 \cos 0° = 10 \text{ kN}$$
$$F_{1y} = F_1 \sin 0° = 10 \sin 0° = 0 \text{ kN}$$
$$F_{2x} = F_2 \cos 60° = -5 \cos 60° = -2.5 \text{ kN}$$
$$F_{2y} = F_2 \sin 60° = 5 \sin 60° = 4.330 \text{ kN}$$
$$F_{3x} = -F_3 \cos 45° = -8 \cos 45° = -5.657 \text{ kN}$$
$$F_{3y} = -F_3 \sin 45° = -8 \cos 45° = -5.657 \text{ kN}$$

Notice that F_{2x}, F_{3x}, and F_{3y} are *negative* quantities to reflect the proper directions of the vectors with respect to the positive x- and y-axes. The forces may now be written in terms of the unit vectors $\mathbf{i}$ and $\mathbf{j}$:

$$F_1 = F_{1x}\mathbf{i} + F_{1y}\mathbf{j} = 10\mathbf{i} + 0\mathbf{j} = 10\mathbf{i} \text{ kN}$$
$$F_2 = F_{2x}\mathbf{i} + F_{2y}\mathbf{j} = -2.5\mathbf{i} + 4.330\mathbf{j} \text{ kN}$$
$$F_3 = F_{3x}\mathbf{i} + F_{3y}\mathbf{j} = -5.657\mathbf{i} - 5.657\mathbf{j} \text{ kN}$$

Earlier in this chapter we learned that a resultant is the sum of two or more vectors. Here we define a **resultant force** as the sum of two or more forces. Therefore, the resultant force $\mathbf{F}_R$ is the vector sum of the three forces. Adding corresponding components, we obtain

$$\mathbf{F}_R = (10 - 2.5 - 5.657)\mathbf{i} + (0 + 4.330 - 5.657)\mathbf{j}$$
$$= 1.843\mathbf{i} - 1.327\mathbf{j} \text{ kN}$$

The signs on the x and y components of $\mathbf{F}_R$ are significant. A positive sign on the x component and a negative sign on the y component means that the resultant force lies in the fourth quadrant. The magnitude of $\mathbf{F}_R$ is

$$F_R = \sqrt{1.843^2 + (-1.327)^2}$$
$$= 2.271 \text{ kN}$$

The direction of $\mathbf{F}_R$ with respect to the positive x-axis is

$$\phi = \tan^{-1}(-1.327/1.843) = -35.8°$$

where the minus sign on the angle is consistent with the fact that $\mathbf{F}_R$ lies in the fourth quadrant, as shown in Figure 4.15.

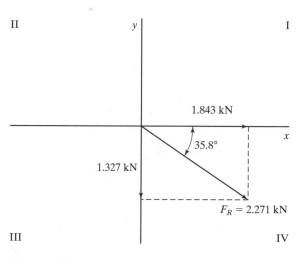

Figure 4.15.   Resultant force for Example 4.4.

PRACTICE!

1. Find the resultant force for the forces shown by (a) using the parallelogram law and (b) by resolving the forces into their x and y components.
   Answer: 178 N, $-15.1°$

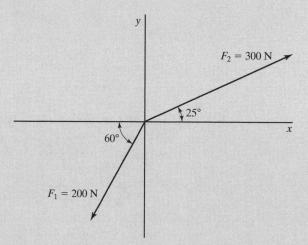

2. Find the resultant force for the forces shown by (a) using the parallelogram law and (b) by resolving the forces into their x and y components.
   Answer: 166 N, 5.5°

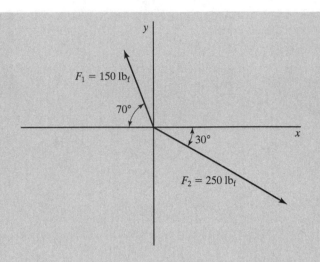

3. Find the resultant force for the forces shown by (a) using the parallelogram law and (b) by resolving the forces into their x and y components.
   Answer: 26.0 N, 75.0°

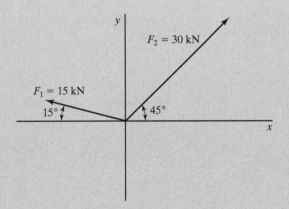

4. Consider the three forces, $\mathbf{F}_1 = 5\mathbf{i} + 2\mathbf{j}$ kN, $\mathbf{F}_2 = -2\mathbf{i} - 5\mathbf{j}$ kN, and $\mathbf{F}_3 = \mathbf{i} - \mathbf{j}$ kN. Find the resultant force, its magnitude, and direction with respect to the positive x-axis.
   Answer: $4\mathbf{i} - 4\mathbf{j}$ kN, 5.66 kN, −45.0°

5. Consider the three forces, $\mathbf{F}_1 = 2\mathbf{i} - 7\mathbf{j}$ lb$_\mathrm{f}$, $\mathbf{F}_2 = 5\mathbf{i} + 8\mathbf{j}$ lb$_\mathrm{f}$, and $\mathbf{F}_3 = 3\mathbf{i} + 4\mathbf{j}$ lb$_\mathrm{f}$. Find the resultant force, its magnitude and direction with respect to the positive x-axis.
   Answer: $10\mathbf{i} + 5\mathbf{j}$ lb$_\mathrm{f}$, 11.2 lb$_\mathrm{f}$, 26.6°

6. Consider three forces, $\mathbf{F}_1 = 3\mathbf{i} + 5\mathbf{j} - 2\mathbf{k}$ N, $\mathbf{F}_2 = -\mathbf{i} - 4\mathbf{j} + 3\mathbf{k}$ N, and $\mathbf{F}_3 = 2\mathbf{i} - 2\mathbf{j} + 6\mathbf{k}$ N. Find the resultant force and its magnitude.
   Answer: $4\mathbf{i} - \mathbf{j} + 7\mathbf{k}$ N, 8.12 N

APPLICATION: STABILIZING A COMMUNICATIONS TOWER WITH CABLES

Tall slender structure often incorporate cables to stabilize them. The cables, which are connected at various points around the structure and along its length, are connected to concrete anchors buried deep in the ground. Shown in Figure 4.16(a) is a typical communications tower that is stabilized with several cables. On this particular tower each ground anchor facilitates two cables that are connected at a common point, as shown in Figure 4.16(b). The upper and lower cables exert forces of 15 kN and 25 kN, respectively, and their directions are

(a)

(b)

(c)

Figure 4.16.   A communications tower stabilized with cables.

45° and 32°, respectively, as measured from the ground [Figure 4.16(c)]. What is the resultant force exerted by the cables on the ground anchor?

Any two forces in three-dimensional space lie in a single plane, so we may arbitrarily locate our two cable forces in the x–y plane. Thus, we have two coplanar forces that act concurrently at the origin. We let $F_1 = 15$ kN and $F_2 = 25$ kN. We resolve the forces into their x and y components:

$$F_{1x} = F_1 \cos 45° = 15 \cos 45° = 10.607 \text{ kN}$$

$$F_{1y} = F_1 \sin 45° = 15 \sin 45° = 10.607 \text{ kN}$$

$$F_{2x} = F_2 \cos 32° = 25 \cos 32° = 21.201 \text{ kN}$$

$$F_{2y} = F_2 \sin 32° = 25 \sin 32° = 13.248 \text{ kN}$$

The forces may now be written in terms of the unit vectors $\mathbf{i}$ and $\mathbf{j}$:

$$\mathbf{F}_1 = F_{1x}\mathbf{i} + F_{1y}\mathbf{j} = 10.607\mathbf{i} + 10.607\mathbf{j} \text{ kN}$$

$$\mathbf{F}_2 = F_{2x}\mathbf{i} + F_{2y}\mathbf{j} = 21.201\mathbf{i} + 13.248\mathbf{j} \text{ kN}$$

The resultant force $\mathbf{F}_R$, is the vector sum of the two forces. Adding corresponding components, we obtain

$$\mathbf{F}_R = (10.607 + 21.201)\mathbf{i} + (10.607 + 13.248)\mathbf{j}$$

$$= 31.808\mathbf{i} + 23.855\mathbf{j} \text{ kN}$$

The magnitude of $\mathbf{F}_R$ is

$$F_R = \sqrt{R_x^2 + R_y^2}$$

$$= \sqrt{31.808^2 + 23.855^2}$$

$$= 39.76 \text{ kN}$$

and the direction of $\mathbf{F}_R$ with respect to the ground is

$$\phi = \tan^{-1}(23.855/31.808)$$

$$= 36.9°$$

What does our answer mean, and how would it be used? The resultant force would be used by an engineer (probably a civil engineer) to design the concrete anchor. A force of nearly 40 kN directed at an angle of about 37° with respect to the ground would have a tendency to pull the anchor out of the ground. If not designed properly, the anchor could become loose or break under the load, thereby causing an unbalanced force on the tower. Look carefully at Figure 4.16(b). Notice that the two cables connect via turnbuckles at a ring assembly connected to a single rod that goes into the concrete anchor, which is not shown. The resultant force would also be used to ascertain the structural integrity of the ring assembly and rod.

4.4 FREE-BODY DIAGRAMS

One of the most important steps in the general analysis procedure is to construct a diagram of the system being analyzed. In engineering mechanics, this diagram is referred to as a free-body diagram. A **free-body diagram** is a *diagram that shows all external forces acting on the body*. As the term implies, a free-body diagram shows only the body in question, being isolated or "free" from all other bodies. The body is conceptually removed from all supports, connections, and regions of contact with other bodies. All forces produced by these external influences are schematically represented on the free-body diagram. In a free-body diagram, only the *external* forces acting on the body in question are considered in the analysis. There may be **internal forces** (i.e., forces originating from inside the body that act on other parts of the body), but it can be shown that these forces cancel one another and therefore do not contribute to the overall mechanical state of the body. The free-body diagram is one of the most critical parts of a mechanical analysis. It focuses the engineer's attention on the body being analyzed and helps to identify all the external forces acting on the body. The free-body diagram also helps the engineer to write the correct governing equations.

Free-body diagrams are used in statics, dynamics, and strength of materials, but their application to statics and strength of materials will be emphasized here. **Statics** is the branch of engineering mechanics that deals with bodies in static equilibrium. If a body is in static equilibrium, the external forces cause the body to be in a state of balance. Even though the body does not move, it experiences stresses and deformations that must be determined if its performance as a structural member is to be evaluated. In order to determine the forces that act on the body, a free-body diagram must be properly constructed.

Procedure for Constructing Free-Body Diagrams

The following procedure should be followed when constructing free-body diagrams:

1. *Identify* the body you wish to isolate and make a *simple drawing* of it.
2. Draw the appropriate *force vectors* at all locations of supports, connections and contacts with other bodies.
3. Draw a force vector for the *weight* of the body, unless the gravitational force is to be neglected in the analysis.
4. *Label* all forces that are known with a numerical value and those that are unknown with a letter.
5. Draw a *coordinate system* on, or near, the free-body to establish directions of the forces.
6. Add *geometric data* such as lengths, angles, etc., as required.

Free-body diagrams for some of the most common force configurations are illustrated in Figure 4.17.

PROFESSIONAL SUCCESS: DON'T BEGIN IN THE MIDDLE OF A PROBLEM

It's human nature to want to finish a job in the least amount of time. Sometimes, we take shortcuts without taking enough time to assure that the job is done thoroughly. Like everyone else, engineers are only human and may sometimes take shortcuts in the solution of a problem. Engineers may take shortcuts for a variety of reasons. Perhaps the engineer is simply overloaded with work, and the only way to meet deadlines is to spend less time on each problem. Perhaps the engineers manager has unrealistic expectations and does not budget enough time for each project. Time and budget-related reasons, while serious enough to warrant corrective action, are not usually the reasons that engineers take shortcuts in their analytical work. They take shortcuts because they either have become lax in their problem-solving practices or have forgotten how to perform a thorough analysis of a problem. Perhaps they have forgotten some of the steps in the general analysis procedure, or even worse, never learned them at all.

Regardless of the underlying reasons, the practice of taking problem-solving shortcuts may provoke an engineer to begin an analysis "in the middle of the problem." How does this happen? In an attempt to solve the problem more efficiently, the engineer may want to get right to the equations and calculations. By going directly to the *governing equations* and *calculations* steps of the analysis procedure, three crucial steps are omitted: problem statement, diagram, and assumptions. How can an engineer solve a problem if he or she does not even state what the problem is? The engineer may defensively exclaim. "But I know what the problem statement is. It's in my head." A problem statement not written is a not a problem statement! Others who will review the analysis cannot read minds. A good engineer documents everything in writing, including problem statements. The engineer further retorts, "Everyone knows exactly what the component looks like, and the forces acting on it are straightforward. A free-body diagram is unnecessary." Everyone may be intimately familiar with the component's configuration and loading today, but 18 months from now, when the analysis is reevaluated because the component failed in its first year of service, everyone, including the engineer who did the analysis, may not remember all the details. Once again, written documentation is essential. The formulation of good assumptions is as much an art as it is a science. A hurried engineer may declare, "The assumptions are obvious. It's no big deal," The assumptions may or may not be obvious, but they are critical to the outcome of the problem. Assumptions must be explicitly stated, and the governing equations and calculations must be consistent with those assumptions. If the component failed in its first year of service, it is perhaps because the engineer *thought* the assumptions were obvious, and they were not, resulting in a flawed analysis and a failed component.

While you are in school, develop the habit of conscientiously applying the general analysis procedure to all your analytical problem-solving work. Then, as you make the transition from student to engineering professional, you will not experience the pitfalls of beginning "in the middle of a problem."

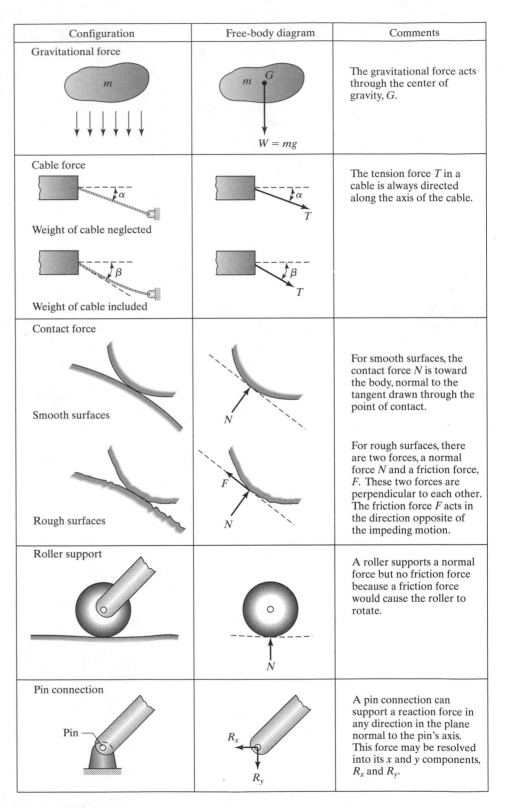

Figure 4.17.   Free-body diagrams for some common force configurations.

PRACTICE!

1. A crate hangs by a rope as shown. Construct a free-body diagram of the crate.

2. Two crates hang by ropes from a ceiling as shown. Construct a free-body diagram of (a) crate *A* and (b) crate *B*.

3. A wooden block rests on a rough inclined plane as shown. Construct a free-body diagram of the block.

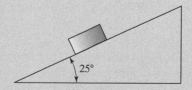

4. An obliquely loaded I-beam is supported by a roller at *A* and a pin at *B* as shown. Construct a free-body diagram of the beam. Include the weight of the beam.

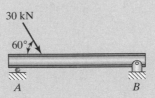

5. Two pipes rest in a long V-shaped channel as shown. Construct a free-body diagram of each pipe.

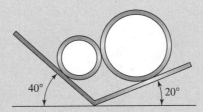

6. A box is held in position on the bed of a truck by a cable as shown. The surface of the truck bed is rough. Construct a free-body diagram of the box.

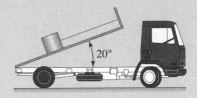

4.5   EQUILIBRIUM

Equilibrium is a state of balance between or among opposing forces, and it is one of the most important concepts in engineering mechanics. There are two types of equilibrium in engineering mechanics: static and dynamic. If a body is in static equilibrium, the body does not move, whereas if a body is in dynamic equilibrium, the body moves with a constant velocity. In this book, we will restrict our discussion to static equilibrium. Furthermore, we will confine our treatment of static equilibrium to *concurrent* force systems. In a concurrent force system, the lines of action of all forces pass through a single point, so the forces do not have a tendency to rotate the body. Therefore, there are no *moments of force* to deal with, only the forces themselves. Because the forces act concurrently, the body effectively becomes a *particle* (i.e., a dimensionless point in space through which the forces act). The actual body may or not be a particle, but is modeled as such for purposes of the analysis. This concept will be demonstrated in some examples later.

A body is in static equilibrium if the vector sum of all external forces is zero. Consistent with this definition, the condition of static equilibrium may be stated mathematically as

$$\Sigma \mathbf{F} = \mathbf{0} \tag{4.20}$$

where the summation symbol Σ denotes a sum of all external forces. Note that the zero is written as a vector to preserve the vector character of the equation across the equal sign. Equation (4.20) is a necessary and sufficient condition for equilibrium according to Newton's second law, which can be written as $\Sigma \mathbf{F} - m\mathbf{a}$. If the sum of the forces is zero, then $m\mathbf{a} = \mathbf{0}$. The quantity m is a scalar that can be divided out, leaving $\mathbf{a} = \mathbf{0}$. Thus, the acceleration is zero, so the body either moves with a constant velocity or remains at rest. Equation (4.20) is a vector equation that may be broken into its scalar components. Writing the equation in terms of the unit vectors $\mathbf{i}$, $\mathbf{j}$, and $\mathbf{k}$, we obtain

$$\Sigma F_x \mathbf{i} + \Sigma F_y \mathbf{j} + \Sigma F_z \mathbf{k} = \mathbf{0} \tag{4.21}$$

where the three terms on the left side are the total *scalar* forces in the x, y, and z directions, respectively. Equation (4.21) can only be satisfied if the sum of the scalar forces in each coordinate direction is zero. Hence, we have three scalar equations

$$\Sigma F_x = 0, \; \Sigma F_y = 0, \; \Sigma F_z = 0 \tag{4.22}$$

These relations are referred to as the *equations of equilibrium for a particle*. Each of these three scalar equations must be satisfied for the particle to be in equilibrium. If *any one* of these scalar equations is not satisfied, the particle is not in equilibrium. For example, if $\Sigma F_x = 0$ and $\Sigma F_y = 0$, but $\Sigma F_z \neq 0$, the particle will be in equilibrium in the x and y directions, but will accelerate in the z direction. Similarly, if $\Sigma F_x = 0$, but $\Sigma F_y \neq 0$ and $\Sigma F_z \neq 0$, the particle will be in equilibrium in the x direction, but will have components of acceleration in the y and z directions.

Equations (4.22) are the governing equations for a particle in static equilibrium. Using those equations and a free-body diagram of the particle, the unknown external forces can be determined. Consider the particle in Figure 4.18(a). A force of 2 kN acts

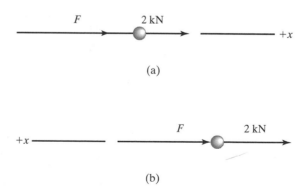

(a)

(b)

Figure 4.18. A force of $F = -2$ kN is required to maintain equilibrium, regardless of the orientation of the coordinate system.

on the particle in the positive x direction. An unknown force F, whose direction is *assumed* to act in the positive x direction, also acts on the particle. Applying the first equation of equilibrium, we have

$$\Sigma F_x = 0 = +F + 2$$

Both forces are positive because they act in the positive x direction. Solving for the unknown force F, we obtain

$$F = -2 \text{ kN}$$

Thus, in order for the particle to be in equilibrium, a 2-kN force acting to the *left* must be applied. The negative sign on the answer is consistent with the direction of the positive x-axis. In mechanics, the orientation of the coordinate system is arbitrary (i.e., does not affect the solution), as long as it is used consistently. Let's rework the example by reversing the direction of the x-axis. As shown in Figure 4.18(b), the positive x-axis is now directed to the *left*, but the forces remain unchanged. Writing the equation of equilibrium, we have

$$\Sigma F_x = 0 = -F - 2$$

Solving yields

$$F = -2 \text{ kN}$$

and we obtain the same answer as before. The direction of the x-axis has no influence on the answer. In both cases, the negative sign indicates that the direction of F required to maintain the particle in equilibrium is *opposite* to the assumed direction.

The examples that follow demonstrate how to find forces acting on a particle. Each example is worked in detail, in accordance with the general analysis procedure of (1) problem statement, (2) diagram, (3) assumptions, (4) governing equations, (5) calculations, (6) solution check, and (7) discussion. For the sake of simplicity, the examples are limited to coplanar force systems.

EXAMPLE 4.4

Problem statement

Two blocks hang from cords as shown in Figure 4.19. Find the tension in each cord.

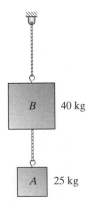

Figure 4.19.   Suspended blocks for Example 4.4.

Diagram

In order to find the tension in each cord, a separate free-body diagram is constructed for each block. The most critical part of a free-body diagram is the inclusion of every external force acting on the body in question. Two forces act on block A, its weight and the tension force in the lower cord. Three forces act on block B, its weight, the tension force in the lower cord, and the tension force in the upper cord. All forces are concurrent, so we treat the boxes as particles. The free-body diagrams for the blocks are shown in Figure 4.20(a).

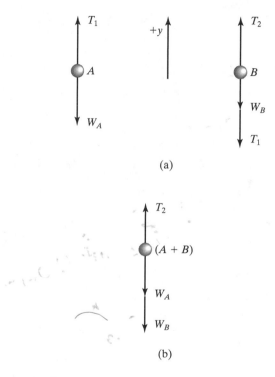

(a)

(b)

Figure 4.20. Free-body diagrams for Example 4.4.

Assumptions

1. All forces are concurrent.
2. The weights of the cords are negligible.
3. The cords are sufficiently flexible to hang straight down.

Governing equations

Because forces act in one direction only, there is only one governing equation: the equation of equilibrium for the vertical direction. Thus, for both blocks, we have

$$\sum F_y = 0$$

Calculations

To solve the problem, the equation of equilibrium must be written for both blocks. Noting the direction of the positive y-axis and using the free-body diagrams in Figure 4.20(a), we have

Block A:

$$\sum F_y = 0 = T_1 - W_A$$
$$= T_1 - (25 \text{ kg})(9.81 \text{ m/s}^2)$$

Block B:

$$\sum F_y = 0 = T_2 - T_1 - W_B$$
$$= T_2 - T_1 - (40 \text{ kg})(9.81 \text{ m/s}^2)$$

Solving the first equation for T_1, we obtain

$$T_1 = 245.25 \text{ N}$$

Substituting this value of T_1 into the second equation and solving for T_2, we obtain

$$T_2 = 637.65 \text{ N}$$

We commonly express engineering answers in three significant figures, so our answers are reported as

$$T_1 = \underline{245 \text{ N}}, T_2 = \underline{638 \text{ N}}$$

Solution Check

No mathematical or calculation-related errors are detected. Do the answers seem reasonable? The lower cord supports block A only, so tension, T_1, is simply the weight of block A. Because the upper cord supports both blocks, the tension T_2 should be the sum of the weights:

$$W_A + W_B = (m_A + m_B)\, g$$
$$= (25 \text{ kg} + 40 \text{ kg})(9.81 \text{ m/s}^2)$$
$$= 637.65 \text{ N}$$

Our solution checks out.

Discussion

An alternative method for finding the tension in the upper cord T_2 is to construct a free-body diagram of both blocks as a *single* particle. The interesting thing about this approach is that the tension forces produced by the lower cord on both blocks are ignored because they are *internal* forces, not external forces. The internal forces exerted on each block by the lower cord are equal in magnitude, but opposite in direction; hence, they cancel, thereby having no overall mechanical effect on the system. There are three external forces acting on the combined blocks, the weights of each block, and the tension T_2. Using the free-body diagram in Figure 4.20(b), we have

$$\Sigma F_y = 0 = T_2 - W_A - W_B$$
$$= T_2 - (25 \text{ kg} + 40 \text{ kg})(9.81 \text{ m/s}^2)$$

which yields

$$T_2 = 637.65 \text{ N}$$

EXAMPLE 4.5

Problem statement

A 200-kg engine block hangs from a system of cables as shown in Figure 4.21. Find the tension in cables AB and AC.

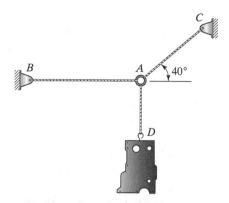

Figure 4.21. Suspended engine block for Example 4.5.

Diagram

We have a coplanar force system in which the force in each cable acts concurrently at *A*, so we construct a free-body diagram for a "particle" at *A*. (See Figure 4.22). The tension force in cable AB acts to the left along the *x*-axis, and the tension force in cable AC acts along a line 40° with respect to the *x*-axis. The tension force in cable AD, which is equivalent to the engine block's weight, acts straight down.

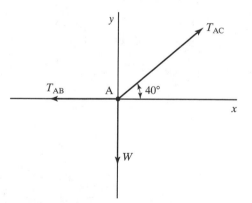

Figure 4.22. Free-body diagram for Example 4.5.

Assumptions

1.   All forces are concurrent at *A*.
2.   The weights of the cables are negligible.
3.   All cables are taut.

Governing equations

The governing equations are the equations of equilibrium in the x and y directions:

$$\Sigma F_x = 0$$
$$\Sigma F_y = 0$$

Calculations

Using the free-body diagram in Figure 4.22, the equations of equilibrium become

$$\sum F_x = 0 = -T_{AB} + T_{AC} \cos 40°$$

$$\sum F_y = 0 = T_{AC} \sin 40° - W$$

where $W = mg = (200 \text{ kg})(9.81 \text{ m/s}^2) = 1962$ N. The second equation can be immediately solved for T_{AC},

$$T_{AC} = \underline{3052 \text{ N}}$$

Substituting this value of T_{AC} into the first equation and solving for T_{AB}, we get

$$T_{AB} = \underline{2338 \text{ N}}$$

Solution check

To verify that our answers are correct, we substitute them back into the equilibrium equations. If they satisfy the equations, they are correct.

$$\Sigma F_x = -2338 \text{ N} + (3052 \text{ N}) \cos 40° = -0.032 \approx 0$$

$$\Sigma F_y = (3052 \text{ N}) \sin 40° - (200 \text{ kg})(9.81 \text{ m/s}^2) = -0.212 \approx 0$$

Within the numerical precision of the calculations, the sum of the forces in the x and y directions is zero. Our answers are therefore correct.

Discussion

Now that we know the tension forces in the cables, what do we do with them? Knowing the forces per se do not really tell us how the cables perform structurally. The next step in the analysis would be to determine the stress in each cable. If the calculated stresses are less than an allowable or design stress, the cables will support the engine block without experiencing failure. In this situation, failure most likely means cable breakage, but may also mean permanent cable strain. Stress and strain would have to be calculated in order to make a full structural assessment of the cables.

PRACTICE!

For the following practice problems, use the general analysis procedure of (1) problem statement, (2) diagram, (3) assumptions, (4) governing equations, (5) calculations, (6) solution check, and (7) discussion.

1. A 30-cm diameter solid steel sphere hangs from cables as shown. Find the tension in cables AB and AC. For the density of steel, use $\rho = 7270 \text{ kg/m}^3$.

   Answer: $T_{AB} = T_{AC} = 712.9 \text{ N}$

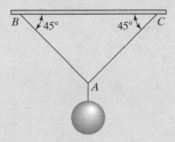

2. A 250-kg cylinder rests in a long channel as shown. Find the forces acting on the cylinder by the sides of the channel.

   Answer: 1999 N, 893 N

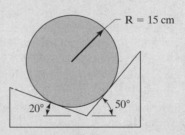

3. Three coplanar forces are applied to a box in an attempt to slide it across the floor, as shown. If the box remains at rest, what is the friction force between the box and the floor?

   Answer: 116.5 N

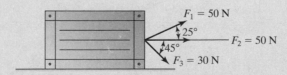

4. A 15-kg flowerpot hangs from wires as shown. Find the tension in wires AB and AC.

   Answer: $T_{AB} = 88.3 \text{ N}, T_{AC} = 117.7 \text{ N}$

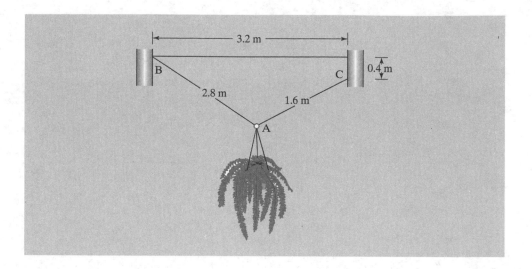

4.6   STRESS AND STRAIN

If the vector sum of the external forces acting on a body is zero, the body is in a state of static equilibrium. There are also internal forces acting on the body. Internal forces are *caused* by external forces, but internal forces do not affect the equilibrium of the body. So, if internal forces do not affect equilibrium, how are they important? To illustrate the importance of internal forces, let's use a familiar example from the sport of weight lifting. (See Figure 4.23.) As a weight lifter holds a heavy set of weights, his body and the weights are momentarily in a state of static equilibrium. The gravitational force of the weights is balanced by the force exerted on the bar by the weight lifter's hands or shoulders, and the gravitational forces of the weights plus his body are balanced by the force exerted by the floor on his feet. Are there internal forces acting on the weight lifter? Most definitely, yes. If the weight he is holding is great, he is painfully aware of those internal forces. The external forces of the weights and the reaction at the floor cause internal forces in his arms, torso, and legs. The magnitude of these internal forces usually limits the time the weight lifter can sustain his position to only a few seconds. Like the weight lifter, engineering structures such as buildings, bridges, and machines experience internal forces when external forces are applied to them. Engineering structures, however, must usually sustain internal forces for long periods of time, perhaps years. Principles of statics alone, which yield the external forces acting on a body, are insufficient to define the mechanical state of the body. In order for an engineer to make a complete assessment of the structural integrity of any body, internal forces must be considered. From the internal forces and the deformations resulting from them, stress and strain are determined.

4.6.1   Stress

The concept of stress is of prime importance in mechanics of materials. **Stress** is the primary physical quantity that engineers use to ascertain whether a structure can withstand the external forces applied to it. By finding stresses, engineers have a standard method of comparing the abilities of given materials to withstand external forces. There are two types of stress: *normal stress* and *shear stress*. In this book, we will confine our attention to normal stress. Normal stress is the *stress that acts normal (perpendicular)*

Figure 4.23. A weight lifter is in equilibrium, but his body is in a state of stress. (Art by Kathryn Hagen).

to a selected plane or axis within a body. Normal stress is often associated with the stress in the axial direction in long slender members such as rods, beams, and columns. Consider the slender bar shown in Figure 4.24. An axial force F acts on each end of the bar, maintaining the bar in equilibrium, as indicated in Figure 4.24(a). Now, suppose that we pass an imaginary plane through the bar perpendicular to its axis, as shown in Figure 4.24(b). Conceptually, we then remove the bottom portion of the bar that was "cut away" by the imaginary plane. In removing the bottom portion of the bar, we also removed the force applied at the bottom end of the bar that balanced the force applied at the top end. To restore equilibrium, we must apply an equivalent force P at the "cut" end. This force, unlike the external force applied at the top of the bar, is an *internal* force because it acts *within* the bar. The internal force P acts perpendicular to the cross-sectional area created by passing the imaginary plane through the bar, as indicated in Figure 4.24(c). The normal stress σ in the bar is defined as the internal force P divided by the cross-sectional area A:

$$\sigma = \frac{P}{A} \tag{4.23}$$

This mathematical definition of normal stress is actually an *average* normal stress, because there may be a variation of stress across the cross section of the bar. Stress variations are normally present only near points where the external forces are applied,

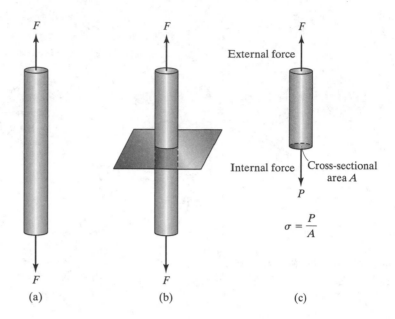

Figure 4.24. Normal stress in a rod.

however, so Equation (4.23) may be used in the majority of stress calculations without regard to stress variations. The cross section of the bar in Figure 4.24 is circular, but the quantity A represents the cross-sectional area of a member of any shape (i.e., circular, rectangular, triangular, etc.). Note that the definition of stress is very similar to that of pressure. Both quantities are defined as a force divided by an area. Accordingly, stress has the same units as pressure. Typical units for stress are kPa or MPa in the SI system and psi or ksi in the English system.

In Figure 4.24, the force vectors are directed away from each other, indicating that the bar is stretched. The normal stress associated with this force configuration is referred to as *tensile stress* because the forces place the body in tension. Conversely, if the force vectors are directed toward each other, the bar is compressed. The normal stress associated with this force configuration is referred to as *compressive stress* because the forces place the body in compression. These two force configurations are illustrated in Figure 4.25. One may think that the type of normal stress, tensile or compressive, does not matter, since Equation (4.23) says nothing about direction. However, some materials can withstand one type of stress more readily than the other. For example, concrete is stronger in compression than in tension. Consequently, concrete

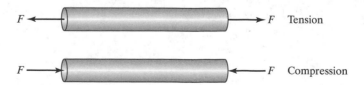

Figure 4.25. External forces for tension and compression.

is typically used in applications where the stresses are compressive, such as columns that support bridge decks and highway overpasses. When concrete members are designed for applications that involve tensile stresses, reinforcing bars are used.

4.6.2   Strain

External forces are responsible for producing stress, and they are also responsible for producing deformation. Deformation may also be caused by temperature changes. **Deformation** is defined as a *change in the size or shape of a body*. No material is perfectly rigid; hence, when external forces are applied to a body, the body changes its size or shape according to the magnitude and direction of the external forces applied to it. We have all stretched a rubber band and noticed that its length changes appreciably under a small tensile force. All materials—steel, concrete, wood, and other structural materials—deform to some extent under applied forces, but the deformations are usually too small to detect visually, so special measuring instruments are employed. Consider the bar shown in Figure 4.26. Prior to applying an external force, the bar has a length L. Now, the bar is placed in tension, applying an external force F at each end. The tensile force causes the bar to increase in length by an amount δ. The quantity δ is called the *normal deformation* or *axial deformation*, since the change in length is normal to the di-

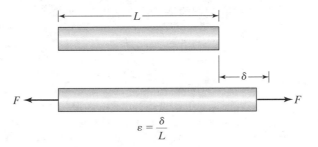

Figure 4.26.   Normal strain in a rod.

rection of the force, which is along the axis of the rod. Depending on the bar's material and the magnitude of the applied force, the normal deformation may be small, perhaps only a few thousandths of an inch. In order to normalize the change in size or shape of a body with respect to the body's original geometry, engineers use a quantity called **strain**. There are two types of strain, *normal strain* and *shear strain*. In this book, we confine our attention to normal strain. Normal strain ε is defined as the *normal deformation δ divided by the original length L*:

$$\varepsilon = \frac{\delta}{L} \tag{4.24}$$

Because strain is a ratio of two lengths, it is a dimensionless quantity. It is customary, however, to express strain as a ratio of two length units. In the SI unit system, strain is usually expressed in units of μm/m because, as mentioned before, deformations are typically small. In the English unit system, strain is usually expressed in units of in/in. Since strain is a dimensionless quantity, it is sometimes expressed as a percentage. The normal strain illustrated in Figure 4.26 is for a body in tension, but the definition given by Equation (4.24) also applies to bodies in compression.

4.6.3   Hooke's Law

About three centuries ago, the English mathematician Robert Hooke (1635–1703) discovered that the force required to stretch or compress a spring is proportional to the displacement of a point on the spring. The law describing this phenomena, known as Hooke's law, is expressed mathematically as

$$F = kx \qquad (4.25)$$

where F is force, x is displacement, and k is a constant of proportionality called the spring constant. Equation (4.25) applies only if the spring is not deformed beyond its ability to resume its original length after the force is removed. A more useful form of Hooke's law for engineering materials has the same mathematical form as Equation (4.25), but is expressed in terms of stress and strain:

$$\sigma = E\varepsilon \qquad (4.26)$$

Hooke's law, given by Equation (4.26), states that the stress σ in a material is proportional to the strain ε. The constant of proportionality E is called the **modulus of elasticity** or *Young's modulus*, after the English mathematician Thomas Young (1773–1829). Like the spring equation, the engineering version of Hooke's law applies only if the material is not deformed beyond its ability to resume its original size after the force is removed. A material that obeys Hooke's law is said to be *elastic* because it returns to its original size after the removal of the force. As strain ε is a dimensionless quantity, the modulus of elasticity E has the same units as stress. Equation (4.26) describes a straight line with E as the slope. The modulus of elasticity is an experimentally derived quantity. A sample of the material in question is subjected to tensile stresses in a special apparatus that facilitates a sequence of stress and strain measurements in the elastic range of the material. The **elastic range** is the distance or extent a material can be deformed and still be capable of returning to its original shape. Stress-strain data points are plotted on a linear scale, and a best fit straight line is drawn through the points. The slope of this line is the modulus of elasticity E.

A useful relationship may be obtained by combining Equations (4.23), (4.24), and (4.26). The axial deformation δ may be expressed directly in terms of the internal force P and the geometrical and material properties of the member. This is done by substituting the definition of strain ε given by Equation (4.24) into Equation (4.26), Hooke's law, and noting that normal stress is the internal force divided by the cross-sectional area given in Equation (4.23). Thus, the resulting expression is

$$\delta = \frac{PL}{AE} \qquad (4.27)$$

Equation (4.27) is useful because the strain does not have to be calculated first to find the deformation of the member. However, this equation is valid only over the linear region of the stress-strain curve.

4.6.4   Stress-Strain Diagram

A **stress-strain diagram** *is a graph of stress as a function of strain in a given material*. The shape of this graph varies somewhat with material, but stress-strain diagrams have some common features. A typical stress-strain diagram is illustrated in Figure 4.27. The upper stress limit of the linear relationship described by Hooke's law is called the *proportional limit*, labeled point A. At any stress between point A and the *elastic limit*, labeled point B, stress is not proportional to strain, but the material will still return to its original size after the force is removed. For many materials, the proportional and elastic limits are very close

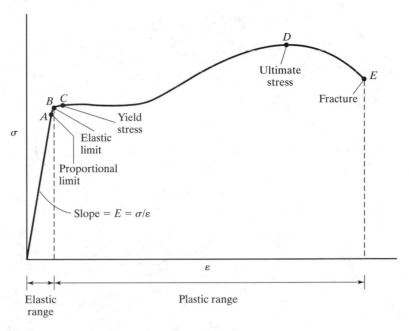

Figure 4.27. A typical stress-strain diagram.

together. Point C is called the **yield stress** or *yield strength*. Any stress above the yield stress will result in *plastic* deformation of the material (i.e., the material will not return to its original size, but will deform permanently). As the stress increases beyond the yield stress, the material experiences a large increase in strain for a small increase in stress. At about point D, called the **ultimate stress** or *ultimate strength*, the cross-sectional area of the material begins to decrease rapidly until the material experiences *fracture* at point E.

In the next example, we use the general analysis procedure of (1) problem statement, (2) diagram, (3) assumptions, (4) governing equations, (5) calculations, (6) solution check, and (7) discussion.

EXAMPLE 4.6

Problem statement

A 200-kg engine block hangs from a system of cables as shown in Figure 4.28. Find the normal stress and axial deformation in cables AB and AC. The cables are 0.7 m long and have a diameter of 4 mm. The cables are steel with a modulus of elasticity of $E = 200$ GPa.

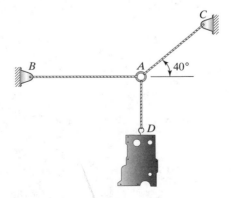

Figure 4.28. Suspended engine block for Example 4.6.

Diagram

We will presume that the statics portion of the problem has been solved, so a free-body diagram of the entire system is unnecessary. Diagrams showing a cross section of the cables and the corresponding internal forces are sufficient. (See Figure 4.29.)

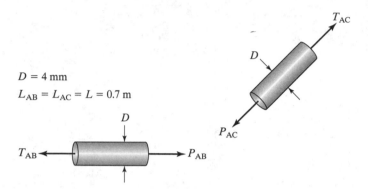

$D = 4$ mm
$L_{AB} = L_{AC} = L = 0.7$ m

Figure 4.29. Cables for Example 4.6.

Assumptions

1. Cables are circular in cross section.
2. Cables have the same modulus of elasticity.
3. Stress is uniform in the cables.

Governing equations

Cross-sectional area:

$$A = \frac{\pi D^2}{4}$$

Normal stress:

$$\sigma = \frac{P}{A}$$

Axial deformation:

$$\delta = \frac{PL}{AE}$$

Calculations

The cross-sectional area of the cables is

$$A = \frac{\pi D^2}{4}$$

$$= \frac{\pi(0.004 \text{ m})^2}{4} = 1.2566 \times 10^{-5} \text{ m}^2$$

From a prior statics analysis, the tensions in cables AB and AC are 2338 N and 3052 N, respectively. Hence, the normal stress in each cable is

$$\sigma_{AB} = \frac{P_{AB}}{A}$$

$$= \frac{2338 \text{ N}}{1.2566 \times 10^{-5} \text{ m}^2} = 186.1 \times 10^6 \text{ N/m}^2 = \underline{186.1 \text{ MPa}}$$

$$\sigma_{AC} = \frac{P_{AC}}{A}$$

$$= \frac{3052 \text{ N}}{1.2566 \times 10^{-5} \text{ m}^2} = 242.9 \times 10^6 \text{ N/m}^2 = \underline{242.9 \text{ MPa}}$$

The deformation in each cable is

$$\delta_{AB} = \frac{P_{AB}L}{AE}$$

$$= \frac{(2338 \text{ N})(0.7 \text{ m})}{(1.2566 \times 10^{-5} \text{ m}^2)(200 \times 10^9 \text{ N/m}^2)} = 6.51 \times 10^{-4} \text{ m} = \underline{0.651 \text{ mm}}$$

$$\delta_{AC} = \frac{P_{AC}L}{AE}$$

$$= \frac{(3052 \text{ N})(0.7 \text{ m})}{(1.2566 \times 10^{-5} \text{ m}^2)(200 \times 10^9 \text{ N/m}^2)} = 8.50 \times 10^{-4} \text{ m} = \underline{0.850 \text{ mm}}$$

Solution check

One way to check the validity of the results is to compare the relative magnitudes of the stress and deformation in each cable. The internal force in cable *AC* is greater than the normal stress in cable *AB*. Consequently, the normal stress and axial deformation in cable *AC* must also be greater because the cables are geometrically and materially identical. Our calculations show that this is indeed the case.

Discussion

The deformations are small, less than a millimeter in both cables. These deformations would probably not be significant in an engine hoist application and would not be perceptible by the naked eye. Are the stresses excessive? Will they plastically deform the cables? To answer these questions, we must know something about the yield stress of the cable material and the stresses for which the cables were designed.

PRACTICE!

In the following practice problems, use the general analysis procedure of (1) problem statement, (2) diagram, (3) assumptions, (4) governing equations, (5) calculations, (6) solution check, and (7) discussion.

1. A solid rod of stainless steel ($E = 190$ GPa) is 50 cm in length and has a 4 mm $\times$ 4 mm cross section. The rod is subjected to an axial tensile force of 8 kN. Find the normal stress, strain, and axial deformation.

    Answer: 500 MPa, 1.32 mm, 0.00263

2. A 25-cm-long 10-gauge wire of yellow brass ($E = 105$ GPa) is subjected to an axial tensile force of 1.75 kN. Find the normal stress and deformation in the wire. A 10-gauge wire has a diameter of 2.588 mm.
   Answer: 333 MPa, 0.792 mm

3. An 8-m-high granite column sustains an axial compressive load of 500 kN. If the column shortens 0.12 mm under the load, what is the diameter of the column? For granite, $E = 70$ GPa.
   Answer: 0.779 m

4. A solid rod with a length and diameter of 1 m and 5 mm, respectively, is subjected to an axial tensile force of 20 kN. If the axial deformation is measured as $\delta = 1$ cm, what is the modulus of elasticity of the material?
   Answer: 1018 GPa

5. A plastic ($E = 3$ GPa) tube with an outside and inside diameter of 6 cm and 5.4 cm, respectively, is subjected to an axial compressive force of 12 kN. If the tube is 25 cm long, how much does the tube shorten under the load?
   Answer: 1.86 mm

4.7 DESIGN STRESS

Most engineering structures are not designed to deform permanently or fracture. Every member in a structure must maintain a certain degree of dimensional control to assure that it does not plastically deform, thereby losing its size or shape, interfering with surrounding structures or other members in the same structure. Obviously, the members must not fracture either, because this would lead to a catastrophic failure that would result in material and financial loss and perhaps the loss of human life. Therefore, members in most structures are designed to sustain a maximum stress that is *below* the yield stress on the stress-strain diagram for the particular material used to construct that member. This maximum stress is called the *design stress* or **allowable stress**. When a properly designed member is subjected to a load, the stress in the member will not exceed the design stress. Because the design stress is within the elastic range of the material, the member will return to its original dimensions after the load is removed. A bridge, for example, sustains stresses in its members while traffic passes over it. When there is no traffic, the members in the bridge return to their original size. Similarly, while a boiler is operating, the pressure vessel sustains stresses that deform it, but when the pressure is reduced to atmospheric pressure, the vessel returns to its original dimensions.

If a structural member is designed to carry stresses below the yield stress, how does an engineer choose what the allowable stress should be? And why choose a stress below the yield stress in the first place? Why not design the member by using the yield stress itself, since that would allow the member to carry the maximum possible load? Engineering design is not an exact science. If it was, structures could be designed with ultimate precision by using the yield stress, or any other stress for that matter, as the design stress, and the design stress would never be exceeded while the structure was in service. Because design is not an exact science, engineers incorporate an allowance in their designs that takes into account the following uncertainties:

1. *Loadings* The design engineer may not anticipate every type of loading or the number of loadings that may occur. Vibration, impact, or accidental loadings may occur that were not accounted for in the design of the structure.

2. *Failure modes* Materials can fail by one or more of several different mechanisms. The design engineer may not have anticipated every failure mode by which the structure can possibly fail.

3. *Material properties* Physical properties of materials are subject to variations during manufacture, and there are experimental uncertainties in their numerical values. Properties may also be altered by heating or by deformation during manufacture, handling, and storage.

4. *Deterioration* Exposure to the elements, poor maintenance, or unexpected natural phenomena may cause a material to deteriorate, thereby compromising its structural integrity. Various types of corrosion are the most common forms of material deterioration.

5. *Analysis* Engineering analysis is a critical part of design, and analysis involves making simplifying assumptions. Thus, analytical results are not precise, but are approximations.

To account for the uncertainties listed, engineers use a design or allowable stress based on a parameter called the **factor of safety**. The factor of safety (*F.S.*) is defined as the *ratio of the failure stress to the allowable stress*:

$$F.S. = \frac{\sigma_{\text{fail}}}{\sigma_{\text{allow}}} \tag{4.28}$$

Because the yield stress is the stress above which a material plastically deforms, the yield stress σ_y is commonly used as the failure stress σ_{fail}. The ultimate stress σ_u may also be used. The failure stress is always greater than the allowable stress, so $F.S. > 1$. The value chosen for the factor of safety depends on the type of engineering structure, the relative importance of the member compared with other members in the structure, the risk to property and life, and the severity of the design uncertainties previously listed. For example, to minimize weight, the factor of safety for aircraft and spacecraft structures is typically close to 1. However, the factor of safety for ground-based structures such as dams, bridges and buildings may be higher, perhaps 1.5 or 2. High-risk structures that pose a safety hazard to people in the event of failure, such as certain nuclear power plant components, may have a factor of safety as high as 3. Factors of safety for structural members in specific engineering systems have been standardized through many years of testing and industrial evaluation. Factors of safety are often defined by building codes or engineering standards established by city, state, or federal agencies and professional engineering societies.

APPLICATION: DESIGNING A TURNBUCKLE

Turnbuckles are special mechanical fasteners that facilitate connections between cables, chains, or cords. A basic turnbuckle consists of a slender, cylindrical shaped body threaded on each end to accept an eyebolt, a hook, or other type of tying component. The tension in the cables that are tied to a turnbuckle is adjusted by rotating the body of the turnbuckle. Turnbuckles are designed such that tightening or loosening may be accomplished without twisting the cables. Like the cables that are connected to them, turnbuckles must sustain the tensile stresses to which they are subjected. Consider a turnbuckle used to adjust the tension in a cable that stabilizes a communications tower. From a prior analysis, the tension in the cable is determined to be 25kN. The loaded turnbuckle is shown in Figure 4.30(a). Let us suppose that, as a new engineer, your first job is to select a turnbuckle for this application. Turnbuckles are available in a variety of sizes and materials from several suppliers. Hardware suppliers specify the maximum recommended load that a particular turnbuckle can sustain without failing. It is, therefore, a simple matter for you, the end user, to

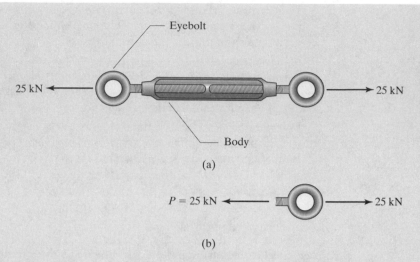

Figure 4.30. A loaded turnbuckle.

select a turnbuckle with a recommended maximum load that is somewhat greater than the actual load of 25kN. But how did the engineers who designed the turnbuckle obtain this load value? The example that follows shows how fundamental concepts of stress and factor of safety may be used to design the eyebolt portion of a turnbuckle. The general analysis procedure of (1) problem statement, (2) diagram, (3) assumptions, (4) governing equations, (5) calculations, (6) solution check, and (7) discussion, is used.

Problem statement

Determine the minimum-diameter eyebolt in a turnbuckle used to stabilize a communications tower. The tensile force in the cable is 25kN. The eyebolt is to be made of AISI 4130 steel, a high-strength forging steel. (AISI is an abbreviation for American Iron and Steel Institute.) To account for potential high wind loads and other uncertainties, use a factor of safety of 2.0.

Diagram

The internal and external forces acting on the eyebolt are shown in Figure 4.30(b).

Assumptions

1. Stress is uniform in the eyebolt.
2. Stress in the eyebolt is purely axial.
3. Consider stress in the main body of the eyebolt only, not the threads.

Governing equations

The governing equations for this, problem are the cross-sectional area for a circular bolt, the definition of normal stress, and the factor of safety.

$$A = \frac{\pi D^2}{4} \qquad \text{(a)}$$

$$\sigma_{\text{allow}} = \frac{P}{A} \qquad \text{(b)}$$

$$F.S. = \frac{\sigma_{\text{fail}}}{\sigma_{\text{allow}}} = \frac{\sigma_y}{\sigma_{\text{allow}}} \qquad \text{(c)}$$

Calculations

In the third governing equation, we have used the yield stress σ_y as the failure stress. The yield stress of AISI 4130 steel is 760 MPa. The objective of the analysis is to find the diameter D of the eyebolt required to sustain the applied load. There are three unknown quantities: σ_{allow}, A, and D. Because the three governing equations are not dependent, we may combine them algebraically to obtain the diameter, D. Upon substituting Equation (a) into Equation (b) and then Equation (b) into Equation (c) we obtain

$$D = \left(\frac{4 \, P.F.S.}{\pi \sigma_y} \right)^{1/2}$$

$$= \left(\frac{4(25 \times 10^3 \, \text{N})(2.0)}{\pi(760 \times 10^6 \, \text{Pa})} \right)^{1/2}$$

$$= 9.15 \times 10^{-3} \, \text{m} = \underline{9.15 \, \text{mm}}$$

Solution check

No errors were found.

Discussion

The minimum eyebolt diameter that will sustain the applied load with a factor of safety of 2.0 is 9.15 mm. In English units, this diameter is

$$D = 9.15 \text{ mm} \times \frac{1 \text{ in}}{25.4 \text{ mm}} = 0.360 \text{ in}$$

Bolts come in standard diameters, and 0.360 in is not a standard size. Bolts are typically available in standard sizes of $\frac{1}{4}$-in, $\frac{5}{16}$-in, $\frac{3}{8}$-in etc. A $\frac{5}{16}$-in (0.3125 in) bolt is too small, so the $\frac{3}{8}$-in (0.375 in) should be chosen, even though it is slightly larger than required. It should be emphasized that this analysis reflects only a part of the analysis that would be required in the total design of a turnbuckle. Stresses in the threads of the eyebolt and the turnbuckle body, as well as the main body of the turnbuckle itself, would also have to be calculated.

PRACTICE!

In the practice problems, use the general analysis procedure of (1) problem statement, (2) diagram, (3) assumptions, (4) governing equations, (5) calculations, (6) solution check, and (7) discussion.

1. A rod of aluminum 6061-T6 has a square cross section measuring 0.25 in × 0.25 *in*. Using the yield stress as the failure stress, find the maximum tensile load that the rod can sustain for a factor of safety of 1.5. The yield stress of aluminum 6061-T6 is 240 MPa.

   Answer: 6.45 kN

2. A concrete column with a diameter of 60 cm supports a portion of a highway overpass. Using the ultimate stress as the failure stress, what is the maximum compressive load that the column can carry for a factor of safety of 1.25? For the ultimate stress of concrete, use $\sigma_u = 40$ MPa.

   Answer: 9.05 MN

3. A column of rectangular cross section constructed from fir timber is subjected to a compressive load of 6 MN. If the width of the column is 12 cm, find the depth required to sustain the load with a factor of safety of 1.6. The ultimate stress of fir is $\sigma_u = 50$ MPa.

   Answer: 1.60 m

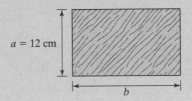

KEY TERMS

allowable stress
Cartesian unit vector
deformation
elastic range
equilibrium
factor of safety
force
force system

free-body diagram
internal force
mechanics
modulus of elasticity
resultant
resultant force
scalar
statics

strain
stress
stress-strain diagram
ultimate stress
vector
yield stress

REFERENCES

Bedford, A. and W. Fowler, *Engineering Mechanics: Statics*, 3d ed., Upper Saddle River, NJ: Prentice Hall, 2002.

Beer, F.P., E.R. Johnston, E.R. Eisenberg, and G.H. Staab, *Vector Mechanics for Engineers: Statics*, 7th ed. NY: McGraw-Hill, 2003.

Johnston, E.R. and J.T. DeWolf, *Mechanics of Materials*, 3d ed., NY: McGraw-Hill, 2002.

Hibbeler, R.C., *Engineering Mechanics: Statics*, 10th ed., Upper Saddle River, NJ: Prentice Hall, 2004.

Hibbeler, R.C., *Mechanics of Materials*, 5th Edition, Upper Saddle River, NJ: Prentice Hall, 2003.

Problems

1. Find the resultant force for the forces shown in Figure P4.1 (a) by using the parallelogram law and (b) by resolving the forces into their x and y components.

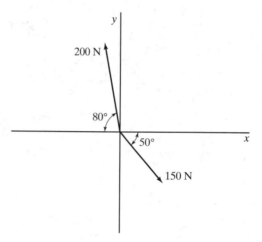

Figure P4.1.

2. Find the resultant force for the forces shown in Figure P4.2 (a) by using the parallelogram law and (b) by resolving the forces into their x and y components.

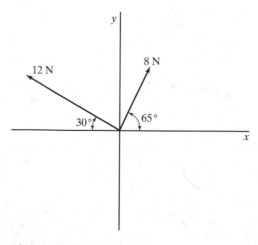

Figure P4.2.

3.   For the three forces shown in Figure P4.3, find the resultant force, its magnitude, and direction with respect to the x-axis.

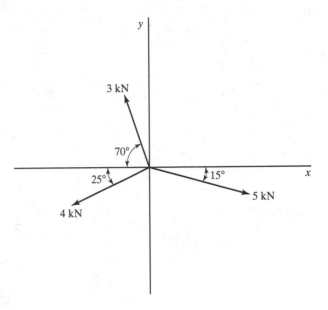

Figure P4.3.

4.   For the three forces shown in Figure P4.4, find the resultant force, its magnitude, and direction with respect to the x-axis.

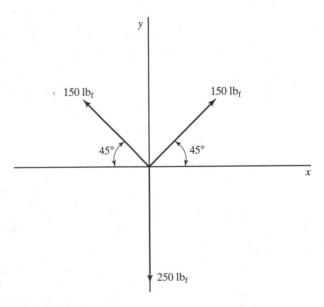

Figure P4.4.

5.   Consider the three forces $F_1 = 3i + 2j − 5k$ N, $F_2 = −4i − 7j + 2k$ N, and $F_3 = i − j − k$ N. Find the resultant force and its magnitude.

6.   A particle is subjected to three forces: $F_1 = 3i + 5j − 8k$ kN, $F_2 = −2i − 3j + 4k$ kN, and $F_3 = −i − 2j + 5k$ kN. Is this particle in equilibrium? Explain.

7. A particle is subjected to three forces: $\mathbf{F}_1 = 6\mathbf{i} - a\mathbf{j} - 7\mathbf{k}$ N, $\mathbf{F}_2 = -4\mathbf{i} - 5\mathbf{j} + b\mathbf{k}$ N, and $\mathbf{F}_3 = c\mathbf{i} - 2\mathbf{j} + 4\mathbf{k}$ N. Find the values of the scalars a, b, and c such that the particle is in equilibrium.

8. Find the magnitudes of forces $\mathbf{F}_1$ and $\mathbf{F}_2$ so that the particle P is in equilibrium. (See Figure P4.8.)

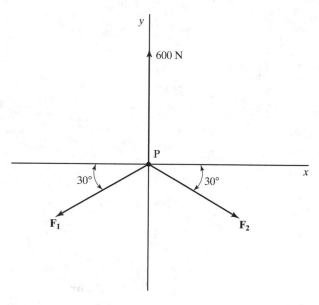

Figure P4.8.

9. Find the magnitude of the force $\mathbf{F}$ and its direction θ so that the particle P is in equilibrium. (See Figure P4.9.)

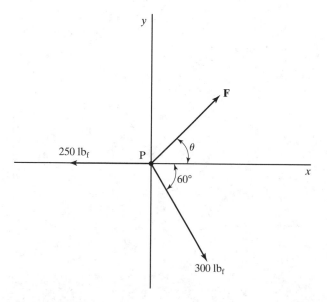

Figure P4.9.

10.   A gusset plate is subjected to the forces shown in Figure P4.10. Find the magnitude and direction θ of the force in member B so that the plate is in equilibrium.

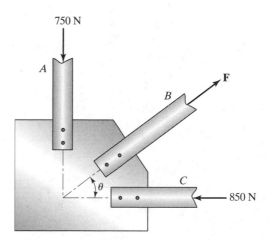

Figure P4.10.

For problems 11 through 27, use the general analysis procedure of (1) problem statement, (2) diagram, (3) assumptions, (4) governing equations, (5) calculations, (6) solution check, and (7) discussion.

11.   A 400-kg crate hangs from ropes as shown in Figure P4.11. Find the tension in ropes AB and AC.

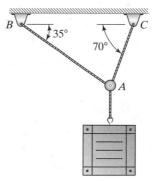

Figure P4.11.

12.   A 250-lb$_m$ box is held in place by a cord with a spring scale on an inclined plane as shown in Figure P4.12. If all surfaces are smooth, what is the force reading on the scale?

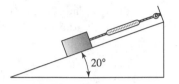

Figure P4.12.

13. A concrete pipe with an inside and outside diameter of 60 cm and 70 cm, respectively, hangs from cables as shown in Figure P4.13. The pipe is supported at two locations, and a spreader bar maintains cable segments AB and AC at 45°. Each support carries half the total weight of the pipe. If the density of concrete is $\rho = 2320 \text{ kg} / \text{m}^3$, find the tension in cable segments AB and AC.

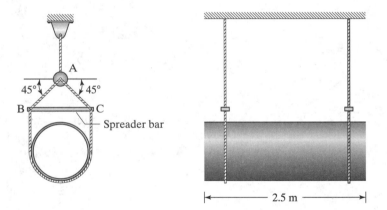

Figure P4.13.

14. A construction worker holds a 600-lb$_f$ crate in the position shown in Figure P4.14. What force must the worker exert on the cable?

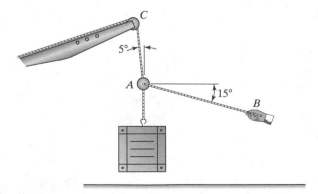

Figure P4.14.

15.   A 10-slug cylinder is suspended by a cord and frictionless pulley system as shown in Figure P4.15. A person standing on the floor pulls on the free end of the cord to hold the cylinder in a stationary position. What is the person's minimum weight for this to be possible?

Figure P4.15.

16.   A 150-N traffic signal is suspended from a symmetrical system of cables as shown in Figure P4.16. Find the tension in all cables. Cable BC is horizontal.

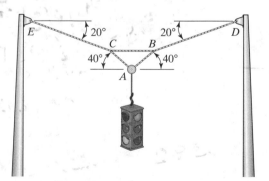

Figure P4.16.

17.   A 1.4-m-diameter wrecking ball hangs motionless from a 1.75-cm diameter cable. The wrecking ball is solid and is constructed of steel ($\rho = 7800 \ \text{kg} \ / \ \text{m}^3$). If the cable is 18 m long, how much does the cable stretch? For the cable, use $E = 175$ GPa.

18.   A 4-m-high wooden column with a rectangular cross section is subjected to a 210 kN axial compressive force. The modulus of elasticity of the wood is 13

GPa. If one side of the column is 25 cm across, find the minimum dimension of the other side to keep the columns deformation under 1.3 mm.

19.   The column is subjected to a 15-kN force as shown in Figure P4.19. Find the average normal stress in the column.

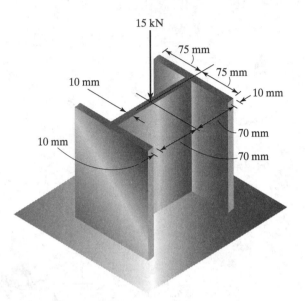

15 kN

75 mm

75 mm

10 mm

10 mm

10 mm

70 mm

70 mm

10 mm

Figure P4.19.

20.   A solid composite shaft is subjected to a 2-MN force as shown in Figure P4.20. Section AB is red brass ($E = 120$ GPa), and section BC is AISI 1010 steel

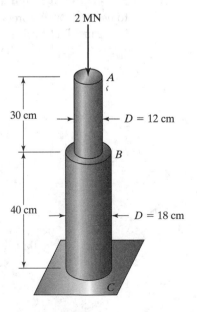

2 MN

A

30 cm

$D = 12$ cm

B

40 cm

$D = 18$ cm

C

Figure P4.20.

$(E = 200 \text{ GPa})$. Find the normal stress in each section and the total axial deformation of the shaft.

21. A tapered column of concrete $(E = 30 \text{ GPa})$ is subjected to a 200-kN force as shown in Figure P4.21. Find the axial deformation of the column. *Hint*: Express the cross-sectional area A as a function of x and perform the integration:

$$\delta = \int_0^L \frac{P \, dx}{A(x)E}$$

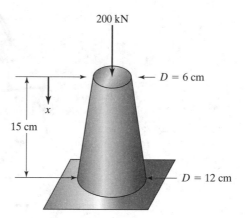

200 kN

$D = 6$ cm

x

15 cm

$D = 12$ cm

Figure P4.21.

22. A 12-cm × 12-cm square plate of AISI 1010 steel $(E = 200 \text{ GPa})$ is subjected to normal tensile forces of 15 kN and 20 kN on the top and right edges as shown in Figure P4.22. The thickness of the plate is 5 mm, and the left and bottom edges of the plate are fixed. Find the normal strain and deformation of the plate in the horizontal and vertical directions.

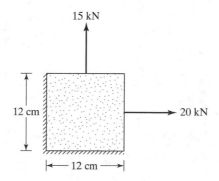

15 kN

12 cm

20 kN

12 cm

Figure P4.22.

23. A tensile test is conducted on a steel specimen with a diameter of 8.0 mm and a test length of 6.0 cm. The data is shown in the table. Plot the stress-strain diagram, and find the approximate value of the modulus of elasticity for the steel.

Load (kN)	Deformation (mm)
2.0	0.0119
5.0	0.0303
10.0	0.0585
15.0	0.0895
20.0	0.122
25.0	0.145

24. A 1-cm-diameter, 0.4-m-long steel rod is to be used in an application where it will be subjected to an axial tensile force of 15 kN. The factor of safety based on yield stress must be at least 1.5, and the axial deformation must not exceed 2 cm. Is the steel whose stress-strain diagram is shown in Figure P4.24 suitable for this application? Explain.

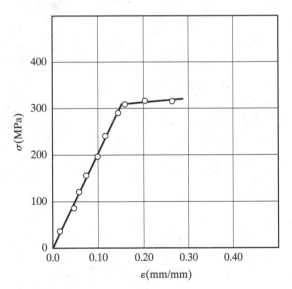

Figure P4.24.

25. A 4-cm-diameter rod of AISI 302 stainless steel is to subjected to an axial tensile force such that the factor of safety based on the yield stress is 1.75. Find the maximum allowable tensile force. The yield stress of AISI 302 stainless steel is $\sigma_y = 520$ MPa.

26.   A 18-in-diameter sandstone column is subjected to an axial compressive force of 2×10^6 lb$_\text{f}$. Find the factor of safety based on the ultimate compressive stress. For the ultimate compressive stress of sandstone, use $\sigma_u = 85$ MPa.

27.   A slender cylindrical member in a child's toy made of polystyrene plastic is to be subjected to an axial tensile force of 400N. Find the minimum diameter of this member for a factor of safety of 1.5 based on yield stress. The yield stress for polystyrene is $\sigma_y = 55$ MPa.

5

Electrical Circuits

5.1   INTRODUCTION

Electrical engineering is one of the most diverse and well-established branches of engineering. Electrical engineers design systems and devices that harness the power of electricity to perform a variety of tasks. Electricity is one of the most useful forms of energy, and it impacts our everyday lives in fundamental ways. Without electricity, commonplace, but important devices such as automobiles, aircraft, computers, household appliances, telephones, television, radio, and electric lights would not exist. The historical roots of electricity can be traced to such notable scientists, engineers and technologists as Alessandro Volta (1745–1827), Andre Ampere (1775–1836), Georg Ohm (1787–1854), Michael Faraday (1791–1867), James Joule (1818–1889), Heinrich Hertz (1857–1894), and Thomas Edison (1847–1931). These individuals, among others, established the fundamental theoretical and practical foundations of electrical phenomena. This chapter deals with a category of electrical engineering referred to as *electrical circuits*. In nearly every electrical engineering curriculum, *electrical circuit analysis* is one of the first courses taken by the engineering student. The principles covered in basic electrical circuit theory are so important that even nonelectrical engineering majors are often required to take at least one course in the subject. Nearly all branches of electrical engineering are fundamentally based on circuit theory. The only subject in electrical engineering that is more fundamental than circuit theory is electromagnetic field theory, which deals with the physics of electromagnetic fields and waves.

As an electrical engineering subject, electrical circuits may be broken down into two general areas: *power* and *signal*. Power may be subdivided further into three categories, *power generation, heating and lighting*, and *motors and generators*. Similarly, signal may be divided into three

OBJECTIVES

After reading this chapter, you will have learned

- The relationship between charge and current
- The concept of voltage
- The concept of resistance
- How to combine resistances in series and parallel
- How to use Ohm's law
- How to analyze simple DC circuits
- How to use Kirchhoff's laws of circuit analysis

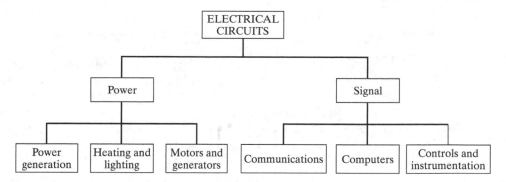

Figure 5.1. Topical structure of electrical circuits.

subcategories: *communications, computers,* and *controls and instrumentation*. This structure is schematically illustrated in Figure 5.1. Power deals with systems designed to provide electrical energy to various mechanical and electrical devices. Power generation refers to the production and transmission of electrical power by power plants. The energy source from which these power stations derive electrical power is typically fossil fuels, nuclear materials, or moving water. To a lesser degree, solar or wind power is also used. Electrical energy is required to run heating and cooling equipment such as furnaces, electric heaters, boilers, and air conditioners. The illumination provided by incandescent and fluorescent lights requires electrical power. Motors are found in numerous systems, including refrigerators, furnaces, fans, CD players, kitchen appliances, and printers. In motors, electrical energy is converted to mechanical energy via a rotating shaft. Unlike a motor, a generator is used to convert mechanical energy to electrical energy. Generators are used in power plants, automobiles, and other power systems. The signal area of electrical circuits deals with systems that transmit and process information. The power transmitted is not a primary consideration in signal applications. Communications refers to the transmission of information via electrical signals. Telephone, television, radio, and computers are types of communication systems. The heart of a computer is its digital circuits, circuits that utilize logic operations for the rapid processing of information. Computers are such a dominant area of engineering that electrical engineering programs in the United States are referred to as electrical and computer engineering to give students the option of focusing on the hardware (electrical) or the software, firmware, and operating systems (computer) aspects of the field. Controls are special circuits that activate or adjust other electrical or mechanical devices. A thermostat that turns a furnace or air conditioner on and off is a simple example. Instrumentation circuits are used to process electrical signals generated by various types of sensors that control a device. For example, an automobile has a circuit that processes an electrical signal generated by a temperature sensor in the cooling system. If the temperature exceeds a certain value, the circuit activates a visual display that warns the driver of an overheating condition.

Now that the basic topical structure of electrical circuits has been defined, what is an electrical circuit? An **electrical circuit** may be defined as *two or more electrical devices interconnected by conductors*. In electrical circuits, there are numerous types of electrical devices such as resistors, capacitors, inductors, diodes, transistors, transformers, batteries, lamps, fuses, switches, and motors. The "conductors" that interconnect these devices is usually a wire or a metal pathway integrated on a printed circuit board. Electrical circuits can be very simple, such as the circuit in a flashlight containing two batteries, a lightbulb, and a switch. Most electrical circuits, however, are much more complex than a flashlight. A standard television contains, among other things, power supplies, amplifiers, speakers, and a cathode ray tube. The microprocessor in a computer may contain the equivalent of millions of transistors interconnected in a single chip that is smaller than a business card. Electrical engineers use principles of electrical circuit theory to analyze and design a wide variety of systems. Look around you. How many devices do you see nearby that utilize electricity for their operation? You are probably not reading this book by the illumination of candles, but by incandescent or fluorescent lights. The room most likely has several electrical outlets on the walls that facilitate the operation of various electrical devices such as computers, vacuum cleaners, clocks, toasters, microwave ovens, etc. Electrical devices are so pervasive that we take them for granted, but our world would be vastly different without them. To anyone born in an industrialized nation during the latter half of the 20th century, a world without television, stereo, cellular phones, and CD players would seem foreign and strange. Electrical devices change rapidly, being driven by the ever increasing need for higher speed, smaller size, and lower cost. This period of time saw gargantuan mainframe computers with thousands of heat-generating vacuum tubes evolve into desktop computers with cracker-sized microprocessors. The second half of the 20th century also witnessed sweeping improvements in telecommunications, automotive, electronics, and automation.

All electrical devices have circuits of one kind or another, and the electrical engineer must know how to design these circuits to perform specific electrical functions. Some familiar examples of devices that have electrical circuits are shown in Figures 5.2 and 5.3. Students must learn the fundamental principles of electrical circuits before proceeding with more advanced study in circuit analysis and other electrical engineering courses.

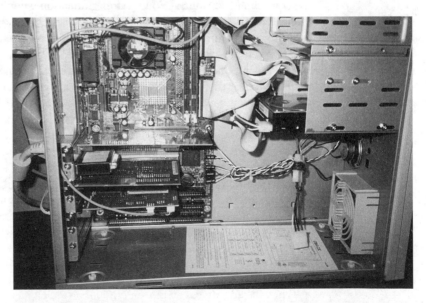

Figure 5.2.   A computer contains electrical circuits that perform a variety of functions.

Figure 5.3. In future space missions, a nano rover for planetary surface exploration will rely on miniature electrical circuits for its operation. (Image courtesy of NASA).

PROFESSIONAL SUCCESS: RETAIN YOUR COURSE MATERIALS

Engineering students may sometimes wonder, "How much of my engineering course materials should I keep after completing a course or after graduation? Should I sell my textbooks back to the bookstore? Should I discard my lecture notes, exams, and laboratory reports? Will I need these materials after I graduate?" The engineering curriculum is a challenging academic road to travel. By the time you graduate, you will have devoted much time and energy and spent a lot of money in the pursuit of your engineering degree. Do not trivialize this great commitment by squandering your textbooks for a few dollars. As you complete each engineering course, keep your books and other course materials for reference in future engineering courses. Engineering courses build on one another, so you will most likely need these resources to help you learn new material. Never sell an engineering text back to the bookstore just because you are a little short of cash. Your engineering texts are a wellspring of information, the backbone of your engineering course work. Will you need your books even

after graduation when you have secured employment as an engineer? Depending on the nature of your engineering position and the company you work for, your college textbooks could be a valuable resource, particularly in engineering design and analysis. Because you do not know exactly what kinds of engineering activities you will be involved in after graduation, keep your textbooks.

At the end of each course, organize your lecture notes, lab reports, homework problems, exams, and other materials into a three-ring binder. Label the binder with the course name and number. Divide the binder into sections with dividers and labeled tabs. You will probably need a section for lecture notes, homework problems, exams, quizzes, and laboratory reports. Depending on the nature of the course, other sections may be required. In addition to your engineering courses, you should probably keep materials from technical support courses such as physics, chemistry, and mathematics. Retaining course materials will help you as a student and as a practicing engineer.

5.2   ELECTRIC CHARGE AND CURRENT

We are familiar with forces caused by bodies in contact with other bodies and gravity. Forces exerted on bodies by other bodies are commonly encountered in a variety of everyday situations and engineering structures. The gravitational force is an attractive force that tends to move objects toward one another, the most common example being the earth's gravitational force that attracts objects toward the center of the earth, thereby maintaining those objects on its surface. Gravitational forces govern the motions of planets, stars, galaxies, and other celestial objects in the universe, and yet it is the weakest of all the natural forces. A type of force that is much stronger than gravity is electrical in nature. Electrical forces are produced by **electric charges**. An electrical force is established between two charged particles when they are in proximity. The force between the particles is attractive if the charges are unlike (i.e., if one charge is positive and the other is negative). The force is repulsive if the charges are alike, that is, if both charges are either positive or negative. (See Figure 5.4.) Charges are created by producing an imbalance in the number of charged particles in the atom. Atoms consist of a nucleus composed of neutrons (neutral particles) and protons (positively charged particles) surrounded by a cloud of electrons (negatively charged particles). An atom with the same number of protons and electrons is electrically neutral (has no charge), because the positive charge of the protons precisely balances the negative charge of the electrons. The atom may become positively charged by losing electrons or negatively charged by gaining electrons from other atoms. For example, rubbing a silk cloth over a glass rod strips some electrons from the surface atoms of the glass, adding them to the atoms of the cloth, thereby creating a negatively charged rod. A negative charge may also be produced on a balloon by rubbing it against our hair. This force is referred to as an *electrostatic* force because the charges are static or stationary. The branch of electrical studies that deals with static charges is called *electrostatics*.

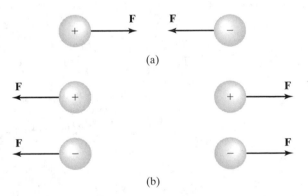

Figure 5.4.   (a) Unlike charges attract and (b) like charges repel.

Electric charges are quantified by means of a physical parameter called the *coulomb* (C). The coulomb, named in honor of the French physicist Charles Coulomb is defined as the *charge possessed by approximately* 6.242×10^{18} *electrons*. Another way to define the coulomb is to state that a single electron has a charge of approximately 1.602×10^{-19} C, the inverse of 6.242×10^{18}. The charge on a single electron is said to be quantized, because it is the smallest amount of charge that can

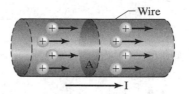

Figure 5.5. Current is the passage of electric charges through a cross-sectional area in a conductor.

exist. Symbols typically used for electric charge are Q or q. The symbol Q usually denotes a constant charge, such as $Q = 2C$, whereas the symbol q usually denotes a charge that is changing with time. In the latter case, charge is sometimes written in the functional form $q(t)$.

When electric charges of the same sign move, an **electric current** is said to exist. To define electric current more precisely, consider the charges moving in a wire perpendicular to a cross-sectional area, A. (See Figure 5.5.) Electric current I is defined as the *rate at which charge flows through the area*. The *average* current that flows through the area may be written in terms of the amount of charge Δq that passes through the area in a given time interval Δt, as

$$I_{av} = \frac{\Delta q}{\Delta t} \qquad (5.1)$$

If the current changes with time, the rate at which charge flows through the area A also changes with time, and the current is an *instantaneous* current expressed as a derivative

$$I = \frac{dq}{dt} \qquad (5.2)$$

The SI unit for electric current is the ampere (A). From its definition, given by Equations (5.1) and (5.2), 1 A of current is equivalent to 1 C of charge passing through the area in 1 s. Hence, 1 A = 1 C/s. Because the ampere is one of the seven base dimensions, electric charge may be alternatively defined as the charge transferred in 1 s by a current of 1 A. To give you a physical feel for current, 1 A is approximately the current that flows through the filament of a 115-V, 100-watt lightbulb. Some electrical devices, such as CD players and radios, may utilize very small currents, on the order of mA or even μA. For example, a typical flashlight utilizes about 300 mA. A toaster may utilize around 8 A, and an electric kitchen range or electric dryer may utilize 15 A or more. The total current supplied to a typical home is around 200 A. Large machines used in heavy industries may utilize hundreds or even thousands of amperes.

In electric circuit theory, current is generally considered to be the movement of *positive* charges. This convention is based on the work of Benjamin Franklin (1706–1790), who conjectured that electricity flowed from positive to negative. Today, we know that electric current in wires and other conductors is due to the drift of free electrons (negatively charged particles) in the atoms of the conductor. When dealing with electric current, we need to distinguish between *conventional current* (the movement of positive charges) and *electron current* (the movement of free electrons). In a real sense, however, it does not matter whether we use conventional current or electron current, because positive charges moving to the right is equivalent to negative charges moving to the left. The only thing that matters is that we use the same sign convention consistently. By adoption, conventional current is generally used in electrical circuit analysis.

There are several types of current in use in various electrical devices, but we will study the two major types. **Direct current (DC)** is a flow of charge in which the direction of flow is always the same. **Alternating current (AC)** is a flow of charge in which the charge flows back and forth, alternating in direction, usually following a sinusoidal pattern. If the current always flows in the same direction, but the magnitude varies somewhat in a periodic fashion, the current is said to be *pulsating* direct current. Power supplies that are poorly filtered generate pulsating direct current. Another type of current is a current that flows in the same direction while increasing or decreasing *exponentially*. Exponentially changing currents are sometimes very short lived, such as when electrical devices are turned on or off. Still another type of current is one that flows in the same direction, while its magnitude varies according to a so-called *sawtooth* function. Sawtooth currents are useful in equipment such as oscilloscopes, which are measurement instruments that display electrical characteristics on a screen. These current types are illustrated in Figure 5.6. It should be noted that the symbol I is usually reserved for DC, whereas the symbol i is generally used for AC or other types of currents that change with time.

Electric current is measured by means of an instrument called an *ammeter*. There are basically two types of ammeters: analog and digital. An analog ammeter

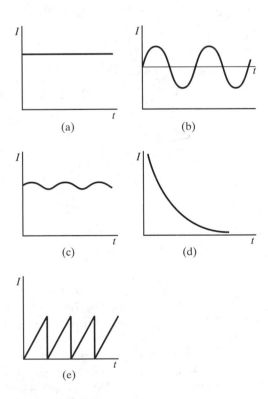

Figure 5.6.   Common types of electric currents: (a) direct current (DC), (b) alternating current (AC), (c) pulsating direct current, (d) exponential current, and (e) sawtooth current.

provides a current reading by means of a needle or pointer that moves across a calibrated scale. Digital ammeters provide a current reading by displaying numbers in a window. Each type of ammeter has two terminals. In order to use an ammeter, the circuit must be *broken* at the location where the current measurement is desired and the ammeter must be inserted directly into the current path. Most ammeters have function switches that enable both direct and alternating current measurements. Most also have manual or automatic range selector functions that facilitate current readings in units of A, mA, or μA.

EXAMPLE 5.1

As an electrical circuit is powered off, the current in a device changes exponentially with time according to the function

$$i(t) = 5\,e^{-kt}\,\text{A}$$

where k is a constant. If $k = 2\,\text{s}^{-1}$, how many coulombs pass through the device during the first second after the power is turned off? What is the current in the device at the instant just prior to turning off the power?

SOLUTION

The current decreases exponentially according to the relation

$$i(t) = 5\,e^{-2t}\,\text{A}$$

The number of coulombs that pass through the device during the first second after the power is turned off may be found by using Equation (5.2),

$$i = \frac{dq}{dt}$$

Multiplying both sides of this equation by dt and integrating, we obtain

$$\int_{q_1}^{q_2} dq = \int_0^1 i(t)\,dt = 5\int_0^1 e^{-2t}\,dt$$

Hence,

$$q_2 - q_1 = \left.\frac{5e^{-2t}}{-2}\right|_0^1 = \frac{5(e^{-2} - e^0)}{-2}$$

$$= 2.16\,\text{C}$$

Thus, 2.16 C pass through the device during the first second after the power is turned off. The current immediately before the power is turned off is the current at $t = 0$ s. Therefore, we have

$$i(0) = 5\,e^{-2(0)} = 5\,e^0$$

$$= 5\,\text{A}$$

APPLICATION: TRANSIENT CURRENT AND THE TIME CONSTANT

Upon turning them on or off, some electrical circuits exhibit exponential variations of current with time. In many cases, these variations are very short in duration, perhaps only a few milliseconds. Such a current variation is referred to as *transient*, because it is very short-lived. A typical transient current has the mathematical form

$$i(t) = C(1 - e^{-t/\tau})$$

where C is a constant, t is time, and τ is the *time constant*. The value of the time constant depends on the specific electrical characteristics of the circuit. For a simple circuit consisting of a resistor in series with an inductor, the time constant is $\tau = L/R$, where L is inductance and R is resistance. By inspection of the equation, the current is zero at $t = 0$, the instant the circuit is turned on. The current then increases exponentially with time until, after a long period of time, the current attains a steady value of C.

The time constant τ is defined as the time it takes for the current to change by 36.8 percent. To see how this works, let's examine the equation more closely. After one time constant ($t = \tau$), the exponential term is e^{-1}, or 0.368, and the current has increased to 0.632 times its steady value of C. After two time constants ($t = 2\tau$), the exponential term is e^{-2}, or 0.135, and the current has increased to 0.865 times its steady value. Extending the analysis to five time constants ($t = 5\tau$), the exponential term is e^{-5}, or 0.00674, and the current has increased to approximately 0.993 times its steady value. (See Table 5.1 and Figure 5.7). Theoretically, the current never reaches a steady value; it asymptotically approaches a steady value. For practical purposes, however, we may say that the current attains a steady value after five time constants because, as shown in Table 5.1, the current comes to within one percent of the steady value. Thus, the "rule of thumb" for transient currents is that it takes five time constants for a steady condition to be achieved.

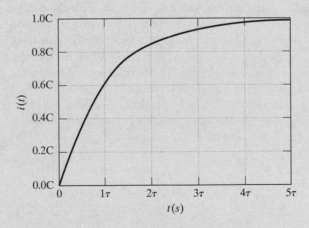

Figure 5.7. After five time constants, the current has practically reached a steady value.

TABLE 5.1

t(s)	$e^{-t/\tau}$	$i(t)$(A)
0	1	0
τ	0.368	0.632 C
2τ	0.135	0.865 C
3τ	0.050	0.950 C
4τ	0.0183	0.9817 C
5τ	0.00674	0.99326 C
∞	0	C

PRACTICE!

1. How many electrons are represented by a charge of 1 μC? 50 pC?
   Answer: 6.242×10^{12}, 3.121×10^{8}

2. The rate at which charge moves through a conductor is given by the relation
   $$q(t) - C \ln(t + 1) + 2t^2 \text{ C}$$
   where C is a constant. Find the current at $t = 0$ s and $t = 2$ s.
   Answer: $(C/3 + 8)$ A

3. The current in a device varies with time according to the function
   $$i(t) = (1 + 2\,e^{-5t}) \text{ A}$$
   How many coulombs pass through the device during the time interval $1 < t < 3$ s? What is the current for large values of time?
   Answer: 2.0027 C, 1 A

4. The current in a device varies sinusoidally with time, according to the function
   $$i(t) = 5 \sin(\pi + 2\pi t) \text{ A}$$
   How many coulombs pass through the device during the time interval $0 < t < 0.5$ s?
   Answer: 1.59 C

5.3   VOLTAGE

In the absence of a controlling force, electric charges in a conductor have a tendency to move about in a random manner. If we want the charges to unitedly move in a single direction so as to constitute an electric current, we must apply an external force to the charges called an *electromotive force* (emf). This force, since it causes a movement of charges through the conductor, does work on the charges. The electromotive force is typically called voltage. We therefore define **voltage** as the *work done in moving a charge of one coulomb*. The unit of voltage is the volt (V), named after the Italian physicist Alessandro Volta, who invented the voltaic battery. Because voltage is defined as the work done in moving a unit charge, one volt is defined as $1 \text{ V} = 1 \text{ J/C}$. *Instantaneous* voltage v is expressed as a derivative,

$$v = \frac{dw}{dq} \tag{5.3}$$

where w is the work measured in joules (J). The symbol V may also be used for voltage. Do not confuse the roman V, which stands for the unit called volts, with italic V, which denotes the variable voltage or potential difference. Voltage is sometimes referred to as *potential difference*. In its technical context, the word *potential* refers to a source of stored energy that is available for doing work. For example, a compressed spring has potential energy, and it performs work when it is allowed to return to its original, undeformed state. A stone that is nudged from the brink of a cliff converts its potential energy to work as it falls to the ground. Voltage is the electrical equivalent of mechanical potential energy. A battery, for example, has potential energy to do electrical work (that is, to drive a current), but does not do so until a closed circuit is connected across the battery. The word *difference* denotes that voltage is always taken *between two points*. To speak of voltage "at a point" is meaningless, unless a second point (reference point) is implied. A voltage exists across the positive and negative terminals of a battery. If we were to place the probes of a voltmeter across the terminals of a standard dry cell, we would measure a voltage of about 1.5 volts. In many circuits, a reference voltage referred to as *ground* is established. Ground may be the actual ground of the earth, referred to as *earth* ground or an arbitrary reference voltage on the chassis or case of the system, referred to as *chassis* ground. In either case, voltage is always taken between two points in the circuit.

We are all familiar with several electrical devices that supply a specified voltage. Batteries supply a voltage by converting chemical energy to electrical energy. Flashlights, lanterns, and electronic devices such as radios, CD players, cameras, and children's toys use batteries as a source of electrical energy. A few of the common types of batteries are illustrated in Figure 5.8. Of all the battery types, the 1.5-volt dry cell [Figure 5.8(a)] is probably the most popular. The 1.5-volt dry cell comes in a variety of sizes, designated by letters such as D, C, A, AA, AAA, etc. Some electronic devices such as radios and digital clocks use 9-volt dry cells [See Figure 5.8(b)]. Automobiles, trucks, and recreational vehicles use large 12-volt or 6-volt batteries for starting and other electrical functions [See Figure 5.8(c).] By connecting a closed circuit across the positive and negative terminals of a battery, DC flows through the circuit. What about the voltage supplied by the electrical outlets in our homes? In the United States, local power utilities supply residential and commercial customers with standard voltages of 110 V and 220 V [See Figure 5.8(d).] Unlike the current supplied by batteries, the current supplied by utility companies is AC that has a frequency of 60 Hz; that is, the current changes direction 60 times per second. Virtually all household appliances and electronic devices—washing machines, ranges, clothes dryers, microwave ovens, toasters, televisions, VCRs, etc.—operate on 110 VAC or 220 VAC. The abbreviation, VAC, means "volts AC," and the abbreviation VDC means "volts DC."

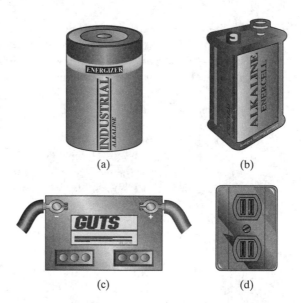

Figure 5.8. Typical voltage sources: (a) 1.5-V dry cell, (b) 9-V dry cell, (c) 6-V or 12-V automotive battery, and (d) standard 110-VAC wall outlet.

Now that voltage has been defined, let's consider the electrical energy that is supplied to, or by, a circuit element. **Circuit element** is a generic term that refers to an *electrical device or component*, such as a resistor, capacitor, inductor, etc. As shown in Figure 5.9, a steady electric current I flows through a circuit element. In order to ascertain whether energy is being supplied *to* the element, or *by*, the element to the rest of the circuit, we must know the direction of current flow and the *polarity* of the voltage across the element. The direction of current flow in Figure 5.9 is from positive to negative, which is consistent with the conventional current standard. Because the current enters the positive terminal of the element, an external electromotive force must be

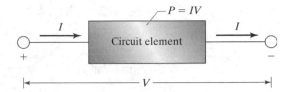

Figure 5.9. A circuit element with the relationship between current I, voltage V, and power P.

driving the current into the element, thereby supplying energy *to* the element. Thus, we say that the circuit element *absorbs* electrical energy. If, on the other hand, the current enters the negative terminal, the element supplies energy to the rest of the circuit. It is important to know the *rate* at which the energy is supplied to, or by, the circuit element. Rearranging Equation (5.3), and denoting voltage by V, we obtain

$$dw = V\,dq \tag{5.4}$$

Dividing both sides of Equation (5.4) by a time interval dt we obtain

$$\frac{dw}{dt} = V\frac{dq}{dt} \tag{5.5}$$

The quantity on the left side of Equation (5.5) is the rate at which work is done to move charge through the circuit element. By definition, the *rate at which work is performed* is **power** P. The quantity dq/dt is defined as electric current I. Hence, the power supplied to or by the circuit element is given by the relation

$$P = VI \tag{5.6}$$

The dimensional consistency of Equation (5.6) can be checked by noting that the units of VI are (J/C)(C/s) or J/s, which is defined as the watt (W), the SI unit for power.

Voltage is measured by means of an instrument called a *voltmeter*. Like ammeters that measure current, there are basically two types of voltmeters: analog and digital. An analog voltmeter provides a voltage reading by means of a needle or pointer that moves across a calibrated scale. Digital voltmeters provide a voltage reading by displaying numbers in a window. Each type of voltmeter has two terminals. Unlike a current measurement, a voltage measurement does not require that the circuit be broken at the location where the measurement is desired. In order to use a voltmeter, the terminals of the meter are connected *across* the device for which the potential difference is to be measured. Most voltmeters have function switches that enable both DC and AC voltage measurements. Most voltmeters also have manual or automatic range-selector functions that facilitate voltage readings in units of V, mV, or μV.

As a final comment on voltage, it may be instructive to invoke a physical analogy to voltage and its relationship to current. Voltage has been defined as the work required to move charge. We also stated that voltage is often referred to as a potential difference. To understand how voltage relates to current, it may be helpful to think of voltage as an electrical "pressure" or, more precisely, a pressure difference. Voltage is the "pressure difference" that drives electric current through a circuit element. In a pipe that carries water or some other fluid, a pressure difference between one end of the pipe and the other is the "potential" that drives the fluid through the pipe. Thus, we may consider voltage across a circuit element to be analogous to the pressure difference across a length of pipe and the flow of charge (current) through the circuit element to be analogous to the flow of fluid in the pipe. In a pipe, if there is no pressure difference, there is no fluid flow. In a circuit element, if there is no potential difference (voltage), there is no current flow.

PRACTICE!

1. A circuit element absorbs 2 W of power due to the passage of a steady current of 250 mA. What is the voltage across the element?

   Answer: 8 V

2. Resistors are devices that absorb electrical energy. If a steady current of 500 mA passes through a resistor with a voltage of 6 V across it, how much power must the resistor be able to absorb? What happens to this absorbed energy? What physical change does the resistor exhibit as it absorbs this energy?

   Answer: 3 W, The energy is transformed to heat, which causes the temperature of the resistor to increase.

3. A 12-V automobile lamp is rated at 40 W. What is the total charge that flows through the filament of the lamp in 1 minute? How many electrons does this represent?

   Answer: 200 C, 1.248×10^{21}

4. A battery-operated radio requires a current of 200 mA at 12 V. Find the power required to run the radio and the energy consumed in 2 hours of operation.

   Answer 2.4 W, 17.3 kJ

5. Borrow a voltmeter from your instructor or the electrical engineering department at your school. Measure the voltage across a 1.5-V and a 9-V dry cell. What are your voltage readings?

5.4   RESISTANCE

In addition to current and voltage, resistance is a very important electrical quantity. Electrical **resistance** may be defined as an *impedance to current flow through a circuit element*. All circuit elements, including even the **conductors** (wires) that connect them, impede the flow of current to some extent. When current flows in a conductor, free electrons collide with the lattices of the atoms inside the conductor. These collisions tend to retard or impede the organized motion of electrons through the conductor. Resistance in the wires connecting circuit elements is generally undesirable, but there are numerous situations where resistance is needed in electrical circuits to control other electrical quantities. The circuit element specifically designed for providing resistance in circuits is the **resistor**. Of all the circuit elements used in electrical circuits, the resistor is the most common. When electrical engineers design circuits, the circuit elements and their connections are drawn as a schematic diagram. The schematic symbol for a resistor is a zig-zag line, as shown in Figure 5.10(a). A very popular type of resistor uses carbon as the resistive material. A *carbon-composition* resistor consists of carbon particles mixed

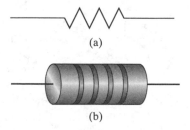

(a)

(b)

Figure 5.10.   Resistance. (a) Schematic symbol and (b) actual resistor.

with a binder and molded into a cylindrical shape. A *carbon-film* resistor consists of carbon powder that is deposited on an insulating substrate. The wires connected to the body of the resistor, or any type of circuit element for that matter, are called *leads*. A typical carbon composition resistor is illustrated in Figure 5.10(b).

There are other types of resistors in addition to the carbon devices. Some resistors employ a wire wrapped around a central core of ceramic or other insulating material. These resistors are referred to as wire-wound resistors. Wire-wound resistors are generally larger than carbon resistors and can handle more power. Other resistors use a combination of ceramic and metal for their resistive material. These resistors are referred to as CERMETS. The resistive material in some resistors is a metal or metal oxide. Resistors are manufactured in a variety of package styles, sizes, and power capabilities. An assortment of resistors used in various electrical circuit applications is shown in Figure 5.11.

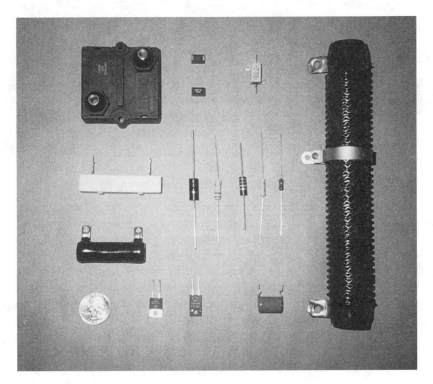

Figure 5.11.   Assortment of resistors for various applications. (Resistors for photograph provided courtesy of Ohmite Manufacturing Co., Skokie, IL.)

The unit for electrical resistance is the *ohm* (Ω), in honor of Georg Ohm, who is credited with formulating the relationship between current, voltage, and resistance, based on experiments performed in 1826. A resistor with a very small resistance has a low ohm value, whereas a resistor with a very high resistance has a high ohm value. Because resistances of various magnitudes are needed in specific circuit applications, resistors are manufactured in a wide range of ohm values. For example, some manufacturers supply carbon-composition resistors in the range 2.2 Ω to 1 MΩ. Some precision resistors are available in very small resistances, such as 0.008 Ω. It is interesting to note that 0.008 Ω is about the same resistance as a 1.5-m length of 12-gauge copper wire, the size of wire typically used in the electrical systems of homes. The resistance of most resistors is fixed, but some resistors are adjustable by means of a sliding or rotating electrical contact. This type of adjustable resistor is known as a potentiometer, or rheostat.

In circuit analysis, it is often necessary to determine the *total* or *equivalent* resistance of two or more resistors connected together. There are two ways in which circuit elements can be connected to each other. If the circuit elements are connected *end to end*, the elements are said to be connected in **series**. If the circuit elements are connected *across* each other, the elements are said to be connected in **parallel**. Figure 5.12 shows, in schematic form, three resistors connected in series and three resistors connected in parallel. The total resistance R_t for resistors connected in series is simply the arithmetic sum of the resistances for each resistor. Thus,

$$R_t = R_1 + R_2 + R_3 + \cdots + R_N \text{ (series)} \tag{5.7}$$

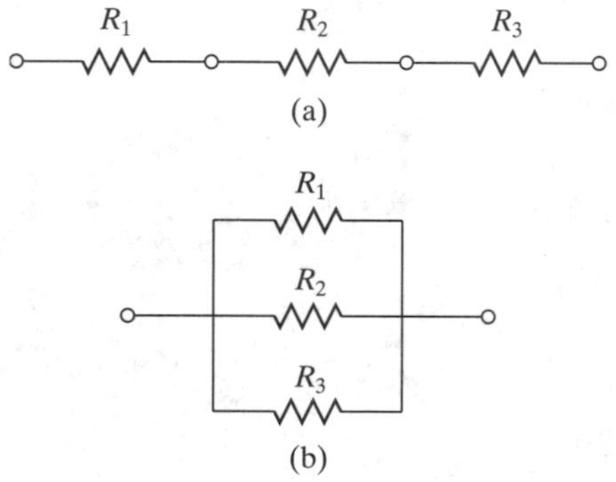

Figure 5.12. Resistors connected in (a) series and (b) parallel.

where N is the total number of resistors connected in series. The total resistance for resistors connected in parallel is given by the relation

$$\frac{1}{R_t} = \frac{1}{R_1} + \frac{1}{R_2} + \frac{1}{R_3} + \cdots + \frac{1}{R_N} \text{ (parallel)} \tag{5.8}$$

where, as before, N is the total number of resistors. To obtain the total resistance, R_t, we simply find the reciprocal of the total resistance by using Equation (5.8) and then invert it.

Resistance is measured by means of an instrument called an *ohmmeter*. Like ammeters and voltmeters that measure current and voltage, there are basically two types of ohmmeters: analog and digital. An analog ohmmeter provides a resistance reading by means of a needle or pointer that moves across a calibrated scale. Digital ohmmeters provide a resistance reading by displaying numbers in a window. Each type of ohmmeter has two terminals. The terminals are connected across the resistor for which the measurement is desired. Ohmmeters supply a current to the resistor, so the current in the circuit itself must be turned off while the measurement is being made. Ohmmeters have manual or automatic range selector functions that facilitate resistance readings in units of Ω, kΩ, or MΩ.

EXAMPLE 5.2

Find the total resistance for the resistor circuit shown in Figure 5.13.

SOLUTION

The resistor configuration in Figure 5.13 is a series–parallel combination. The 1-kΩ, 500-Ω, and 20-kΩ resistors are connected in parallel, and the 200-Ω resistor is connected in series with the set of parallel resistors. To find the total resistance, we must first find the equivalent resistance for the three resistors that are connected in parallel by using Equation (5.8). We then add that equivalent resistance to the 200-Ω resistor by using Equation (5.7). We assign the resistors the variable names

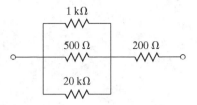

Figure P5.13. Resistor circuit
for Example 5.2.

$$R_1 = 1\ \text{k}\Omega,\ R_2 = 500\ \Omega,\ R_3 = 20\ \text{k}\Omega,\ \text{and } R_4 = 200\ \Omega$$

The equivalent resistance R_P of the three resistors in parallel is given by

$$\frac{1}{R_p} = \frac{1}{R_1} + \frac{1}{R_2} + \frac{1}{R_3}$$

$$= \frac{1}{1000\ \Omega} + \frac{1}{500\ \Omega} + \frac{1}{20,000\ \Omega} = 3.050 \times 10^{-3}\ \Omega^{-1}$$

Thus,

$$R_p = \frac{1}{3.050 \times 10^{-3}\ \Omega^{-1}} = 328\ \Omega$$

We now add R_p to R_4 in series to obtain the total resistance, R_t:

$$R_t = R_p + R_4$$

$$= 328\ \Omega + 200\ \Omega = 528\ \Omega$$

Hence, the series–parallel combination of resistors has a total resistance of 528 Ω. This means that the resistor configuration is equivalent to a *single* resistor with a resistance of 528 Ω.

PRACTICE!

1. What is the total resistance of five resistors, each with a resistance of $R\ \Omega$, if the resistors are connected in series? In parallel?
    Answer: series: $5R\ \Omega$, parallel: $R/5\ \Omega$

2. Consider two resistors connected in parallel. The resistance R_1 of the first resistor is very large, and the resistance R_2 of the second resistor is very small. What is the approximate total resistance?
    Answer: R_2

3. Find the total resistance for the resistor circuit shown in the accompanying figure.
    Answer: 59.5 Ω

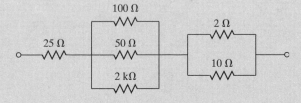

4. Find the total resistance for the resistor circuit shown in the accompanying figure.

   Answer: 13.3 kΩ

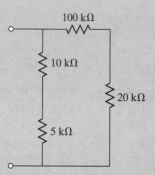

5. Find the total resistance for the resistor circuit shown in the accompanying figure.

   Answer: 1.998 Ω

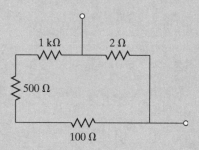

6. For the resistor circuit shown in the accompanying figure, what resistance must R_1 have to give a total resistance of 250 Ω?

   Answer: 525 Ω

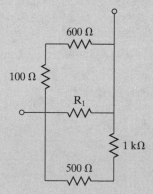

5.5   OHM'S LAW

In a series of experiments performed in 1826, the German physicist Georg Ohm discovered a relationship between the voltage across a conductor and the current flow through it. This relationship, known as **Ohm's law**, states that *the potential difference across a* *conductor is directly proportional to the current.* Stated mathematically, Ohm's law is

$$V \propto I \tag{5.9}$$

where V is potential difference (voltage) and I is current. Equation (5-9) may be written as an equality by introducing a constant of proportionality R, denoting resistance:

$$V = RI \tag{5.10}$$

Ohm's law, given by Equation (5.10), is one of the simplest, but most important laws in electrical circuit theory. Because the unit of voltage is volt (V) and the unit of current is ampere (A), a resistance of one ohm (Ω) is defined as $1\ \Omega = 1$ V/A. Hence, a resistor with a resistance of $1\ \Omega$ carrying a current of 1 A will have a voltage across it of 1 V. Unlike the law of universal gravitation or Newton's laws of motion, Ohm's law is not a fundamental law of nature. Ohm's law is an empirical (experimental) relationship that is valid only for certain materials. The electrical properties of *most* materials is such that the ratio of voltage to current is a constant and, according to Ohm's law, that constant is the resistance of the material. Ohm's law applies to wires and other metal conductors and, of course, resistors. A resistor, showing the relationship between voltage V, current I, and resistance R, is depicted in Figure 5.14.

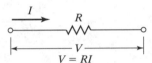

Figure 5.14. Ohm's law.

A resistance absorbs electrical energy. As current flows through a resistor, the absorbed electrical energy is transformed into thermal energy (heat), which is transferred to the surroundings. The rate at which the absorbed electrical energy is transformed into heat is referred to as *power dissipation*. All resistive circuit elements dissipate energy in the form of heat. We can calculate the power dissipation by combining Ohm's law given by Equation (5.10) with the relationship between power P, voltage V, and current I:

$$P = VI \tag{5.11}$$

Upon substituting Ohm's law, $V = RI$, into Equation (5.11), we obtain

$$P = I^2 R \tag{5.12}$$

A second relationship for power dissipation may be obtained by substituting Ohm's law in the form $I = V/R$ into Equation (5.11), yielding

$$P = \frac{V^2}{R} \tag{5.13}$$

Equations (5.12) and (5.13) are useful for finding the power dissipation from a resistive circuit element when the resistance and either the current or the voltage is known.

APPLICATION: SIZING A RESISTOR FOR A POWER-SUPPLY CIRCUIT

Resistors are available in a variety of ohm values and power ratings. The power rating of a resistor is the maximum number of watts of absorbed electrical power that the resistor is capable of dissipating as heat. If a resistor is used in a circuit where the actual power exceeds the power rating specified by the resistor supplier, the resistor may overheat. Resistance is a function of temperature, so if a resistor overheats, its resistance may vary significantly, thereby altering the electrical characteristics of the circuit. In extreme cases, an overheated resistor may even cause a complete failure of the device and perhaps a fire. The physical size of a resistor is usually an indication of its power rating. Large resistors have a lot of surface area and are therefore able to transfer more heat to the surroundings. Some resistors have ridges or fins to increase their surface area, whereas others, in order to minimize the resistor's temperature, have built-in heat sinks or provisions for mounting to heat sinks. Large resistors designed for high-power applications are called *power resistors*. A typical chassis-mounted power resistor is illustrated in Figure 5.15.

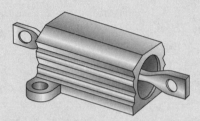

Figure 5.15. A power resistor.

Suppose that we are designing a power-supply circuit. Our circuit design calls for a resistor that carries a direct current of 800 mA and has a voltage drop of 24 V. What is the resistance of the resistor? What power rating must the resistor have? The resistance may be calculated by using Ohm's law,

$$R = \frac{V}{I}$$
$$= \frac{24 \text{ V}}{0.800 \text{ A}} = 30 \ \Omega$$

The power absorbed by the resistor may be found by using Equations (5.12) or (5.13). Let's calculate the power by using both equations to verify that we obtain the same result. Using Equation (5.12), we have

$$P = I^2R$$
$$= (0.800 \text{ A})^2(30 \ \Omega) = 19.2 \text{ W}$$

Using Equation (5.13), we have

$$R = \frac{V^2}{R}$$
$$= \frac{(24 \text{ V})^2}{30 \ \Omega} = 19.2 \text{ W}$$

Hence, we need a power resistor with a resistance of 30 Ω and the resistor must be capable of dissipating 19.2 W of power. It turns out that 30 Ω is a common resistance value for power resistors supplied by many manufacturers, but can we buy a resistor with a 19.2-W power rating? Resistors are available only in certain sizes and therefore only in certain power ratings. One supplier has power resistors in power ratings of 5, 10, 15, and 25 W. A power rating of 15 W is too low, so we choose a 25-W resistor even though it will handle more power than the design value. The additional 5.8 W may be considered a "factor of safety" for the resistor.

PRACTICE!

1. A 100-Ω carbon-composition resistor has a voltage of 12 V across it. What is the current? How much power does the resistor dissipate?
   Answer: 120 mA, 1.44 W

2. A circuit design calls for a resistor that will produce a voltage drop of 15 V where the current is 200 mA. How much power does the resistor dissipate? What resistance is required?

    Answer: 3.0 W, 75 Ω

3. A portable, 1320-W forced-air heater runs on standard 110-V residential voltage. The heating element is a nichrome ribbon that crosses in front of a polished metal plate. What is the current drawn by the heater? What is the resistance of the nichrome heating element?

    Answer: 12.0 A, 9.17 Ω

4. Two 75-Ω resistors connected in parallel dissipate 2.5 W each. What is the voltage across the resistors? What is the current in each resistor?

    Answer: 13.7 V, 183 mA

5.6 SIMPLE DC CIRCUITS

With any subject, we learn by studying the basic principles first and then progress toward more complex concepts. The study of electrical circuits works the same way. Beginning engineering students must first acquire a solid grasp of the fundamentals before proceeding to more advanced material. Hence, this section deals with some basic circuit concepts. Because the current changes direction in AC circuits, the analysis of AC circuits can be quite complex. We therefore focus our attention on DC circuits. A simple electrical circuit consists of two or more electrical devices interconnected by conductors. In many electrical circuits, there are numerous types of electrical devices such as resistors, capacitors, inductors, diodes, transistors, transformers, batteries, lamps, fuses, switches, and motors. Because the resistor is the most common circuit device and the analysis of resistors is the most straightforward, our coverage is limited to resistive circuits (i.e., circuits that have resistors as the only circuit element other than a source of constant voltage, such as a battery).

Consider the electrical circuit for a common household flashlight. As shown in Figure 5.16(a), a basic flashlight contains two 1.5-V dry cells, a lamp, and a switch. The conductor that interconnects these devices is normally a metal strip that helps hold the batteries in place and serves as a spring member in the switch mechanism. When the switch is closed, a direct current flows in a closed loop through the dry cells, switch, and lamp filament. Because the dry cells are connected in series, the voltages of each dry cell add, providing a total voltage of 3 V. Based on an arbitrarily selected standard, the direction for conventional current flow is *from* the positive terminal of the voltage source to the external circuit. The electrical schematic diagram that represents the flashlight circuit is shown in Figure 5.16(b). A **schematic diagram** is a *symbolic representation of the devices and interconnections in the circuit*. The schematic diagram may be loosely considered as the electrical equivalent of the free-body diagram used in engineering mechanics. A free-body diagram schematically shows a mechanical system with all the external forces acting on it as well as the other physical features of the system. To the mechanical and civil engineer, the free-body diagram is an indispensable analytical tool. Likewise, the schematic diagram is an indispensable tool to the electrical engineer. A schematic diagram shows how all the electrical devices are interconnected and also shows their numerical values. For example, a schematic diagram consisting of a voltage source, a resistor, a capacitor, and an inductor would show how these circuit elements are interconnected and would indicate the potential difference of

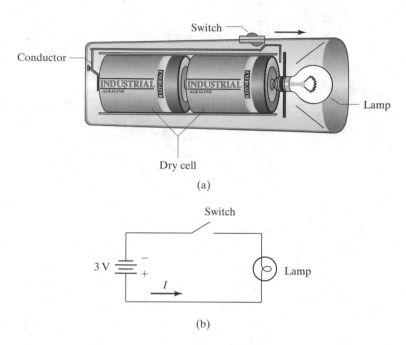

Figure 5.16. A common flashlight. (a) The actual device. (b) The schematic diagram.

the voltage source in volts (V), the resistance of the resistor in ohms (Ω), the capacitance of the capacitor in farads (F), and the inductance of the inductor in henrys (H). The schematic diagram contains all the pertinent information an engineer needs to evaluate the electrical functions of the circuit.

Every electrical device (circuit element) has a unique schematic symbol. Illustrated in Figure 5.17 are the schematic symbols for a few common circuit elements. By examining the schematic symbols in the figure, we note that the schematic symbol resembles the electrical or mechanical characteristics of the actual electrical device. The schematic symbol for a battery, for example, is a series of short parallel lines of alternating lengths. Batteries consist of at least two elements or terminals, a positive and a negative, separated by a substance that participates in a chemical reaction. The schematic symbol for a resistor is a zig-zag line. Resistors retard the flow of current through them, so a zig-zag line, indicative of an impeded electrical path, is used. A switch is an electrical "gate" that is either open or closed, allowing the current to flow or not. Capacitors consist of two plates separated by a dielectric (nonconducting) material. Inductors are coils of wire wrapped around a core. The major element in lamps is a filament in which a portion of the absorbed electrical energy is converted into visible light. Obviously, there are many more electrical devices used in electrical circuits than those shown in Figure 5.17. During the course of the electrical engineering program, the engineering student will become familiar with numerous electrical devices and their corresponding schematic symbols.

Circuit elements are broadly classified into two categories: *active* elements and *passive* elements. An active circuit element is a device that *supplies* energy to an external circuit. Common examples of active elements are batteries and generators. A passive circuit element, therefore, is any device that is not active. Resistors, capacitors, and inductors are common examples of passive elements. The two most important types of active circuit elements are referred to as the **independent voltage source** and the **independent current source**. An independent voltage source is a two-terminal circuit

Circuit element	Schematic symbol	Actual device
Battery		
Resistor		
Switch		
Capacitor		
Inductor		
Lamp		

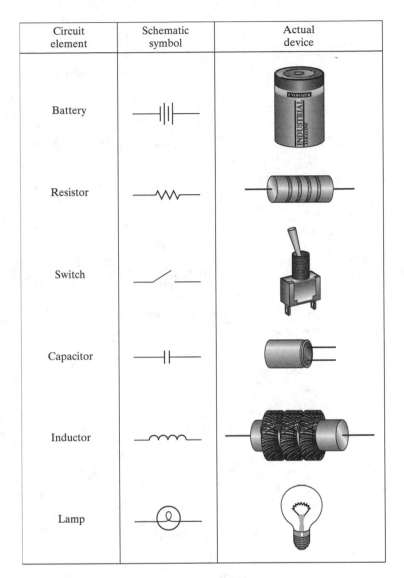

Figure 5.17. Common circuit elements and their schematic symbols.

element, such as a battery or generator, that maintains a specified voltage between its terminals. The voltage is independent of the current through the element. Because the voltage is independent of current, the *internal* resistance of the independent voltage source is zero. Actual voltage sources such as batteries do not have a zero internal resistance, but the internal resistance can be neglected if the resistance of the external circuit is large. Thus, the independent voltage source is an idealization that simplifies circuit analysis. The schematic symbol for the independent voltage source is illustrated in Figure 5.18(a). An independent current source is a two-terminal circuit element through which a specified current flows. The current is independent of the voltage across the element. Hence, like the independent voltage source, the independent current source is an idealization. The schematic symbol for the independent current source is shown in Figure 5.18(b).

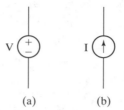

Figure 5.18. Schematic symbols for (a) independent voltage source and (b) independent current source.

In the next two examples, we demonstrate how to analyze simple DC circuits by using Ohm's law and other fundamental electrical relationships. Each example is worked in detail, following the general analysis procedure of (1) problem statement, (2) diagram, (3) assumptions, (4) governing equations, (5) calculations, (6) solution check, and (7) discussion.

EXAMPLE 5.3

Problem statement

The DC circuit shown in Figure 5.19 consists of a 10-V independent voltage source connected to two resistors in series. Find (a) the current, (b) the voltage across each resistor, and (c) the power dissipated by each resistor.

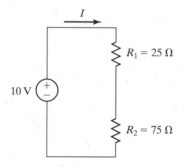

Figure 5.19. DC circuit for Example 5.3.

Diagram

The diagram for this problem is the schematic diagram shown in Figure 5.19.

Assumptions

1. The voltage source is ideal.
2. The resistance of the connecting wires is negligible.
3. The resistances of the resistors are constant.

Governing equations

Three equations are needed to solve this problem. There are two resistors in the circuit, so we need a formula for the total resistance. We also need Ohm's law and a relation for the power dissipation. The three equations are

$$R_t = R_1 + R_2$$
$$V = IR$$
$$P = I^2 R$$

Calculations

(a) All the elements, voltage source, and resistors in this simple DC circuit are connected in series, so the current through each element is the same. The total resistance is found by adding the values of each resistor and then by using Ohm's law to calculate the current. The total resistance is

$$R_t = R_1 + R_2$$
$$= 25\ \Omega + 75\ \Omega = 100\ \Omega$$

We have effectively combined two resistors into one resistor with an equivalent resistance. The voltage across this equivalent resistor is 10 V. The current is found by using Ohm's law:

$$I = \frac{V}{R_t}$$
$$= \frac{(10\ \text{V})}{100\ \Omega} = 0.1\ \text{A} = \underline{100\ \text{mA}}$$

(b) Now that the current is known, the voltage across each resistor can be calculated. Once again, we use Ohm's law:

$$V_1 = IR_1$$
$$= (0.1\ \text{A})(25\ \Omega) = \underline{2.5\ \text{V}}$$
$$V_2 = IR_2$$
$$= (0.1\ \text{A})(75\ \Omega) = \underline{7.5\ \text{V}}$$

(c) The power dissipated as heat by each resistor is

$$P_1 = I^2 R_1$$
$$= (0.1\ \text{A})^2(25\ \Omega) = \underline{0.25\ \text{W}}$$
$$P_2 = I^2 R_2$$
$$= (0.1\ \text{A})^2(75\ \Omega) = \underline{0.75\ \text{W}}$$

Solution check

After a careful review of our solution, no errors are found.

Discussion

Note that resistors R_1 and R_2 dissipate the same fractions of the total power as their fractions of the total resistance: 25 percent and 75 percent, respectively. Note also that the sum of the voltages across the resistors equals the voltage of the independent voltage source and that the voltage across each resistor is proportional to the resistance of that resistor. This type of resistor circuit is known as a *voltage divider*, because it divides the total voltage into two or more specified voltages.

EXAMPLE 5.4

Problem statement

The DC circuit shown in Figure 5.20 consists of a 200-mA independent current source connected to two resistors in parallel. Find the voltage across the resistors and the current in each resistor.

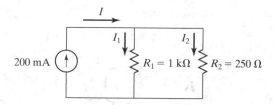

Figure 5.20. DC circuit for Example 5.4.

Diagram

The diagram for this problem is the schematic diagram shown in Figure 5.20.

Assumptions

1. The current source is ideal.
2. The resistance of the connecting wires is negligible.
3. The resistances of the resistors are constant.

Governing equations

Two equations are needed to solve this problem. There are two resistors in the circuit, so we need a formula for the total resistance. We also need Ohm's law. These equations are

$$\frac{1}{R_t} = \frac{1}{R_1} + \frac{1}{R_2}$$

$$V = IR$$

Calculations

All the elements, current source, and resistors in this simple DC circuit are connected in parallel, so the voltage across each element is the same. We can calculate the voltage across the resistors by finding the total resistance. Two resistors connected in parallel add according to the formula

$$\frac{1}{R_t} = \frac{1}{R_1} + \frac{1}{R_2}$$

$$= \frac{1}{1000\ \Omega} + \frac{1}{250\ \Omega} = 0.005\ \Omega^{-1}$$

Inverting to obtain the total resistance R_t, we have

$$R_t = \frac{1}{0.005\ \Omega^{-1}} = 200\ \Omega$$

Using Ohm's law, we find that the voltage across the resistors is

$$V = IR_t$$

$$= (0.2\ \text{A})(200\ \Omega) = \underline{40\ \text{V}}$$

Examine the circuit closely. When the 200-mA current reaches the junction of the first resistor R_1, part of the current flows into R_1 and the remainder flows into R_2. Hence, the total current, I is "split" in some fashion between the two resistors. The current in each resistor can be calculated by applying Ohm's law for each resistor. Thus,

$$I_1 = \frac{V}{R_1}$$

$$= \frac{40 \text{ V}}{1000 \ \Omega} = 0.040 \text{ A} = \underline{\underline{40 \text{ mA}}}$$

$$I_2 = \frac{V}{R_2}$$

$$= \frac{40 \text{ V}}{250 \ \Omega} = 0.160 \text{ A} = \underline{\underline{160 \text{ mA}}}$$

Solution check

After a careful review of our solution, no errors are found.

Discussion

The total current 200 mA equals the sum of the currents in resistors R_1 and R_2: 40 mA and 160 mA, respectively. It is important to note that the currents in R_1 and R_2 are inversely related to the resistance values. Resistor R_1 is larger than R_2, so it carries a smaller current. Most of the current is in resistor R_2, because the current "prefers" to take the path of least resistance. The resistance of R_2 is one-fourth the resistance of R_1, so the current in resistor R_1 is one-fourth the current in resistor R_2. This type of resistor circuit is known as a *current divider*, because it divides the total current into two or more specified currents.

PRACTICE!

1. For the resistor circuit shown, find (a) the current, (b) the voltage across each resistor, and (c) the power dissipated by each resistor.
    Answer: (a) 250 mA (b) 25 V, 5.0 V, 20 V (c) 6.25 W, 1.25 W, 5.0 W

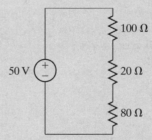

2. For the resistor circuit shown, find the current in each resistor and the voltage across each resistor.
    Answer: 20 mA, 80 mA; 15.0 V, 4.0 V, 1.0 V, 18.0 V, 2.0 V

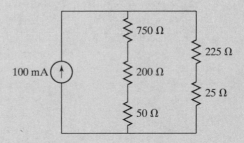

5.7 KIRCHHOFF'S LAWS

Ohm's law $V = IR$ is a fundamental and powerful principle for calculating current, voltage and power associated with a single resistor or a simple combination of resistors. However, Ohm's law alone cannot be used to analyze the majority of simple DC circuits. In addition to Ohm's law, two additional laws, stated by the German physicist Gustav Kirchhoff (1824–1887), are required. These two laws are known as **Kirchhoff's current law (KCL)** and **Kirchhoff's voltage law (KVL)**. Let us first consider Kirchhoff's current law, for which we will hereafter use the abbreviation KCL.

5.7.1 Kirchhoff's Current Law

Kirchhoff's current law states that the *algebraic sum of the currents entering a node is zero*. To understand the physical meaning of KCL, we must first understand what a *node* is. An electrical circuit consists of circuit elements (i.e., resistors, capacitors, inductors, etc.) interconnected by conductors. A node is defined as a *point of connection of two or more circuit elements*. The actual node may or may not be a physical "point" where the conductors from two or more circuit elements come together, but rather a general region in which all points on the conductor are electrically the same. Consider the circuit shown in Figure 5.21(a).

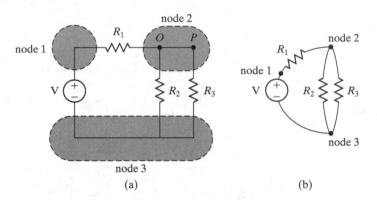

Figure 5.21. A circuit with three nodes. (a) Standard schematic. (b) Schematic redrawn to emphasize that there are only three nodes.

Node 1 is not a single point, but a collection of points, indicated by the shaded region, anywhere along the conductor that connects the independent voltage source to resistor R_1. One may be tempted to define two separate nodes, one node at point O and another node at point P, but points O and P are electrically identical, since they are separated only by pure conductors, not circuit elements. Hence, the entire shaded region surrounding points O and P is node 2. Similarly, node 3 is the entire shaded region shown, because all points on the conductors in this region are electrically identical. The understanding of the node concept may be facilitated by redrawing the schematic in a different form, as shown in Figure 5.21(b). The conductors' lengths have been "shrunk" and the ends of the circuit elements brought together into common points, which are the nodes of the circuit.

Having provided a verbal definition of KCL and defined the term node, we are now ready to give a mathematical definition of KCL. The mathematical expression for KCL is

$$\Sigma I_{\text{in}} = 0 \qquad (5.14)$$

where I_{in} is a single current, entering or leaving, a specified node. If the current enters the node, I_{in} is positive, whereas if the current leaves the node, I_{in} is negative. Consider

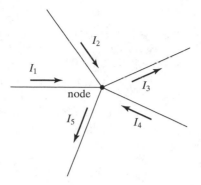

Figure 5.22. A node with five currents, three entering and two leaving.

the node shown in Figure 5.22. Five currents are entering or leaving the node. Kirchhoff's current law for this configuration is written as

$$\Sigma I_{in} = 0$$
$$= I_1 + I_2 - I_3 + I_4 - I_5$$

where minus signs are used for currents I_3 and I_5 because these currents are leaving the node. Conservation of charge in a perfect conductor is the physical principle on which KCL is based. Suppose that the right side of Equation (5.14) were replaced by a nonzero constant Δ. A positive value of Δ would imply that the node accumulates charges, and a negative value of Δ would imply that the node is a source of charges. A node consists of perfect conductors and therefore cannot accumulate or generate charges. Another way of saying this is that whatever current enters the node must exit the node.

5.7.2 Kirchhoff's Voltage Law

Kirchhoff's voltage law, hereafter abbreviated as KVL, states that the *algebraic sum of the voltages around a loop is zero*. The mathematical form of KVL law is

$$\Sigma V = 0 \tag{5.15}$$

A loop is defined as a *closed path in a circuit*. Kirchhoff's voltage law applies for any closed loop, regardless of the number of circuit elements contained in the loop. Consider the simple series circuit shown in Figure 5.23. A 10-V ideal voltage source is connected in series with two resistors, forming a closed loop. Kirchhoff's voltage law for this circuit is written as

$$\Sigma V = 0$$
$$= +10 - V_1 - V_2$$

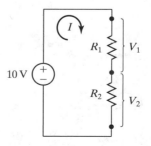

Figure 5.23. For this circuit, Kirchhoff's voltage law states that $\Sigma V = 0 = 10 - V_1 - V_2$.

where V_1 and V_2 are the voltages across resistors R_1 and R_2, respectively. The negative signs on the voltages mean that the voltage *drops* as we proceed around the loop in a clockwise sense, following the direction for conventional current. The voltage across the ideal voltage source is 10 V. A portion of this voltage is dropped by resistor R_1, and the remaining voltage is dropped by resistor R_2, bringing the total voltage drop to 10 V. Stated another way, the voltage rises equal the voltage drops, so KVL may also be written as $V_1 + V_2 = 10$. Alternatively, we may proceed around the loop in a counterclockwise direction, in which case the signs of all the voltages change, and KVL is expressed as

$$\Sigma V = 0$$
$$= -10 + V_1 + V_2$$

Regardless of which direction is used, the same result is obtained; that is, the sum of the voltages across the resistors equals the voltage of the independent voltage source. Conservation of energy is the physical principle on which KCL is based. Kirchhoff's voltage law states that there is no gain or loss of electrical potential energy for the charges that are traveling around a closed loop. Suppose that the right side of Equation (5.15) were replaced by a nonzero constant Δ. A nonzero value of Δ would imply that electrical energy is either being created or destroyed in the loop. This is contrary to the first law of thermodynamics, which states that energy is conserved.

In the next example, using KCL and KVL, we demonstrate how to analyze a simple DC circuit. The example is worked in detail, with the general analysis procedure of (1) problem statement, (2) diagram, (3) assumptions, (4) governing equations, (5) calculations, (6) solution check, and (7) discussion.

EXAMPLE 5.5

Problem statement

For the DC circuit shown in Figure 5.24, find the voltage across each resistor and the current in each resistor.

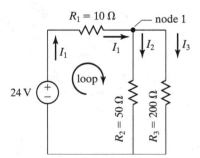

Figure 5.24. Circuit for Example 5.5.

Diagram

The diagram for this problem is the schematic diagram shown in Figure 5.24.

Assumptions

1. The voltage source is ideal.
2. The resistance of the connecting wires is negligible.
3. The resistances of the resistors are constant.

Governing equations

Three governing equations are needed to solve this problem:

$$\Sigma I_{in} = 0 (\text{KCL})$$
$$\Sigma V = 0 (\text{KVL})$$
$$V = IR (\text{Ohm's law})$$

Calculations

We designate the current through the ideal voltage source and resistor R_1 as I_1. At node 1, the current splits into two currents that flow through resistors R_2 and R_3. Applying KCL to node 1, we have

$$\Sigma I_{in} = 0$$
$$= I_1 - I_2 - I_3$$

Invoking Ohm's law, we can rewrite the relation as

$$\frac{V_1}{R_1} = \frac{V_2}{R_2} + \frac{V_3}{R_3}$$

Resistors R_2 and R_3 are connected in parallel, so the voltage across them is the same. Because $V_2 = V_3$, we can simplify the KCL relation further as

$$\frac{V_1}{R_1} = V_2 \left(\frac{1}{R_2} + \frac{1}{R_3} \right) \tag{a}$$

Kirchhoff's voltage law, written for the loop containing the voltage source as well as R_1 and R_2 is

$$\Sigma V = 0$$
$$= 24 - V_1 - V_2 \tag{b}$$

Solving Equation (b) for V_1 and substituting the result into Equation (a), we obtain a relation in terms of V_2 only. Thus, we have

$$\frac{24 - V_2}{R_1} = V_2 \left(\frac{1}{R_2} + \frac{1}{R_3} \right) \tag{c}$$

After a little algebra, we solve Equation (c) for V_2 and obtain

$$V_2 = V_3 = \underline{19.2 \text{ V}}$$

To find the voltage across resistor R_1, we substitute the calculated value of V_2 into Equation (c), which yields

$$V_1 = 24 - V_2$$
$$= 24 \text{ V} - 19.2 \text{ V} = \underline{4.8 \text{ V}}$$

Now that all voltages have been calculated, it is a straightforward matter to calculate the current in each resistor, using Ohm's law:

$$I_1 = \frac{V_1}{R_1} = \frac{4.8 \text{ V}}{10 \ \Omega} = 0.48 \text{ A} = \underline{480 \text{ mA}}$$

$$I_2 = \frac{V_2}{R_2} = \frac{19.2 \text{ V}}{50 \ \Omega} = 0.384 \text{ A} = \underline{384 \text{ mA}}$$

$$I_3 = \frac{V_3}{R_3} = \frac{19.2 \text{ V}}{200 \ \Omega} = 0.096 \text{ A} = \underline{96 \text{ mA}}$$

Solution check

After a careful review of our solution, no errors are found.

Discussion

The total current I_1 that flows through the ideal voltage source and resistor R_1 can be found by first calculating the total resistance and then using Ohm's law. Resistors R_2 and R_3 add in parallel and that resistance adds in series to R_1. Thus, the total resistance is

$$R_t = \frac{1}{\dfrac{1}{R_2} + \dfrac{1}{R_3}} + R_1$$

$$= \frac{1}{\dfrac{1}{50\ \Omega} + \dfrac{1}{200\ \Omega}} + 10\ \Omega = 50\ \Omega$$

The total current I_1 is

$$I_1 = \frac{V}{R_t}$$

$$= \frac{24\ \text{V}}{50\ \Omega} = 0.48\ \text{A}$$

which is in agreement with our previous result.

PRACTICE!

1. For the node shown in the accompanying figure, find the current I_4. Does current I_4 enter or leave the node?
   Answer: 13 A, leave

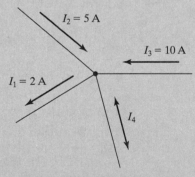

2. For the DC circuit shown in the accompanying figure, find the voltage across each resistor and the current through each resistor.
   Answer: 8.077 V, 1.923 V, 1.923 V; 0.269 A, 0.0769 A, 0.1923 A

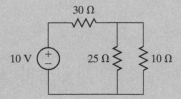

3. For the DC circuit shown in the accompanying figure, find the voltage across each resistor and the current through each resistor.

Answer: 100 V, 7.69 V, 3.85 V, 11.55 V; 100 mA, 76.9 mA, 23.1 mA

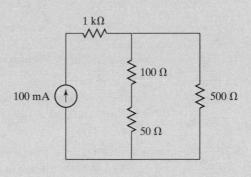

KEY TERMS

alternating current (AC)	independent current source	resistance
circuit element	independent voltage source	resistor
conductor	Kirchhoff's current law (KCL)	schematic diagram
direct current (DC)	Kirchhoff's voltage law (KVL)	series
electric charge	Ohm's law	voltage
electrical circuit	parallel	
electric current	power	

REFERENCES

Bird, J.O., *Electrical Circuit Theory and Technology*, 2d ed., Boston, MA: Butterworth-Heinemann, 2003.

Johnson, D.E., J.L. Hilburn, and J.R. Johnson, *Basic Electric Circuit Analysis*, 5th ed., Upper Saddle River, NJ: Prentice Hall, 2000.

Nilsson, J.W. and S.A. Riedel, *Electric Circuits*, 6th ed., Reading, MA: Addison-Wesley, 2000.

Roadstrum, W.H. and D.H. Wolaver, *Electrical Engineering for All Engineers*, 2d ed., NY: John Wiley & Sons, 1993.

Smith, R.J. and R.C. Dorf, *Circuits, Devices and Systems*, 5th ed., NY: John Wiley & Sons, 1992.

Problems

1. The flow of charge in a conductor varies with time according to the function

$$q(t) = (1 - 4e^{-kt}) \text{ C}$$

If $k = 0.2 \text{ s}^{-1}$, find the current at $t = 5$ s. What is the current for very large values of time?

2. For a period of 1 s immediately after the power is turned on, the current in an electrical device varies with time according to the function

$$i(t) = 3t^{\frac{1}{2}} \text{ A}$$

How many coulombs have passed through the device during the first 0.25 s? 0.75 s? What is the current at the instant the power is turned on?

3. After the power is turned off, the current in an electrical device varies with time according to the function

$$i(t) = 4e^{-kt} \text{ A}$$

If $k = 0.1 \text{ s}^{-1}$, how many coulombs have passed through the device during the first 2 s? 5 s? What is the current at the instant the power is turned off? What is the current a long time after the power is turned off?

4. The current in a device varies with time according to the function

$$i(t) = 3e^{-t/\tau} \text{ A}$$

where τ is the time constant. How many time constants are required for the current to drop to 250 mA? 10 mA?

5. A miniature incandescent lamp is connected to a 6-V lantern battery. If the current flow through the filament of the lamp is 85 mA, how much power does the lamp absorb?

6. A standard power value for a household incandescent lightbulb is 60 W. What is the current through the filament of such a lightbulb if the voltage is 110 V? Is the entire 60 W of electrical power converted into visible light?

7. A standard voltage for homes in the United States is 110 V. Each circuit in the home is protected by a circuit breaker, a safety device designed to break the flow of current in the event of an electrical overload. A particular circuit must provide power to a baseboard electric heater, lights, and two televisions. If the total power required for these devices is 1.8 kW, what is the minimum required amperage of the circuit breaker?

8. Using an order-of-magnitude analysis, estimate the amount of electrical energy (J) used per person in the United States each year. What is the corresponding power (W)?

9. A common resistance of carbon resistors is 33 Ω. How many 33-Ω resistors, connected in parallel, are needed to give a total resistance of 5.5 Ω?

10. Consider two resistors, R_1 and R_2. The resistance of R_1 is lower than the resistance of R_2. If these two resistors are connected in parallel, which of the following statements about the total resistance is true?

    A. The total resistance is higher than the resistance of R_2.
    B. The total resistance is between the resistances of R_1 and R_2.
    C. The total resistance is lower than the resistance of R_1.

11. Without doing the calculation, what is the approximate total resistance of a 10-Ω resistor and a 1-M Ω resistor connected in parallel?

12. Find the total resistance for the resistor circuit shown in Figure P5.12.

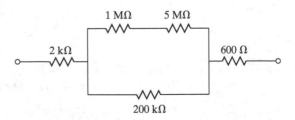

Figure P5.12.

13. Find the total resistance for the resistor circuit shown in Figure P5.13

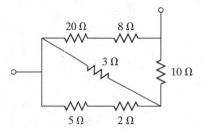

Figure P5.13.

14. Find the total resistance for the resistor circuit shown in Figure P5.14

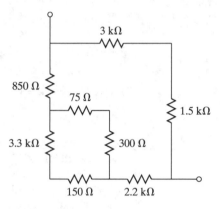

Figure P5.14.

15. Find the total resistance for the resistor circuit shown in Figure P5.15

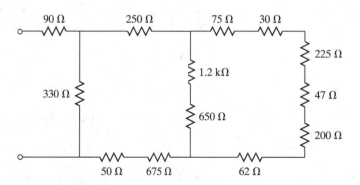

Figure P5.15.

16. For the resistor circuit shown in Figure P5.16, what resistance must resistor R_1 have to give a total resistance of 20 Ω?

17. For the resistor circuit shown in Figure P5.17, what resistance must resistor R_3 have to give a total resistance of 500 Ω?

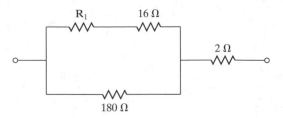

Figure P5.16.

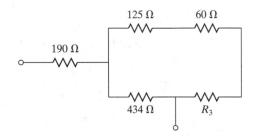

Figure P5.17.

18. Precision resistors are resistors whose resistance is known to within a tolerance of ±1 percent or less. These resistors are typically used in *current-sensing* applications. In this application, a precision resistor with a very low resistance is connected in a circuit where a measurement of current is desired. Because the resistance is low, the resistor does not significantly affect the electrical attributes of the circuit. Current is measured, not by using an ammeter, but by placing a voltmeter across the resistor. By knowing the resistance of the resistor, the current can be readily calculated by using Ohm's law. Furthermore, by a judicious selection of the resistance of the resistor, the voltmeter can be made to read the current directly. If the voltmeter is to read current directly, what should the resistance of the precision resistor be?

19. A 47-Ω power resistor carries a current of 300 mA. What is the voltage across the resistor? What is the power dissipation? If power resistors are available in power ratings of 1, 2, 5, and 10 W, which power rating should probably be selected?

20. Borrow an ohmmeter from your instructor or the electrical engineering department at your school. Measure the resistance of a 40-W incandescent lightbulb. What is the resistance? If this type of lightbulb operates on 110 V, what is the current through the filament? Is the resistance of the lightbulb the same as your measured value when the filament is hot?

For problems 5.21 through 5.30, use the general analysis procedure of (1) problem statement, (2) diagram, (3) assumptions, (4) governing equations, (5) calculations, (6) solution check, and (7) discussion.

21. A simple DC circuit consists of a 12-V independent voltage source and three resistors as shown in Figure P5.21. Find (a) the current, (b) the voltage across each resistor, and (c) the power dissipated by each resistor.

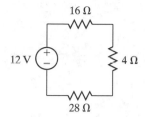

Figure P5.21.

22.   A 50-V independent voltage source supplies power to three resistors in the DC circuit shown in Figure P5.22. For each resistor, find the voltage drop and current.

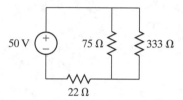

Figure P5.22.

23.   Four resistors are connected in parallel across a 200-mA independent current source as shown in Figure P5.23. What is the voltage across the resistors and the current in each resistor?

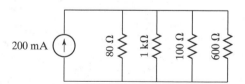

Figure P5.23.

24.   Two power resistors, a 130-Ω fixed-carbon resistor and a variable wire-wound resistor, are connected in series with a 100-V independent voltage source as shown in Figure P5.24. One terminal of the variable resistor is a slider that contacts the wire windings as it moves along the resistor. The maximum resistance of the variable resistor is 1.5 kΩ. If 30 percent of the resistor's windings carry current, find (a) the current, (b) the voltage across the variable resistor, and (c) the power dissipated by both resistors.

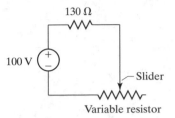

Figure P5.24.

25.   For the DC circuit shown in Figure P5.25, find the voltage across each resistor and the current in each resistor.

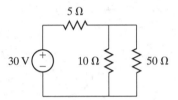

Figure P5.25.

26.   For the DC circuit shown in Figure P5.26, find the voltage across each resistor and the current in each resistor. Find the power dissipations in the 20-Ω and 100-Ω resistors.

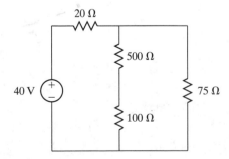

Figure P5.26.

27.   For the DC circuit shown in Figure P5.27, find the voltage across each resistor and the current in each resistor.

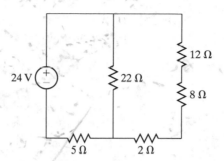

Figure P5.27.

28.   For the DC circuit shown in Figure P5.28, find the voltage across each resistor and the current in each resistor.

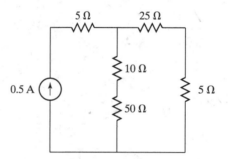

Figure P5.28.

29.   For the DC circuit shown in Figure P5.29, find the voltage across each resistor and the current in each resistor. Find the power dissipations in the 2-Ω, 6-Ω and 22-Ω resistors.

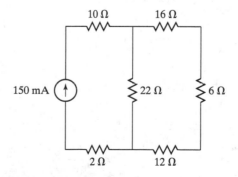

Figure P5.29.

30.   For the DC circuit shown in Figure P5.30, find the voltage across each resistor and the current in each resistor.

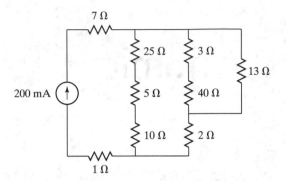

Figure P5.30.

6

Thermodynamics

6.1   INTRODUCTION

One of the most important subjects in the study of engineering is thermodynamics. Because the principles of thermodynamics find applications in virtually every engineering system, many colleges and universities require all engineering majors to take at least one course in the subject. As a specific engineering discipline, thermodynamics typically falls within the domain of mechanical and chemical engineering, because mechanical and chemical engineers have primary responsibility for designing and analyzing energy-based systems. Consistent with this observation, **thermodynamics** may be defined as *the science of energy transformation and utilization.* This is a very broad definition, so it is no wonder that thermodynamics spans all engineering disciplines.

The word *thermodynamics* originates from the Greek words *therm* (heat) and *dynamics* (power). These root words are appropriate, because thermodynamics often deals with systems that convert heat to power. Thermodynamics is the science that describes how energy is converted from one form into another. One of the most important physical laws is the **first law of thermodynamics**, which states that energy can be converted from one form to another, but the total energy remains constant. A popular statement of the first law of thermodynamics is that energy cannot be created or destroyed. For example, a boulder poised on the brink of a cliff has potential energy by virtue of its height above the ground. As the boulder falls toward the ground, its speed increases, thereby converting its potential energy to kinetic energy, but the total energy at any point is constant. Thermodynamics is also the science that reveals whether a given energy conversion is physically possible. This concept is revealed by the **second law of thermodynamics**, which states that energy conversions occur in the direction of decreasing quality of energy. For example, a

OBJECTIVES

After reading this chapter, you will have learned

- The importance of thermodynamics in engineering
- The relationships between the various types of pressures
- The thermodynamic meaning of temperature
- The various forms of energy
- How to determine various forms of work
- The difference between heat and temperature
- How to use the first law of thermodynamics to analyze basic energy systems
- What a heat engine is
- How to analyze a basic heat engine using the first and second laws of thermodynamics

hot beverage on a table eventually cools by itself to the temperature of the surroundings, thereby degrading the energy of the high temperature beverage into a less useful form.

Interestingly, working steam engines were developed prior to the emergence of thermodynamics as a science. Two Englishmen, Thomas Savery and Thomas Newcomen, constructed crude steam engines in 1697 and 1712, respectively. Practical improvements in these first steam engines were made in the ensuing years, but the fundamental thermodynamic principles on which they operated were not fully understood until much later. The first and second laws of thermodynamics were not pronounced until the 1850s. The laws of thermodynamics and other thermodynamics concepts were pioneered by scientists and mathematicians such as Gabriel Fahrenheit (1686–1736), Sadi Carnot (1796–1832), Rudolph Clausius (1822–1888), William Rankine (1820–1872), and Lord Kelvin (1824–1907). These individuals, and many others, laid the theoretical foundations for modern thermodynamics.

Because thermodynamics is the science of energy, it would be difficult to find any engineering system that does not embody thermodynamic principles in some way. Principles of thermodynamics are at work all around us. The illumination by which you are reading this page is produced by converting electrical energy to visible light. Our homes are maintained at comfortable living temperatures by furnaces, heat pumps, and air conditioners. These devices utilize the energy contained in fossil fuels or electrical energy to heat and cool our homes. Common household appliances such as dishwashers, microwave ovens, refrigerators, humidifiers, clothes dryers, toasters, water heaters, irons, and pressure cookers rely on principles of thermodynamics for their operation. Industrial systems that utilize thermodynamics include internal combustion and diesel engines, turbines, pumps and compressors, heat exchangers, cooling towers, and solar panels, to name a few. Shown in Figures 6.1, 6.2, and 6.3 are some engineering systems that utilize thermodynamic processes.

Figure 6.1. Household refrigerators utilize basic principles of thermodynamics for removing thermal energy from food items. (Courtesy of KitchenAid, Benton Harbor, MI.)

Figure 6.2. Wind turbines convert wind energy to electrical energy. The wind power plant shown here is owned and operated by the Sacramento Municipal Utility District. (Courtesy of the U.S. Department of Energy/National Renewable Energy Laboratory.)

Figure 6.3. The Vogtle Electric Generating Plant, located in eastern Georgia, converts nuclear energy to electrical energy. The plant is capable of producing over 2000 MW of power. (Courtesy of Southern Nuclear, Birmingham, AL.)

6.2   PRESSURE AND TEMPERATURE

Systems that use thermodynamic processes for their operation can be described by certain physical characteristics. Any such characteristic of a system is called a *property*. In its broadest engineering context, a property can refer to any physical aspect of a system, such as length, density, velocity, modulus of elasticity, viscosity, etc. In thermodynamics, a property usually refers to a characteristic that relates directly to the energy of the system. Two of the most important properties in thermodynamics are *pressure* and *temperature*.

6.2.1   Pressure

When a fluid (liquid or gas) is confined by a solid boundary, the fluid exerts a force on the boundary. The direction of the force is normal (perpendicular) to the boundary. From a microscopic point of view, the force is the result of a change in momentum experienced by the fluid molecules as they collide with the solid surface. Molecules collide with the surface in many directions, but the overall effect of the collisions is a net force that is normal to the surface. **Pressure** is defined as the *normal force exerted by a fluid per unit area*. Thus, the formula for pressure is

$$P = \frac{F}{A} \qquad (6.1)$$

where P is pressure (N/m^2), F is the normal force (N), and A is area (m^2). For a liquid at rest, pressure increases with depth as a consequence of the weight of the liquid. For example, the pressure exerted by sea water on the hull of a submarine at a depth of 200 m is greater than the pressure exerted by seawater on a scuba diver at a depth of 10 m. The pressure in a tank containing a gas is essentially constant, however, because the weight of the gas is usually negligible compared with the force required to compress the gas. For example, consider a gas enclosed in a piston-cylinder device, as illustrated in Figure 6.4. A force F is applied to the piston, compressing the gas in the

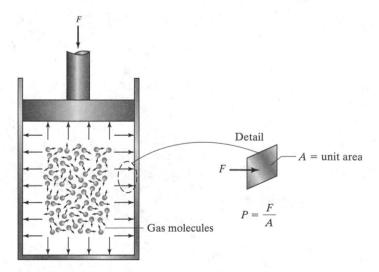

Figure 6.4.   An enclosed gas exerts a pressure on the walls of its container.

cylinder. A constant pressure, whose magnitude is given by Equation (6.1), acts on *all* interior surfaces of the enclosure. If the force F increases, the pressure P increases accordingly.

The unit of pressure in the SI system is N/m^2, which is defined as the pascal (Pa). Hence, $1 \text{ Pa} = 1 \text{ N/m}^2$. The pascal is a very small unit of pressure, so it is customary to use the standard SI multiples kPa (kilopascal) and MPa (megapascal), which stand for 10^3 Pa and 10^6 Pa, respectively. Pressure is sometimes expressed in terms of the bar $(1 \text{ bar} = 10^5 \text{ Pa})$. The most commonly used unit of pressure in the English system is pound-force per square inch (lb_f/in^2), typically written in shorthand notation as psi.

When doing calculations involving pressure, care must be taken to specify the reference on which the pressure is based. Pressure referenced to a perfect vacuum is called *absolute pressure* (P_{abs}). *Atmospheric pressure* (P_{atm}) is the pressure exerted by the atmosphere at a specified location. At sea level, atmospheric pressure is $P_{atm} = 101,325$ Pa = 14.696 psi. At higher elevations, the atmospheric pressure is lower, due to decreasing air density. The pressure that uses atmospheric pressure as the reference is called *gauge pressure* (P_{gauge}). Gauge pressure is the difference between the absolute pressure and the local atmospheric pressure. Most pressure-measuring instruments, such as an automobile tire gauge, measure gauge pressure. A pressure below atmospheric pressure is called *vacuum pressure* (P_{vac}). Vacuum pressure is measured by vacuum gauges that indicate the difference between the local atmospheric pressure and absolute pressure. Gauge, absolute, and vacuum pressures are all positive quantities and are related to one another by the relations

$$P_{gauge} = P_{abs} - P_{atm} \text{ (for pressures above } P_{atm}) \tag{6.2}$$

$$P_{vac} = P_{atm} - P_{abs} \text{ (for pressures below } P_{atm}) \tag{6.3}$$

The majority of thermodynamic equations and data tables use absolute pressure. Sometimes, the letter "a" is used to specify absolute pressure and the letter "g" is used to specify gauge pressure. For example, absolute, and gauge pressures are sometimes written in English units as psia and psig, respectively.

6.2.2 Temperature

Our physiological sense of temperature tells us how hot or how cold something is, but does not provide a quantitative definition of temperature for engineering use. A scientific definition, based on microscopic considerations, is that temperature is a measure of atomic and molecular kinetic energy of a substance. Thus, at a temperature of absolute zero, all translational, rotational, and vibrational motions of atoms and molecules cease. A practical engineering definition is that **temperature**, or, more specifically, a temperature *difference*, is an *indicator of heat transfer*. As illustrated in Figure 6.5, heat flows from a region of high temperature to a region of low temperature. This engineering definition of temperature is consistent with our common experiences. For example, a hot beverage will gradually cool until it reaches the temperature of the surroundings. Conversely, a cold beverage will eventually warm until it reaches the temperature of the surroundings. In either case, when the beverage attains the temperature of the surroundings, heat transfer stops, and the beverage and surroundings are said to be in *thermal equilibrium*, because their temperatures are equal. We can therefore state that when any two bodies have the same temperature, the bodies are in thermal equilibrium.

The **zeroth law of thermodynamics** states that *if two bodies are in thermal equilibrium with a third body, they are also in thermal equilibrium with each other.* This law is analogous to the arithmetic axiom which states that if $A = C$ and $B = C$, then $A = B$. The

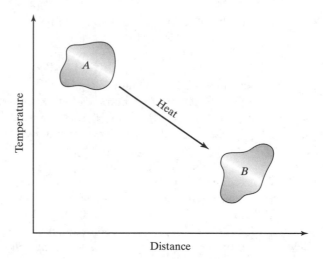

Figure 6.5. Heat is transferred from a high-temperature region to a low-temperature region.

zeroth law, as obvious as it sounds, cannot be derived from the first or second laws of thermodynamics. The zeroth law of thermodynamics is the underlying physical basis for a key element of thermodynamics: temperature measurement. By the zeroth law of thermodynamics, if body A and body B are in thermal equilibrium with body C, then body A and body B are in thermal equilibrium with each other. By letting body C be a thermometer, the zeroth law of thermodynamics infers that bodies A and B are in thermal equilibrium if their temperatures, as measured by the thermometer, are equal. The interesting aspect of the zeroth law is that bodies A and B do not even have to be in physical contact with each other. They only have to have the same temperature to be in thermal equilibrium.

Like length, mass, time, electrical current, luminous intensity, and amount of substance, temperature is a base dimension. As a base dimension, temperature is predicated on a measurable physical standard. Temperature *scales* enable engineers to make temperature measurements on a common basis. International temperature scales have been adopted that are based on fixed reproducible thermodynamic states of matter. The ice point and boiling point of water at 1 atmosphere pressure are defined as 0°C and 100°C, respectively, on the *Celsius* temperature scale. On the *Fahrenheit* temperature scale, these points have the values 32°F and 212°F, respectively. The *Kelvin* and *Rankine* temperature scales are absolute temperature scales that have 0 K and 0°R as their lowest possible temperature values. Thus, we say that *absolute zero* temperature refers to either 0 K or 0°R. By convention, the degree symbol "°" is used for the Celsius, Fahrenheit, and Rankine temperature scales, but not the Kelvin scale. A comparison of these four temperature scales is shown in Figure 6.6.

Because engineers use four different temperature scales in the analysis of thermodynamic systems, it is important to know how to convert from one temperature scale to another. The Kelvin scale is related to the Celsius scale by the formula

$$T(\text{K}) = T(°\text{C}) + 273.15 \tag{6.4}$$

and the Rankine scale is related to the Fahrenheit scale by the formula

$$T(°\text{R}) = T(°\text{F}) + 459.67 \tag{6.5}$$

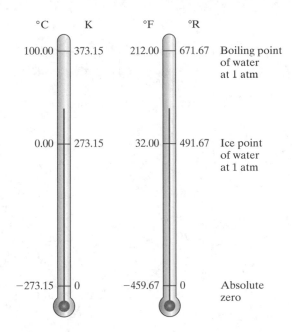

Figure 6.6. The Celsius, Kelvin, Fahrenheit, and Rankine temperature scales.

In the majority of practical applications, temperature precision beyond the decimal point is not required, so the constants in Equations (6.4) and (6.5) are typically rounded to 273 and 460, respectively. The Rankine and Kelvin scales are related by the formula

$$T(°R) = 1.8\,T(K) \tag{6.6}$$

and the Fahrenheit and Celsius scales are related by the formula

$$T(°F) = 1.8\,T(°C) + 32 \tag{6.7}$$

Equations (6.4) through (6.7) are used to convert one temperature value or measurement to another. Using Figure 6.6, we find that the validity of these relations can be readily checked by converting the boiling point and ice point of water from one temperature scale to the other three scales.

We mentioned earlier in this section that temperature difference is an indicator of heat transfer. When calculating a temperature *difference*, it is important to note that the size of the temperature divisions for the Kelvin and Celsius scales are equal and that the size of the temperature divisions for the Rankine and Fahrenheit scales are also equal. In other words, increasing the temperature of a substance by 1 K is the same as increasing the temperature by 1°C. Similarly, increasing the temperature of a substance by 1°R is the same as increasing the temperature by 1°F. Thus, we write the relations for temperature differences as

$$\Delta T(K) = \Delta T(°C) \tag{6.8}$$

and

$$\Delta T(°R) = \Delta T(°F) \tag{6.9}$$

where the Greek symbol Δ refers to a difference or change. When doing thermodynamics calculations involving temperature differences in the SI system, it does not matter whether K or °C is used. Similarly, when doing thermodynamics calculations involving temperature differences in the English system, it does not matter whether °R or °F is used. In analysis work, care must be taken to distinguish between a single temperature value T and a temperature difference ΔT. If the thermodynamic relation is of the form $x = y\Delta T$, it does not matter whether ΔT is expresses in K or °C. If the thermodynamic relation is of the form $x = yT$, however, the temperature T must be expressed in K. The same rules apply for the corresponding English temperature units °R and °F.

EXAMPLE 6.1

The atmospheric pressure in Denver, Colorado (elevation 1 mile) is approximately 83.4 kPa. If we were to inflate the tire of an automobile in Denver to a gauge pressure of 35 psi, what is the absolute pressure in units of kPa?

SOLUTION

In order to work in a consistent set of units, we convert the gauge pressure to units of kPa:

$$35 \text{ psi} \times \frac{1 \text{ kPa}}{0.14504 \text{ psi}} = 241.3 \text{ kPa}$$

Solving for absolute pressure from Equation (6.2), we have

$$P_{abs} = P_{gauge} + P_{atm}$$
$$= 241.3 \text{ kPa} + 83.4 \text{ kPa}$$
$$= 325 \text{ kPa}$$

EXAMPLE 6.2

Steam in a boiler has a temperature of 300°C. What is this temperature in units of K, °R, and °F? If the temperature drops to 225 °C, what is the temperature change in units of K, °R, and °F?

SOLUTION

Using Equation (6.4), we find that the temperature in K is

$$T(\text{K}) = T(\text{°C}) + 273$$
$$= 300\text{°C} + 273$$
$$= 573 \text{ K}$$

Now that the temperature in K is known, we use Equation (6.6) to find the temperature in °R:

$$T(\text{°R}) = 1.8 \, T(\text{K})$$
$$= 1.8(573 \text{ K})$$
$$= 1031\text{°R}$$

Using Equation (6.7), we find that the temperature in °F is

$$T(\text{°F}) = 1.8 \, T(\text{°C}) + 32$$
$$= 1.8(300\text{°C}) + 32$$
$$= 572\text{°F}$$

The temperature change is

$$\Delta T = 300\text{°C} - 225\text{°C}$$
$$= 75\text{°C} = 75 \text{ K}$$

The Rankine and Kelvin temperature scales are related through Equation (6.6). Because these temperature scales are absolute scales, we can write Equation (6.6) in terms of temperature differences as

$$\Delta T(°R) = 1.8 \, \Delta T(K)$$

Hence,

$$\Delta T = 1.8(75 \text{ K})$$
$$= 135°R = 135°F$$

PRACTICE!

1. A pressure gauge on the discharge side of an air compressor reads 260 kPa. What is the absolute pressure at this point in units of psi if the local atmospheric pressure is 95 kPa?

   Answer: 51.5 psi

2. A force of 1.2 kN is applied to the piston of a cylinder, compressing the gas within the cylinder. The piston has a radius of 4 cm. If the local atmospheric pressure is 100 kPa, what is the pressure inside the cylinder?

   Answer: 339 kPa

3. A vacuum gauge connected to a tank reads 30 kPa. If the local atmospheric pressure is 13.5 psi, what is the absolute pressure in units of psi?

   Answer: 8.42 psi

4. A boiler at sea level contains superheated steam at 0.4 MPa absolute pressure and 300°C. Find the gauge pressure in the boiler and the steam temperature in units of K, °R, and °F.

   Answer: 298.7 kPa, 573 K, 1031°R, 572°F

5. A hard-boiled egg removed from a pot of boiling water at 96°C is placed in a 40°F refrigerator to cool. Find the temperature of the egg in units of K, °C, and °R after the egg has attained thermal equilibrium with the refrigerator. What is the temperature change of the egg in units of °F, °C, and K?

   Answer: 164.9°F, 91.6°C, 91.6 K

6.3 FORMS OF ENERGY

The concept of *energy* is central to the study of engineering in general and thermodynamics in particular. A concise definition of **energy** is the *capacity to do work*. If a system has the capacity to do work, it possesses at least one form of energy that is available for transformation to another form of energy. For example, a compressed spring possesses a type of energy referred to as potential energy. As the term implies, potential energy is a type of stored energy that has the potential for producing some useful external effect. Consider a mass attached to a compressed spring, as illustrated in Figure 6.7. When the compressed spring is released, the stored energy in the spring will begin to resume its original undeformed length, imparting a velocity to the mass. As the spring elongates, the potential energy in the spring is converted to kinetic energy. Actually, a small portion of the potential energy in the spring is converted to thermal energy (heat), because there is friction between the mass and the surface, as well as within the spring itself. The important thing to realize is that *all* the potential energy in the compressed spring is converted to other forms of energy (i.e., the total energy of the transformation is constant). In accordance with the first law of thermodynamics, no energy is produced or destroyed during the energy transformation.

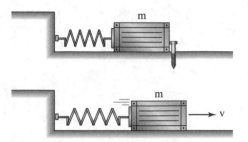

Figure 6.7. The potential energy in a compressed spring is converted to kinetic energy.

Energy can exist in many forms. For purposes of thermodynamic analysis, energy is classified into two broad categories, *macroscopic* energy and *microscopic* energy. Macroscopic forms of energy are those that a whole system possesses with respect to a fixed external reference. In thermodynamics, the macroscopic forms of energy are potential energy and kinetic energy. Potential and kinetic energy are based on external position and velocity references, respectively. Microscopic forms of energy are those that relate to the system on a molecular or atomic level. There are several types of microscopic energies, so we conveniently group them together into a single category referred to as internal energy. Internal energy is the sum of all the various forms of microscopic energies possessed by the molecules and atoms in the system. Potential, kinetic, and internal energy warrant further discussion.

6.3.1 Potential Energy

Potential energy is the stored energy of position possessed by an object. In thermodynamics, there are primarily two forms of potential energy, *elastic* potential energy and *gravitational* potential energy. **Elastic potential energy** is the *energy stored in a deformable body such as an elastic solid or a spring*. **Gravitational potential energy** is the *energy that a system possesses by virtue of its elevation with respect to a reference in a gravitational field*. Elastic potential energy is usually of minor importance in most thermodynamics work, so gravitational potential energy is emphasized here. Gravitational potential energy, abbreviated PE, is given by the relation

$$\mathrm{PE} = mgz \qquad (6.10)$$

where m is the mass of the system (kg), g is gravitational acceleration (m/s^2), and z is the elevation (m) of the center of mass of the system with respect to a selected reference plane. The location of the reference plane is arbitrary, but is usually selected on the basis of mathematical convenience. For example, consider a boulder poised on the edge of a cliff, as illustrated in Figure 6.8. The center of mass of the boulder is 20 m above the ground. A reasonable reference plane is the ground because it is a convenient origin. If the boulder's mass is 1500 kg, the gravitational potential energy of the boulder is

$$\begin{aligned} \mathrm{PE} &= mgz \\ &= (1500 \ \mathrm{kg})(9.81 \ \mathrm{m/s}^2)(20 \ \mathrm{m}) \\ &= 2.94 \times 10^5 \ \mathrm{N{\cdot}m} = 2.94 \times 10^5 \mathrm{J} = 294 \, \mathrm{kJ} \end{aligned}$$

What happens to the boulder's potential energy as it falls from the cliff?

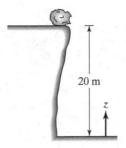

Figure 6.8. A boulder elevated above the ground has gravitational potential energy.

6.3.2 Kinetic Energy

Kinetic energy is the *energy that a system possess as a result of its motion with respect to a reference frame*. Kinetic energy, abbreviated KE, is given by the relation

$$KE = \tfrac{1}{2}\,mv^2 \tag{6.11}$$

where m is the mass of the system (kg) and v is the velocity of the system (m/s). When the boulder in Figure 6.8. is pushed off the cliff, it begins to fall toward the ground. As the boulder falls, its velocity increases, and its potential energy is converted to kinetic energy. If the velocity of the 1500-kg boulder is 10 m/s at a point between the cliff and ground, the boulder's kinetic energy at this point is

$$\begin{aligned}
KE &= \tfrac{1}{2}\,mv^2 \\
&= \tfrac{1}{2}(1500 \text{ kg})(10 \text{ m/s})^2 \\
&= 7.50 \times 10^4 \text{ J} = 75.0 \text{ kJ}
\end{aligned}$$

Immediately before the boulder impacts the ground, all the boulder's potential energy has been converted to kinetic energy. What happens to the boulder's kinetic energy during the impact with the ground?

6.3.3 Internal Energy

Internal energy is the *sum of all the microscopic forms of energy of a system*. Unlike potential energy and kinetic energy, which relate to the energy of a system with respect to external references, internal energy relates to the energy *within* the system itself. Internal energy, denoted by the symbol U, is a measure of the kinetic energies associated with the molecules, atoms, and subatomic particles of the system. Suppose the system under consideration is a polyatomic gas. [A polyatomic gas is a gas that consists of two or more atoms that form a molecule, such as carbon dioxide (CO_2). A monatomic gas consists of only one atom, such as helium (He) or argon (Ar).] Because gas molecules move about with certain velocities, the molecules possess kinetic energy. The movement of the molecules through space is called translation, so we refer to their kinetic energy as *translational* energy. As the gas molecules translate, they also rotate about their center of mass. The energy associated with this rotation is referred to as *rotational* energy. In addition to translating and rotating, the atoms of polyatomic gas molecules oscillate about their center of mass, giving rise to *vibrational* energy. On a subatomic scale, the electrons of atoms "orbit" the nucleus. Furthermore, electrons spin about their own axis, and the nucleus also possesses a spin. The sum of the translational, rotational, vibrational, and subatomic energies constitutes a *fraction* of the internal energy of the system called the *sensible* energy. Sensible energy is the *energy required to change the temperature of a system*. As an example of sensible energy, suppose that we wish to boil a pan of water on the stove. The water is initially at a temperature of about 20°C. The stove burner imparts energy to the water, increasing the kinetic energy of the water molecules. The increase in kinetic energy of the water molecules is manifested as an increase in temperature of the water. As the burner continues to supply energy to the water, the sensible energy of the water increases, thereby increasing the temperature, until the boiling point is reached.

If sensible energy is only a fraction of the internal energy, what kind of energy constitutes the other fraction? To answer this question, we must recognize the various forces that exist between molecules, between atoms, and between subatomic particles. From basic chemistry, we know that various *binding forces* exist between the molecules of a substance. When these binding forces are broken, the substance changes from one *phase* to another. The three phases of matter are solid, liquid, and gas. Binding forces are strongest in solids, weaker in liquids, and weakest in gases. If enough energy is supplied to a solid substance, ice for example, the binding forces are overcome and the substance changes to the liquid phase. Hence, if enough energy is supplied to ice (solid water), the ice changes to liquid water. If still more energy is supplied to the substance, the substance changes to the gas phase. The amount of energy required to produce a phase change is referred to as *latent* energy. In most thermodynamic processes, a phase change involves the breaking of molecular bonds only. Therefore, the atomic binding forces responsible for maintaining the chemical identity of a substance are not usually considered. Furthermore, the binding energy associated with the strong nuclear force—the force that binds the protons and neutrons in the nucleus—is relevant only in fission reactions.

6.3.4 Total Energy

The *total energy* of a system is the sum of the potential, kinetic, and internal energies. Thus, the total energy, *E*, is expressed as

$$E = \text{PE} + \text{KE} + U \tag{6.12}$$

As a matter of convenience, it is customary in thermodynamics work to express the energy of a system on a *per unit mass* basis. Dividing Equation (6.12) by mass *m*, and noting the definitions of potential and kinetic energies from Equations (6.10) and (6.11), we obtain

$$e = gz + \frac{v^2}{2} + u \tag{6.13}$$

where $e = E/m$ and $u = U/m$. The quantities e and u are called the *specific total energy* and *specific internal energy*, respectively.

In the analysis of many thermodynamic systems, the potential and kinetic energies are zero or are sufficiently small that they can be neglected. For example, a boiler containing high-temperature steam is stationary, so its kinetic energy is zero. The boiler has potential energy with respect to an external reference plane (such as the floor on which it rests), but the potential energy is irrelevant, since it has nothing to do with the operation of the boiler. If the potential and kinetic energies of a system are neglected, internal energy is the only form of energy present. Hence, the total energy equals the internal energy, and Equation (6.12) reduces to $E = U$.

The analysis of thermodynamic systems involves the determination of the *change* of the total energy of the system because this tells us how energy is converted from one form to another. It does not matter what the absolute value of the total energy is, as we are interested only in the change of the total energy. This is paramount to saying that it does not matter what the energy reference value is because the change of energy is the same regardless of what reference value we choose. The reference value is arbitrary. Returning to our falling boulder example, the change of potential energy of the boulder does not depend on the location of the reference plane. We could choose the ground as the reference plane or some other location, such as the top of the cliff or any other elevation for that matter. The change of potential energy of the boulder depends only on the elevation change. Hence, if potential and kinetic energies are neglected, the change of total energy of a system equals the change of internal energy, and Equation (6.12) is written as $\Delta E = \Delta U$.

PROFESSIONAL SUCCESS: DEALING WITH ENGINEERING PROFESSORS

As a new engineering student, you may believe that engineering professors are probably not that much different from professors in other disciplines on campus. Perhaps you think that they are not even that much different from people outside higher education who work in nonteaching occupations. However, after you have taken a few engineering courses, your opinion will probably change. Engineering professors are unique. It may even be said that they are somewhat odd. Some engineering professors are overly serious, while others may seem rather light minded. Some engineering professors dress very neatly, wearing a suit, tie, polished shoes, etc., while others come to school looking more like a student, wearing jeans, a sweatshirt, and sneakers. Regardless of their personalities and personal appearances, engineering professors are genuinely interested in their students and desire to see them succeed in their engineering studies. Engineering professors are very knowledgeable people in their disciplines, and they want to share that knowledge with students. They were students once, so they understand what you are going through. Professors are *teachers*, and quality instruction is what students expect from them. However, as a student, you should realize that most professors are involved in numerous activities outside the classroom that may or may not relate directly to teaching. Much of your professor's time is spent developing and improving the engineering curriculum. Depending on the availability of graduate teaching assistants, grading may also occupy a considerable fraction of the professor's time. Some colleges and universities, particularly the larger ones, are referred to as *research* institutions. At these schools, engineering professors are expected to conduct research and publish the results of their research. In addition to publishing research papers, some engineering professors write textbooks. Because most engineering professors specialize in a certain aspect of their discipline, some professors work part time as consultants to private or governmental agencies. Most colleges and universities expect their faculty to render

service to the institution by serving on various campus committees. Some professors, in addition to their research, writing, and service activities, serve as department or program advisors to students. Professors may even be involved with student recruitment, fund raising, professional engineering societies, and a host of other activities.

What does all this mean to you, the engineering student? It means that there are right ways and wrong ways of dealing with your professors. Here are a few suggestions:

- Be an active member of your professor's class. Attend class, arrive on time, take notes, ask questions, and participate. Being actively engaged in the classroom not only helps you learn, but it also helps the professor teach!

- If you need to obtain help from your professor outside of class, schedule an appointment during an office hour, and *keep* the appointment. Unless your professor has an "open door" policy, scheduling appointments during regular office hours is preferred, because your professor is probably involved in research or other activities.

- Engineering professors appreciate students who give their best efforts in solving a problem *before* asking for help. Before you go to the professor's office, be prepared to tell your professor how you approached the problem and where the potential errors are. Many engineering professors become irritated when the first thing a student says is "Look at this problem, and tell me what I'm doing wrong" or "I just can't get the answer in the back of the book." Preparing to ask the right questions before the visit will enable your professor to help you more fully.

- Unless instructed to do otherwise, do not call professors at home. If you need assistance with homework, projects, etc., contact your professor at school during regular office hours if possible or by special appointment. Like students, professors try to have a personal life apart from their day-to-day academic work. How would you like it if your professors called *you* at home to assign additional homework?

- Unless instructed to do otherwise, address professors by their appropriate titles. Do not call them by their first names. Most engineering professors have a PhD degree, so it is appropriate to address these individuals as "Dr. Jones" or "Professor Jones." If the professor does not have a doctorate, the student should address the professor by "Professor Jones," "Mr. Jones," or "Ms. Jones."

6.4   WORK AND HEAT

Work, like energy, is a word that is commonly used in our everyday language and a word that has many meanings. As a student, you know that studying engineering is a lot of "work." When we participate in sports or exercise at the gym, we get a "workout." A person who travels to a place of employment goes to "work," and when a mechanical device stops functioning, we say that it doesn't "work." While various day-to-day usages of this term are thrown about quite casually, engineering defines "work" precisely, with no ambiguity. **Work** is defined as a *form of energy that is transferred across the boundary of a system*. A *system* is a quantity of matter or a region in space chosen for study, and the *boundary* of a system is a real or imaginary surface that separates the system from the surroundings. For example, propane in a fuel tank is a thermodynamic system, and the boundary of the system is the inside surface of the tank wall. Besides work, there is a second form of energy that can be transferred across the boundary of a system. The second form of energy is heat. Heat is a special kind of energy transfer that is easily recognizable and differentiated from work. **Heat** is defined as the *form of energy that is transferred across the boundary of a system by virtue of a temperature difference*. A system with both work and heat crossing the boundary is illustrated in Figure 6.9. Depending on the nature of the interactions of the system with the surroundings, work and heat can be transferred across the boundary in either direction. The only requirement for heat transfer is a temperature difference between the system and the surroundings. If there is no temperature difference between the system and the surroundings, heat cannot be transferred, thus, the only form of energy transfer is work. Because work and heat are forms of energy, both quantities have the same units. Work and heat have units of J in the SI system and Btu in the English system. The most commonly used symbols for work and heat in thermodynamics are W and Q, respectively.

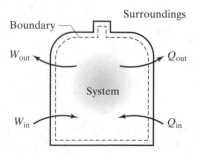

Figure 6.9.   Energy in the form of work or heat can be transferred across the boundary of a system.

Now that work and heat have been defined in general terms, let us examine these forms of energy transfer more closely. In thermodynamics, work is usually categorized as *mechanical* work or *nonmechanical* work. The nonmechanical forms of work include electrical, magnetic, and electrical polarization work. Mechanical forms of work are generally the most important, so we will consider these in some detail.

6.4.1   Mechanical Work

There are several types of mechanical work. From basic physics, the work W done by a force F acting through a displacement s in the *same direction* of the force is given by the relation

$$W = Fs \qquad (6.14)$$

Equation (6.14) is valid only if the force is constant. If the force is not constant (i.e., if the force is a function of displacement), the work is obtained by integration. Thus, Equation (6.14) becomes

$$W = \int_1^2 Fds \qquad (6.15)$$

where the limits 1 and 2 denote the initial and final positions of the displacement, respectively. Equation (6.15) is a general mathematical definition from which equations for the various types of mechanical work are derived. Consider, for example, a vehicle that climbs a rough hill, as shown in Figure 6.10. As the vehicle climbs the hill, it encounters two forces that tend to oppose its motion. Gravity exerts a downward force on the vehicle that retards its upward motion, and friction between the wheels and the rough surface retards its motion along the surface. The vehicle does work against these two forces, and the magnitude of that work is found by integrating the total force from position s_1 to s_2, which is graphically interpreted as the area under the force-displacement curve. The various types of mechanical work are now considered.

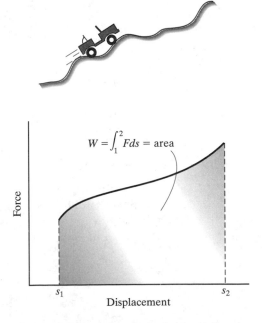

$$W = \int_1^2 Fds = \text{area}$$

Figure 6.10.   As a vehicle climbs a hill, gravitational and friction forces act on it.

Gravitational work

Gravitational work is defined as the *work done on an object by a gravitational force*. In a gravitational field, the force acting on a body is the *weight* of the body, and is given by

$$F = mg \tag{6.16}$$

where m is mass (kg) and g is the local gravitational acceleration (m/s^2). Consider a vehicle that climbs a hill from elevation z_1 to a higher elevation z_2, as shown in Figure 6.11. Substituting Equation (6.16) into Equation (6.15) and integrating, we obtain the gravitational work

$$W_g = \int_1^2 F dz = mg \int_1^2 dz = mg(z_2 - z_1) \tag{6.17}$$

Note that the displacement in Equation (6.17) is in terms of elevation z, because work is defined as a force acting through a distance in the *same direction* of the force. Gravity acts in the *vertical* direction, so Equation (6.17) is written in terms of a vertical distance (elevation) and not a horizontal distance. The gravitational work for the vehicle in Figure 6.11 is *negative*, as the direction of the displacement (upward) is opposite to the direction of the gravitational force (downward). If the vehicle descends the hill, the gravitational work is positive, because the displacement is in the same direction as the force. Note also that gravitational work is equivalent to a change in potential energy, since PE = mgz.

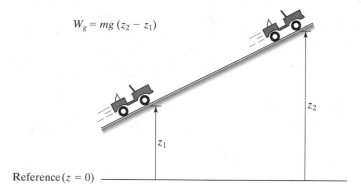

Figure 6.11. Gravitational work is done on a body as it changes elevation.

Acceleration work

Acceleration work is the *work associated with a change in velocity of a system*. Newton's second law states that the force acting on a body equals the product of the body's mass and acceleration. But acceleration a is the time derivative of velocity v, so Newton's second law may be written as

$$F = ma = m\frac{dv}{dt} \tag{6.18}$$

Velocity is the time derivative of displacement,

$$v = \frac{ds}{dt} \tag{6.19}$$

so the differential displacement ds in Equation (6.15) is $ds = v\,dt$. Thus, acceleration work is

$$W_a = \int_1^2 F\,ds = \int_1^2 \left(m\frac{dv}{dt}\right)(v\,dt) = m\int_1^2 v\,dv = \tfrac{1}{2}\,m(v_2^2 - v_1^2) \qquad (6.20)$$

As shown in Figure 6.12, a vehicle traveling along a horizontal road increases its velocity from 10 mi/h to 65 mi/h. In doing so, the vehicle does acceleration work because its velocity changes. We note that the acceleration work is equivalent to a change in kinetic energy, since $\mathrm{KE} = \tfrac{1}{2}\,mv^2$.

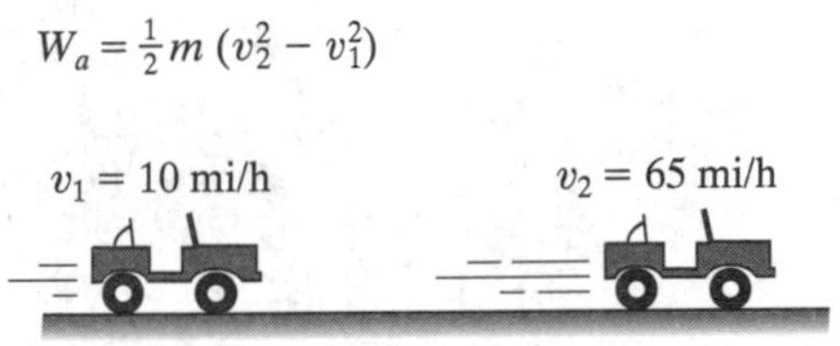

$$W_a = \tfrac{1}{2}m\,(v_2^2 - v_1^2)$$

$v_1 = 10$ mi/h $\qquad\qquad v_2 = 65$ mi/h

Figure 6.12. A body does acceleration work as its velocity changes.

Boundary work

Boundary work is the *work associated with the movement of a solid boundary*. The most common instance of boundary work is the compression or expansion of a gas within a piston-cylinder device, as illustrated in Figure 6.13. A force F is applied to the piston, compressing the gas within the cylinder. Because the cylinder is a closed vessel, the pressure increases as the gas volume decreases. As the gas volume decreases from V_1 to

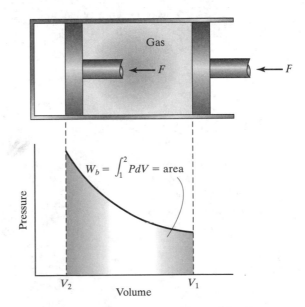

$$W_b = \int_1^2 P\,dV = \text{area}$$

Figure 6.13. Boundary work is performed by a piston as it compresses a gas.

V_2, the pressure increases along a path that depends on certain physical characteristics of the compression process. Pressure is defined as a force divided by area, so the force causing the compression is given by the relation

$$F = PA \qquad (6.21)$$

where A is the surface area of the face of the piston. A differential change in volume, dV is the product of the piston's differential displacement ds and the surface area of the piston A. Hence, $dV = Ads$ and the boundary work becomes

$$W_b = \int_1^2 Fds = \int_1^2 PA\frac{dV}{A} = \int_1^2 PdV \qquad (6.22)$$

Because the product PdV appears in the definition, boundary work is sometimes referred to as "PdV" work. As indicated in Figure 6.13, the magnitude of the boundary work is the area under the pressure-volume curve. In order to evaluate the integral in Equation (6.22), we would have to know the functional relationship between pressure P and volume V. This relationship may be an analytical expression for P as a function of V or a graph that shows the variation of P with V.

Shaft work

Shaft work is *energy transfer by a rotating shaft*. Numerous engineering systems transfer energy by means of a rotating shaft. The drive shaft of an automobile, for example, transfers energy from the transmission to the axle. Energy is transferred from a boat motor to the propeller by a shaft. Even the mixing blades of a food blender perform shaft work on the food. As a shaft rotates, a constant torque is usually applied to the shaft that tends to retard its rotation. As illustrated in Figure 6.14, the torque τ is produced by a force F acting through a moment arm r according to the relation

$$\tau = Fr \qquad (6.23)$$

The force acts through a distance s equal to the circumference times the number of revolutions of the shaft n. Thus,

$$s = (2\pi r)n \qquad (6.24)$$

Upon substituting Equations (6.23) and (6.24) into Equation (6.14), the shaft work, becomes

$$W_{sh} = 2\pi n\tau \qquad (6.25)$$

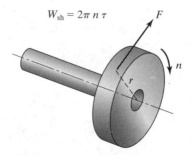

Figure 6.14. Work is produced by a rotating shaft.

Spring work

Spring work is the *work done in deforming a spring.* A force is required to compress or stretch a spring, so work is done. From elementary physics, we know that the force required to deform a linear elastic spring is proportional to the deformation. This principle is known as Hooke's law and is expressed as

$$F = kx \qquad (6.26)$$

where F is force, x is displacement (change in spring length), and k is the spring constant. Substituting Equation (6.26) into Equation (6.15) and noting that $ds = dx$, the spring work becomes

$$W_{sp} = \int_1^2 F\,ds = \int_1^2 (kx)\,dx = \tfrac{1}{2}\,k(x_2^2 - x_1^2) \qquad (6.27)$$

As indicated in Figure 6.15, the initial and final spring displacements are x_1 and x_2, respectively, as measured from the rest (undeformed) position of the spring.

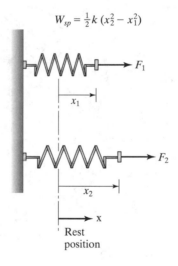

Figure 6.15. Work is done by stretching or compressing a spring.

6.4.2 Heat

As we defined it earlier in this chapter, you now know that *heat* is the transfer of energy across the boundary of a system by virtue of a temperature difference. In order for heat transfer to occur, there must be a temperature difference between the system and the surroundings. The transfer or flow of heat is not the flow of a material substance, as in the case of the flow of a fluid such as air or water. Rather, there is an exchange of internal energy across the system boundary by atomic or molecular motion or by electromagnetic waves. Heat transfer can occur by three distinct mechanisms: *conduction, convection,* and *radiation.* Conduction is the transfer of internal energy in solids and fluids at rest. The actual mechanism of conduction involves kinetic energy exchange between molecules in contact or, in the case of metals, movement of free electrons. Convection is the mechanism by which internal energy is transferred to or from a fluid near a solid surface. Convection is basically conduction at the solid surface with the added complexity of energy transfer by moving fluid molecules. Radiation is the mechanism by which energy is transferred by electromagnetic waves. Unlike conduction and convection, radiation does not require a medium. A familiar example of radiation is the thermal energy that we receive from the sun across the vacuum of space. Regardless of the heat transfer mechanism involved, the *direction* of heat transfer is always from a high-temperature region to a low-temperature region.

Heat transfer occurs all around us. As a familiar example, consider the hot beverage shown in Figure 6.16. Heat is transferred from the beverage to the surroundings by all three heat transfer mechanisms. A portion of the energy is transferred by convection from the liquid to the solid cup wall where the heat is subsequently conducted through the cup wall. That energy is then transferred to the surroundings by convection and radiation. The portion of the energy conducted into the bottom portion of the cup is transferred directly into the tabletop by conduction. The remaining energy is transferred from the surface of the liquid directly to the surroundings by convection and radiation.

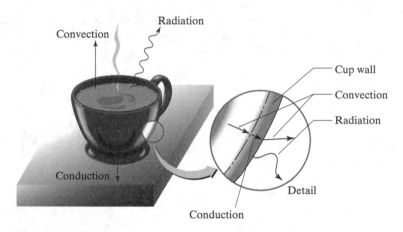

Figure 6.16. A hot beverage resting on a table transfers thermal energy to the surroundings by conduction, convection and radiation.

In the next example, we use the general analysis procedure of (1) problem statement, (2) diagram, (3) assumptions, (4) governing equations, (5) calculations, (6) solution check, and (7) discussion.

EXAMPLE 6.3

Problem statement

A 1200-kg automobile accelerates up a hill, increasing its speed from 5 mi/h to 45 mi/h, along a straight 100-m stretch of road. If the hill makes an angle of 6° with respect to the horizontal, find the total work.

Diagram

The diagram for this problem is shown in Figure 6.17.

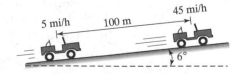

Figure 6.17. Example 6.3.

Assumptions

1. Neglect friction between wheels and road.
2. Neglect aerodynamic friction.
3. Mass of automobile is constant.

Governing equations

Two forms of work, gravitational and acceleration, are involved as the automobile ascends the hill, so we have two governing equations:

$$W_g = mg(z_2 - z_1)$$
$$W_a = \tfrac{1}{2} m(v_2^2 - v_1^2)$$

Calculations

The quantities in the problem statement are given in a mixed set of units, so we first convert the units of all quantities to SI units. Converting the velocities, we obtain

$$5\frac{mi}{h} \times \frac{5280\text{ ft}}{1\text{ mi}} \times \frac{1\text{ m}}{3.2808\text{ ft}} \times \frac{1\text{ h}}{3600\text{ s}} = 2.235\text{ m/s}$$

and

$$45\frac{mi}{h} \times \frac{5280\text{ ft}}{1\text{ mi}} \times \frac{1\text{ m}}{3.2808\text{ ft}} \times \frac{1\text{ h}}{3600\text{ s}} = 20.12\text{ m/s}$$

The vertical position, z_2, of the automobile when it attains a speed of 45 mi/h is

$$z_2 = (100\text{ m})\sin 6° = 10.45\text{ m}$$

Assigning the position of the ground as $z_1 = 0$ m, the gravitational work is

$$W_g = -mg(z_2 - z_1)$$
$$= -(1200\text{ kg})(9.81\text{ m/s}^2)(10.45\text{ m} - 0\text{ m})$$
$$= -1.231 \times 10^5 \text{ J}$$

The acceleration work is

$$W_a = \tfrac{1}{2} m(v_2^2 - v_1^2)$$
$$= \tfrac{1}{2}(1200\text{ kg})[(20.12\text{ m/s})^2 - (2.235\text{ m/s})^2]$$
$$= 2.399 \times 10^5 \text{ J}$$

The total work is the sum of the gravitational and acceleration work.

$$W_t = W_g + W_a$$
$$= -1.231 \times 10^5 \text{ J} + 2.399 \times 10^5 \text{ J} = 1.168 \times 10^5 \text{ J} = \underline{117\text{ kJ}}$$

Solution check

No errors are found.

Discussion

Even though the gravitational work is negative, the total work is positive because the acceleration work is larger in magnitude. We must remember that gravitational work is the work done *by gravity* on the body and not the work done *by the body* in overcoming gravity. The work done by the automobile's engine in order to climb the hill is 123 kJ, but the gravitational work is −123 kJ.

APPLICATION: BOUNDARY WORK DURING A CONSTANT PRESSURE PROCESS

In some thermodynamic systems, boundary work is performed while the pressure remains constant. A common example is the heating of a gas contained in a piston-cylinder device, as illustrated in Figure 6.18(a). As heat is transferred to the gas within the cylinder, the internal energy of the gas increases, as exhibited by an increase in the gas temperature, and the piston moves up. If we assume that the piston-cylinder device is frictionless, the pressure of the gas remains constant, but, since the piston moves, boundary work is still done. Suppose that the frictionless piston-cylinder device shown in Figure 6.18(a) contains 2.5 L of nitrogen at 120 kPa. Heat is then transferred to the nitrogen until the volume is 4 L. Find the boundary work done by the nitrogen during this process.

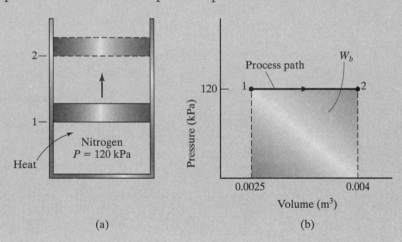

(a)   (b)

Figure 6.18.   A constant pressure process.

Boundary work W_b is given by the relation

$$W_b = \int_1^2 P dV$$

where P is pressure and V is volume. Because the process occurs at constant pressure, P can be brought outside the integral, giving the relation

$$W_b = P \int_1^2 dV$$

The initial and final volumes of the nitrogen are

$$V_1 = 2.5 \, \text{L} = 0.025 \, \text{m}^3 \qquad V_2 = 4 \, \text{L} = 0.004 \, \text{m}^3$$

Thus, the boundary work is

$$W_b = P \int_1^2 dV = P(V_2 - V_1)$$

$$= (120 \times 10^3 \, \text{Pa})(0.004 \, \text{m}^3 - 0.025 \, \text{m}^3)$$

$$= 180 \, \text{J}$$

The boundary work calculated here is the work done *by* the nitrogen on the piston, not the work done *on* the nitrogen by the piston. Figure 6.18(b) shows the *process path* for the constant pressure process that occurs in the piston-cylinder device. The boundary work of 180 J is the shaded area under the process path.

PRACTICE!

1. As a 2500-kg truck climbs a hill, it changes speed from 20 mi/h to 50 mi/h along a straight 1600-ft section of road. If the hill is inclined at an angle of 8° with respect to the horizontal, find the total work.

    Answer: −1.14 MJ

2. A 95-slug automobile changes speed from 55 mi/h to 30 mi/h while climbing a 3° hill. If the change in speed occurs over a 1355-ft straight section of road, find the total work.

    Answer: −5605 Btu

3. A shaft rotating at 1200 rpm (revolutions per minute) experiences a constant torque of 60 N · m. How much work does the shaft perform in one hour?

    Answer: 27.1 MJ

4. The pressure inside a frictionless piston-cylinder device varies according to the function $P = a - bV$, where a and b are constants and V is volume. The initial and final volumes for the process are 1 m^3 and 0.1 m^3, respectively. If $a = 500$ Pa and $b = 2000$ Pa/m^3, find the boundary work.

    Answer: 540 J

5. A linear elastic spring is compressed 3.5 cm from its at-rest position. The spring is then compressed an additional 7.5 cm. If the spring constant is 2600 N/cm, find the work done in compressing the spring.

    Answer: 1.57 kJ

6. A frictionless piston-cylinder device has a diameter of 10 cm. As the gas inside the cylinder is heated, the piston moves a distance of 16 cm. If the gas pressure is maintained at 120 kPa, how much work is done?

    Answer: 151 J

6.5   THE FIRST LAW OF THERMODYNAMICS

The first law of thermodynamics is one of the most important laws in science and engineering. The first law of thermodynamics, often referred to as the law of *conservation of energy*, enables engineers to analyze transformations that occur between the various forms of energy. Stated another way, the first law of thermodynamics allows engineers to study how one form of energy is converted to other forms. The most concise definition of the first law of thermodynamics is *energy is conserved*. Another way to state this law is *energy cannot be created or destroyed, but can only change forms*. The first law of thermodynamics, hereafter referred to as simply "the first law," cannot be proved mathematically. Like Newton's laws of motion, the first law is taken as an axiom, a sound physical principle based on countless measurements. No energy transformation, either natural or man made, is known to have violated the first law.

The first law is a very intuitive concept. Consider the system shown in Figure 6.19. The system may represent any substance or region in space chosen for thermodynamic analysis. The boundary of the system is the surface that separates the system from the surroundings. We may construct a mathematical representation of the first law by applying a simple physical argument. If an amount of energy E_{in} is supplied *to* the system, that energy can *leave* the system, *change* the energy of the system, or both. The energy that leaves the system is E_{out}, and the energy change of the system is ΔE. Thus, the energy that enters the system equals the energy that leaves the system plus the energy change

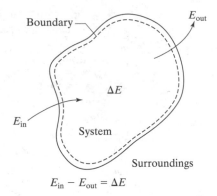

Figure 6.19.   The first law of thermodynamics.

of the system. The first law may therefore be expressed mathematically as

$$E_{\text{in}} = E_{\text{out}} + \Delta E \qquad (6.28)$$

We see that the first law is nothing more than a simple accounting principle that maintains the system's "energy ledger" in balance. In fact, the first law is often referred to as an *energy balance* because that is precisely what it is. In most engineering thermodynamics texts, Equation (6.28) is typically written in the form

$$E_{\text{in}} - E_{\text{out}} = \Delta E \qquad (6.29)$$

As shown in Figure 6.19, E_{in} and E_{out} are energy quantities that are *transferred* across the system boundary, whereas ΔE is the *change* in energy of the system itself. Because E_{in} and E_{out} are transferred forms of energy, these terms can only represent energy in the forms of *heat, work,* and *mass flow*. Heat is the transport of energy across the boundary of a system by virtue of a temperature difference. For heat transfer to occur, there must be a temperature difference between the system and the surroundings. Work may be mechanical in nature, such as the movement of the system boundary or the turning of a shaft inside the system, or electrical in nature, such as the transfer of electrical energy by a wire that penetrates the system boundary. When mass crosses a system boundary, energy crosses the boundary as well, because mass carries energy with it. Thus, the left side of Equation (6.29) becomes

$$E_{\text{in}} - E_{\text{out}} = (Q_{\text{in}} - Q_{\text{out}}) + (W_{\text{in}} - W_{\text{out}}) + (E_{\text{mass, in}} - E_{\text{mass, out}}) \qquad (6.30)$$

where Q denotes heat, W denotes work, and E_{mass} denotes energy transfer by mass flow. The *in* and *out* subscripts refer to energy transferred *in* and *out* of the system, respectively. These energy quantities should always be clearly indicated on a diagram as arrows pointing into or out of the system. The energy change of the system ΔE is the sum of the potential, kinetic, and internal energy changes. Hence, the right side of Equation (6.29) is

$$\Delta E = \Delta \text{PE} + \Delta \text{KE} + \Delta U \qquad (6.31)$$

where PE, KE, and U represent the potential, kinetic, and internal energy, respectively. Most thermodynamic systems of practical interest are stationary with respect to external reference frames, so $\Delta PE = \Delta KE = 0$, leaving $\Delta E = \Delta U$. Furthermore, many thermodynamic systems are *closed*, which means that mass cannot enter or leave the system. For closed systems, the only forms of energy transfer possible are work and heat. The analysis of closed systems is considerably simpler than the analysis of systems that permit mass transfer. In this book, we consider closed systems only. Thus, the first law of thermodynamics for closed systems is

$$(Q_{in} - Q_{out}) + (W_{in} - W_{out}) = \Delta U \tag{6.32}$$

The heat and work transferred across the system boundary causes a change in the internal energy of the system. This change alters the thermodynamic state or condition of the system. A change in the thermodynamic state of a system is called a *process*. The internal energy change ΔU is simply the difference between the internal energies at the end of the process and the beginning of the process. Thus, $\Delta U = U_2 - U_1$, where the subscripts 1 and 2 denote the beginning and end of the process, respectively.

The first law may be expressed in *rate form* by dividing each term in Equation (6.29) by a time interval Δt over which the process occurs. By dividing the energy terms by time, the quantities on the left side of the equation become quantities of *power*, and the quantity ΔE becomes a change in energy that occurs during the specified time interval. Equation (6.29) is then rewritten as

$$\dot{E}_{in} - \dot{E}_{out} = \Delta E/\Delta t \tag{6.33}$$

where $\dot{E}_{in}$ and $\dot{E}_{out}$ denote the *rate* at which energy enters and leaves the system, respectively. The units for $\dot{E}_{in}$ and $\dot{E}_{out}$ are J/s, which is defined as the watt (W). If the problem is stated in terms of energy rates rather than absolute energy quantities, the use of the first law given by Equation (6.33) is preferred over Equation (6.29).

In the following examples, the first law is used to analyze some basic closed thermodynamic systems. We use the general analysis procedure of (1) problem statement, (2) diagram, (3) assumptions, (4) governing equations, (5) calculations, (6) solution check, and (7) discussion.

EXAMPLE 6.4

Problem statement

A closed tank contains a warm liquid whose initial internal energy is 1500 kJ. A paddle wheel connected to a rotating shaft imparts 250 kJ of work to the liquid, while 700 kJ of heat is lost from the liquid to the surroundings. What is the final internal energy of the liquid?

Diagram

A diagram representing the system is illustrated in Figure 6.20. The system is the liquid in the tank. Energy transferred across the system boundary as work and heat is shown.

Assumptions

1. The system is closed.
2. The tank is stationary, so $\Delta PE = \Delta KE = 0$.
3. Energy change in the paddle wheel is negligible.

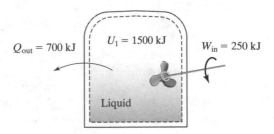

Figure 6.20. System for Example 6.4.

Governing equations

The governing equation for this problem is the first law of thermodynamics for a closed system:

$$(Q_{in} - Q_{out}) + (W_{in} - W_{out}) = \Delta U = U_2 - U_1$$

Calculations

From the diagram, we see that

$$Q_{out} = 700 \text{ kJ}, \qquad W_{in} = 250 \text{ kJ}, \qquad U_1 = 1500 \text{ kJ}$$

but there is no heat input and no work output. Thus,

$$Q_{in} = 0, \qquad W_{out} = 0$$

Substituting known quantities into the first law, we have

$$(0 - 700)\text{kJ} + (250 - 0)\text{kJ} = U_2 - 1500 \text{ kJ}$$

Solving for U_2, the final energy of the liquid, we obtain

$$U_2 = \underline{1050 \text{ kJ}}$$

Solution check

No errors are found.

Discussion

The final internal energy of the liquid is 1050 kJ, a decrease of 450 kJ. The internal energy of the liquid must decrease because more energy (700 kJ) is removed from the system than is supplied (250 kJ) to the system.

EXAMPLE 6.5

Problem statement

The air in a small house is maintained at a constant temperature by an electric baseboard system that supplies 5.6 kW to the house. There are 10 light fixtures in the house and each dissipate 60 W and the major electrical appliances (dishwasher, range, clothes dryer, etc.) have a total dissipation of 2560 W. The house is occupied by four people who each dissipate 110 W. Find the total heat loss from the house to the surroundings.

Diagram

The diagram for this problem is shown in Figure 6.21. The air in the house is the system. Power supplied to the house by the baseboard system, shown as an electrical resistor, is represented on the diagram by electrical power input, $\dot{W}_{in}$. The rate of heat dissipation by lights, people, and appliances is shown on the diagram as $\dot{Q}_{in}$, and the heat loss from the house to the surroundings is shown as $\dot{Q}_{out}$. A careful reading of the problem statement indicates that the change in internal energy of the system is zero, since the baseboard system maintains the air in the house at a constant temperature.

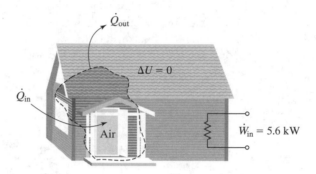

Figure 6.21. System for Example 6.5.

Assumptions

1. The system is closed.
2. Energy change in the contents of the house is zero.
3. All energy transfer rates are constant.

Governing equations

The governing equation for this problem is the first law, in rate form, for a closed system. Because the house is maintained at a constant temperature, $\Delta U = 0$. Thus, we have

$$(\dot{Q}_{in} - \dot{Q}_{out}) + (\dot{W}_{in} - \dot{W}_{out}) = 0$$

Calculations

There are 10 lights that dissipate 60 W each, 4 people who dissipate 110 W each, and appliances that dissipate a total of 2560 W. The total rate of heat transfer into the house is

$$\dot{Q}_{in} = \dot{Q}_{lights} + \dot{Q}_{people} + \dot{Q}_{appliances}$$
$$= 10(60\ W) + 4(110\ W) + 2560\ W = 3600\ W$$

The electrical power supplied to the house by the baseboard system is

$$\dot{W}_{in} = 5600\ W$$

but there is no work output, so $\dot{W}_{out} = 0$. Substituting known quantities into the first law, we have

$$(3600 - \dot{Q}_{out})W + (5600 - 0)W = 0$$

Solving for the heat loss $\dot{Q}_{out}$, we obtain

$$\dot{Q}_{out} = \underline{9.2 \text{ kW}}$$

Solution check

No errors are found.

Discussion

The heat loss of 9.2 kW represents the rate of heat transfer from the house to the surroundings. Heat is lost from the house through the walls, roof, windows, doors, and any other building member that is part of the system boundary. Because the air in the house is maintained at a constant temperature, the rate of energy supplied to the house must equal the rate of energy lost by the house.

PRACTICE!

1. A 2500-kg boulder is pushed off a 75-m high cliff. What is the velocity of the boulder immediately before it strikes the ground? How does the boulder's mass affect the solution?

    Answer: 38.4 m/s, boulder's mass is irrelevant

2. Just before striking the ground, the boulder in practice problem 1 converts all its potential energy to kinetic energy (assuming negligible aerodynamic friction). After colliding with the ground, the boulder comes to rest, converting its kinetic energy into other energy forms. What are these forms?

    Answer: heat, sound, and deformation

3. The fluid in a closed-pressure vessel receives 500 kJ of heat, while a shaft does 250 kJ of work on the fluid. If the final internal energy of the fluid is 1100 kJ, what is the initial internal energy of the fluid?

    Answer: 350 kJ

4. The fluid in a closed tank loses 600 Btu of heat to the surroundings, while a shaft does 850 Btu of work on the fluid. If the initial internal energy of the fluid is 250 Btu, what is the final internal energy of the fluid?

    Answer: 500 Btu

5. A small house is to be air-conditioned. The house gains 18,000 Btu/h of heat from the surroundings, while lights, appliances, and occupants add 6000 Btu/h from within the house. If the house is to be maintained at a constant temperature, what is the required rating of the air conditioner?

    Answer: 24,000 Btu/h

6. A piston-cylinder device containing water is heated. During the heating process, 300 J of energy is supplied to the water, while 175 J of heat is lost through the walls of the cylinder to the surroundings. As a result of the heating, the piston moves, doing 140 J of boundary work. Find the change in the internal energy of the water for this process.

    Answer: −15 J

Engineering is the business of designing and producing devices and systems for the benefit of society. People who practice engineering for a living design and manufacture things—*practical* things that are useful in specific applications. Given the applied nature of engineering, one would assume that engineering education is likewise applied. After all, an engineering education is supposed to prepare students for engineering practice, right? While an engineering education does indeed prepare students for industrial practice, the nature of that preparation may not be what you expect. Generally speaking, engineering courses are very theoretical and mathematical in nature. If you have a few engineering courses under your belt already, you have no doubt discovered this. Engineering courses are usually deep in theory, but shallow in practical aspects. As a result, an electrical engineering student may know how to analyze a circuit with the use of a schematic diagram, but may not be able to recognize an actual electrical component such as a resistor, capacitor, inductor or integrated circuit. Similarly, a mechanical engineering student may be very comfortable with performing a first law analysis of a boiler, compressor, turbine, or heat exchanger, but would not recognize one of these devices if he or she saw one.

So, why is the emphasis placed on theory at the expense of the practical aspects? One of the main reasons is that many professors who are teaching you how to become a practicing engineer have never practiced engineering themselves. This may sound bizarre, but many engineering professors took a teaching position directly out of graduate school after receiving their Ph.D degree, have been teaching ever since, and therefore have little or no industrial experience. This situation is not likely to change significantly in the near future, so it is up to the engineering student to acquire some practical, hands-on experience. Here are some ways:

- Enroll in a vocational or technical course at the university, the local community college, or trade school. Technical programs usually offer a wide variety of very practical courses such as welding, machining, refrigeration repair, auto repair, pipe fitting, electrical wiring, small engine repair, and computer servicing. You should take these courses when they will not interfere with your engineering course work, such as during the summer.

- Take additional laboratory courses. Some engineering courses have laboratories associated with them. The engineering laboratory is a good place to acquire practical engineering skills.

- Read engineering and technical-related magazines and journals. These publications contain articles about real engineering systems that will help you bridge the gap between engineering theory and engineering practice.

- Participate in engineering projects and competitions sponsored by your school and professional engineering societies. The American Society of Mechanical Engineers (ASME), the Society of Automotive Engineers (SAE), the Institute for Electrical and Electronics Engineers (IEEE), the American Institute of Aeronautics and Astronautics (AIAA), and other professional societies sponsor various engineering competitions. Local participation in National Engineers Week, held annually in February, is an excellent opportunity for students to bolster their practical engineering skills.

- Tinker with various mechanical and electrical devices. Find an old electric hand drill and disassemble it. Figure out how it works. Do the same for a telephone, computer hard drive, and a small kitchen appliance. Disassembling, studying, and reassembling things will help you discover how actual devices work. You may even want to perform service on your own automobile, perhaps by replacing the brakes, doing a tune-up, or installing a sound system.

6.6   HEAT ENGINES

The first law of thermodynamics states that energy can be converted from one form to another, but cannot be created or destroyed. The first law is a conservation law, a simple accounting principle that tells us how a system's "energy ledger" is kept in balance. Although the first law tells us what forms of energy are involved in a particular energy conversion, it tells us nothing about whether the conversion is possible or in which direction the conversion process occurs. For example, consider the system in Figure 6.22. A closed tank containing a fluid has a shaft that facilitates the transfer of work to the fluid. When the shaft rotates, work is transferred to the fluid, increasing its internal energy and thereby transferring heat from the fluid to the surroundings, as shown in Figure 6.22(a). During this process, work is converted directly and completely to heat. But when heat is transferred to the fluid, as shown in Figure 6.22(b), the shaft does not rotate, and thus no work is performed. The first law of thermodynamics does not preclude the conversion of heat to work in this system, but we know from experience that such conversion does not occur. Based on this argument, we conclude that work can be converted to heat directly and completely, but heat cannot always be converted to work. The direct conversion from heat to work is impossible without the use of a special device called a heat engine.

A **heat engine** is a *device that converts heat to work*. Before describing how this conversion occurs, we must define an important thermodynamic term: *thermal energy*

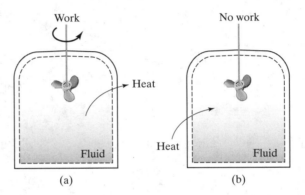

Figure 6.22. Work can always be converted to heat (a), but heat cannot always be converted to work (b).

reservoir. A thermal energy reservoir is a body with a very large thermal capacity. The distinctive characteristic of a thermal energy reservoir is that it can supply or receive large amounts of thermal energy without experiencing any change in temperature. In actual thermodynamic systems, because of their large masses and high heat capacities, expansive bodies of water such as oceans, lakes, or rivers are considered thermal energy reservoirs. The atmosphere is also considered a thermal energy reservoir. Any region or body whose thermal capacity is large compared with the amount of heat it supplies, or receives, may be considered a thermal energy reservoir. There are two types of thermal energy reservoirs: a thermal energy *source* and a thermal energy *sink*. A thermal energy source supplies heat to a system, whereas a thermal energy sink absorbs heat from a system. As illustrated in Figure 6.23, a heat engine receives an amount of heat (Q_{in}) from a high-temperature source and converts a portion of that heat to work (W_{out}). The heat engines rejects the remaining heat (Q_{out}) to a low-temperature sink. There are several thermodynamic systems that qualify as heat engines, but the system that best fits the definition of a heat engine is the steam-power plant. In a steam-power plant, Q_{in} is the heat supplied to a boiler from a combustion process or nuclear reaction. The heat Q_{out}, rejected to a low-temperature sink, is the heat transferred from a heat exchanger to a nearby lake, river, or the atmosphere. The work W_{out} produced by the power plant is the energy generated by a turbine. An electrical generator, which is connected to the turbine via a shaft, generates electrical energy.

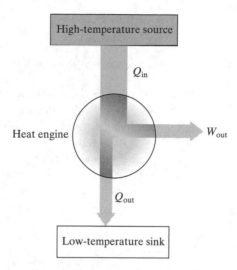

Figure 6.23.   A heat engine converts a portion of the heat it receives from a high-temperature source to work and rejects the remaining heat to a low-temperature sink.

By inspection of Figure 6.23, the first law of thermodynamics for a heat engine is

$$Q_{in} = Q_{out} + W_{out} \qquad (6.34)$$

The work W_{out} is the useful work produced by the heat engine. For a steam-power plant, W_{out} is actually a *net* work because some work has to be supplied to a pump in order to circulate the steam through the boiler and other power-plant components. The heat Q_{out}, rejected to a low-temperature sink, is wasted energy. So why don't we just eliminate Q_{out}, converting all Q_{in} to work? It turns out that, while this idea sounds very attractive, the elimination of Q_{out} violates the second law of thermodynamics. A nonzero amount of waste heat Q_{out} is necessary if the heat engine is to operate at all.

Efficiency is a useful engineering quantity that is used to measure the performance of numerous engineering systems. A general definition of efficiency is

$$\text{efficiency} = \frac{\text{desired output}}{\text{required input}} \qquad (6.35)$$

For heat engines, the desired output is the work output, and the required input is the heat supplied by the high-temperature source. Hence, **thermal efficiency** of a heat engine, denoted η_{th}, is given by the relation:

$$\eta_{th} = \frac{W_{out}}{Q_{in}} \qquad (6.36)$$

In accordance with the first law of thermodynamics, no heat engine (or any other device for that matter) can produce more energy than is supplied to it. Therefore, the thermal efficiency of a heat engine is always less than 1. This fact is apparent from Figure 6.23 because only a portion of the heat supplied to the heat engine is converted to work, the remaining heat being rejected to a low-temperature sink.

EXAMPLE 6.6

A heat engine produces 6 MW of power while absorbing 10 MW from a high-temperature source. What is the thermal efficiency of this heat engine? What is the rate of heat transfer to the low-temperature sink?

SOLUTION

The work output and heat input are given in terms of energy rates, not energy. The first law relation for a heat engine [Equation (6.34),] may be expressed in rate form by dividing each quantity by time. Similarly, the work and heat quantities in Equation (6.36) may be divided by time. Dividing the work and heat quantities by time yields power $\dot{W}_{out}$ and heat transfer rates, $\dot{Q}_{in}$ and $\dot{Q}_{out}$, where the "dot" denotes a rate quantity. Thus, the thermal efficiency of the heat engine is

$$\eta_{th} = \frac{\dot{W}_{out}}{\dot{Q}_{in}}$$

$$= \frac{(6 \text{ MW})}{10 \text{ MW}} = 0.6 \ (60\%)$$

The rate of heat transfer to the low-temperature sink is

$$\dot{Q}_{out} = \dot{Q}_{in} - \dot{W}_{out}$$

$$= 10 \text{ MW} - 6 \text{ MW} = 4 \text{ MW}$$

6.7   THE SECOND LAW OF THERMODYNAMICS

The first law of thermodynamics states that energy is conserved (i.e., energy can be converted from one form to another, but cannot be created or destroyed). The first law tells us what forms of energy are involved in a particular energy conversion, but it does not tell us anything about whether the conversion is possible or in which direction the conversion process occurs. Common experience tells us that a boulder naturally falls from a cliff to the ground, but never jumps from the ground to the top of the cliff by itself. The first law does not preclude the boulder from jumping from the ground to the top of the cliff, because energy (potential and kinetic) is still conserved in this process. We know by experience that a hot beverage naturally cools as heat is transferred from the beverage to the cool surroundings. The energy lost by the beverage equals the energy gained by the surroundings. The hot beverage will not get hotter, however, because heat flows from a high temperature to a low temperature. The first law does not preclude the beverage from getting hotter in a cool room as long as the energy lost by the room equals the energy gained by the beverage. Experience also tells us that if you drop a raw egg on the floor, it breaks and makes a big gooey mess. The reverse process will not occur (i.e., the shell fragments will not automatically reassemble around the egg white and yolk and then rebound from the floor into your hand). Once again, the first law does not preclude the reverse process from occurring; however, the overwhelming experimental evidence tells us that the reverse process does not take place. As a final example, consider the system in Figure 6.24. A closed tank containing a fluid has a shaft that facilitates the conversion between work and heat. Suppose that we wanted to use the apparatus as a heat engine, a device that converts heat to work. If we were to actually build this device and attempt to raise the weight by transferring heat to the fluid, we would discover that the weight would not be raised. As in the previous examples, the first law does not preclude

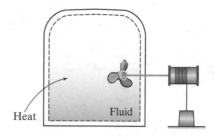

Figure 6.24. Transferring heat to the fluid will not cause the shaft to rotate; therefore, no work will be done to raise the weight.

the conversion of heat to work in this system, but we know from experience that this conversion does not occur.

Based on direct observations of physical systems, it is clear that thermodynamic processes occur only in certain directions. While the first law places no restrictions on the direction in which a thermodynamic process occurs, it does not ensure that the process is *possible*. To answer that question, we need another thermodynamic principle or law that tells us something about the natural direction of thermodynamic processes. That principle is the second law of thermodynamics. In order for a process to occur, *both* the first and second laws of thermodynamics must be satisfied. There are various ways of stating the second law of thermodynamics. One of the most useful forms of the second law of thermodynamics, hereafter referred to as simply "the second law," is that *it is impossible for a heat engine to produce an amount of work equal to the amount of heat received from a thermal energy reservoir.* In other words, the second law states that it is impossible for a heat engine to convert all the heat it receives from a thermal energy reservoir to work. A heat engine that violates the second law is illustrated in Figure 6.25. In order to operate, a heat engine must reject some of the heat it receives from the high-temperature source to a low-temperature sink. A heat engine that violates the second law converts 100 percent of this heat to work. This is physically impossible.

The second law can also be stated as *no heat engine can have a thermal efficiency of 100 percent.* The thermal efficiency of a heat engine, denoted η_{th}, is defined as the ratio of the work output to the heat input:

$$\eta_{th} = \frac{W_{out}}{Q_{in}} \tag{6.37}$$

Clearly, if the thermal efficiency of a heat engine is 100 percent, then $Q_{in} = W_{out}$. If the second law precludes a heat engine from having a thermal efficiency of 100 percent, what is the maximum possible thermal efficiency of a heat engine? As illustrated in Figure 6.26, a heat engine is a device that converts a portion of the heat supplied to it from a high-temperature source into work. The remaining heat is rejected to a low-temperature sink. The thermal efficiency of a heat engine is given by Equation (6.37). Applying the first law to the heat engine, we obtain

$$Q_{in} = Q_{out} + W_{out} \tag{6.38}$$

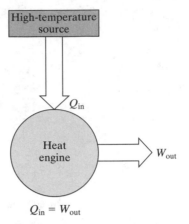

Figure 6.25. This heat engine violates the second law of thermodynamics.

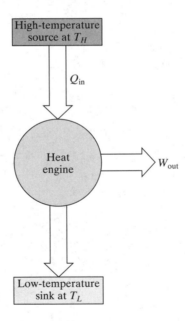

Figure 6.26. A heat engine, operating between thermal energy reservoirs at temperatures T_H and T_L, converts heat to work.

Solving for W_{out} from Equation (6.38) and substituting the result into Equation (6.37), we obtain

$$\eta_{th} = \frac{Q_{in} - Q_{out}}{Q_{in}} = 1 - \frac{Q_{out}}{Q_{in}} \tag{6.39}$$

It can be shown mathematically that, for an *ideal* heat engine operating between source and sink temperatures of T_H and T_L, respectively, the ratio of the heat supplied to the heat rejected equals the ratio of the absolute temperatures of the heat source and heat sink. Thus,

$$\frac{Q_{\text{out}}}{Q_{\text{in}}} = \frac{T_L}{T_H} \tag{6.40}$$

The details of the mathematical proof may be found in most thermodynamics texts. What does it mean for a heat engine to be ideal? The short answer is that a heat engine is considered ideal if the processes within the heat engine itself are reversible. A reversible process is a process that can be reversed in direction without leaving any trace on the surroundings. A simple example of a reversible process is a frictionless pendulum. A frictionless pendulum can swing in either direction without dissipating any heat to the surroundings. A more thorough discussion of this concept may be found in the references at the end of this chapter.

Substituting Equation (6.40) into Equation (6.39), we find that the thermal efficiency for an ideal heat engine becomes

$$\eta_{\text{th, ideal}} = 1 - \frac{T_L}{T_H} \tag{6.41}$$

where T_L and T_H denote the absolute temperatures of the low-temperature sink and high-temperature source, respectively. Because T_L and T_H are absolute temperatures, these quantities must be expressed in units of kelvin (K) or rankine (°R). The thermal efficiency given by Equation (6.41) is the maximum possible thermal efficiency a heat engine can have, and is often referred to as the **Carnot efficiency**, in honor of the French engineer Sadi Carnot. A heat engine whose thermal efficiency is given by Equation (6.41) is a theoretical heat engine only, an idealization that engineers use to compare with real heat engines. No real heat engine can have a thermal efficiency greater than the Carnot efficiency, because no real heat engine is reversible. Hence, the efficiencies of real heat engines, such as steam power plants, should not be compared to 100 percent. Instead, they should be compared to the Carnot efficiency for a heat engine operating between the same temperature limits. The Carnot efficiency is the theoretical upper limit for the thermal efficiency of a heat engine. If a heat engine is purported to have a thermal efficiency greater than the Carnot efficiency, the heat engine is in violation of the second law of thermodynamics.

The first and second laws of thermodynamics are the quintessential governing principles on which all energy processes are based. In summary: *The first law says you can't get something for nothing. The second law says you can't even come close.*

Earlier in this section, we mentioned that there are various ways of stating the second law of thermodynamics. The primary objective of science is to explain the universe in which we live. The second law of thermodynamics, while very useful for analyzing and designing engineering systems, is a scientific principle that has profound consequences. From a scientific standpoint, the second law is considered an "arrow of time," an immutable principle that assigns a natural direction to all physical processes. Stones fall from cliffs, but never the reverse. Heat flows from hot objects to cold objects, but never the reverse. Raw eggs dropped to the floor make a gooey mess, but do not reassemble. Cream mixes with coffee, but once mixed, the coffee and cream do not separate back out. Physical processes are ordered—they follow the arrow of time. Matter spreads and energy spreads, reducing the quality of things. According to the second law, things naturally move from order to disorder, from a higher quality to a lower quality, from a more useful

APPLICATION: EVALUATING A CLAIM FOR A NEW HEAT ENGINE

In a patent application for a new heat engine, an inventor claims that the device produces 1 kJ of work for every 1.8 kJ of heat supplied to it. In the application, the inventor states that the heat engine absorbs energy from a 350°C source and rejects energy to a 25°C sink. Evaluate this claim.

The feasibility of the new heat engine may be checked by ascertaining whether the heat engine violates either the first or second laws of thermodynamics. If the first law is violated, the heat engine would have to produce an amount of work greater than the amount of heat supplied to it. Because $W_{out} < Q_{in}$ (1 kJ < 1.8 kJ) for this heat engine, the first law is satisfied. If the second law is violated, the heat engine would have to have a thermal efficiency greater than the Carnot efficiency for a heat engine operating between the same temperature limits. The actual thermal efficiency of the heat engine is

$$\eta_{th,\,actual} = \frac{W_{out}}{Q_{in}} = \frac{1 \text{ kJ}}{1.8 \text{ kJ}} = 0.556(55.6\%)$$

Noting that the source and sink temperatures must be expressed in absolute units, the Carnot efficiency is

$$\eta_{th,\,Carnot} = 1 - \frac{T_L}{T_H}$$

$$= 1 - \frac{(25°C + 273) \text{ K}}{(350°C + 273) \text{ K}} = 0.522(52.2\%)$$

The actual thermal efficiency of the heat engine is greater than the Carnot efficiency (0.556 > 0.522), so the inventors's claim is invalid. It is physically impossible for this heat engine to produce 1 kJ of work for every 1.8 kJ of heat supplied to it, given the source and sink temperatures specified in the patent application.

state to a less useful state. In short, the second law says that, left to themselves, things get worse. As shown in Figure 6.27, the second law seems to apply to everything, not just energy systems.

Figure 6.27. The second law of thermodynamics has taken its toll on this structure.

PRACTICE!

1. A high-temperature source supplies a heat engine with 25 kJ of energy. The heat engine rejects 15 kJ of energy to a low-temperature sink. How much work does the heat engine produce?
    Answer: 10 kJ

2. A heat engine produces 5 MW of power while absorbing 8 MW of power from a high-temperature source. What is the thermal efficiency of this heat engine? What is the rate of heat transfer to the low-temperature sink?
    Answer: 0.625, 3 MW

3. A heat engine absorbs 20 MW from a 400°C furnace and rejects 12 MW to the atmosphere at 25°C. Find the actual and Carnot thermal efficiencies of this heat engine. How much power does the heat engine produce?
    Answer: 0.400, 0.557, 8 MW

4. Joe, a backyard tinkerer who fancies himself an engineer, tells his engineer neighbor, Jane, that he has developed a heat engine that receives heat from boiling water at 1 atm pressure and rejects heat to a freezer at −5°C. Joe claims that his heat engine produces 1 Btu of work for every 2.5 Btu of heat it receives from the boiling water. After a quick calculation, Jane informs Joe that if he intends to design heat engines, he needs to pursue an engineering education first. Is Jane justified in making this comment? Justify your answer by analysis.
    Answer: Yes, because $\eta_{\text{actual}} = 0.400$ and $\eta_{\text{Carnot}} = 0.282$, which is impossible.

KEY TERMS

Carnot efficiency
elastic potential energy
energy
first law of
    thermodynamics
gravitational potential
    energy
heat

heat engine
internal energy
kinetic energy
potential energy
pressure
second law of
    thermodynamics
temperature

thermal efficiency
thermodynamics
work
zeroth law of
    thermodynamics

REFERENCES

Cengel, Y.A. and M.A. Boles, *Thermodynamics: An Engineering Approach*, 4th ed., NY: McGraw-Hill, 2002.

Sonntag, R.E., C. Borgnakke, and G.J. Van Wylen, *Fundamentals of Thermodynamics*, 6th ed., NY: John Wiley & Sons, 2002.

Moran, M.J. and H.N. Shapiro, *Fundamentals of Engineering Thermodynamics*, 5th ed., NY: John Wiley & Sons, 2003.

Levenspiel, O., *Understanding Engineering Thermo*, Upper Saddle River, NJ: Prentice Hall, 1996.

Hagen, K.D., *Heat Transfer with Applications*, Upper Saddle River, NJ: Prentice Hall, 1999.

Incropera, F.P. and D.P. DeWitt, *Fundamentals of Heat and Mass Transfer*, 5th ed., NY: John Wiley & Sons, 2001.

Problems

1. What gauge pressure would you need to inflate an automobile tire in San Diego, California, to achieve an absolute pressure of 325 kPa?

2. A pressure gauge connected to a tank reads 480 kPa at a location where the atmospheric pressure is 92 kPa. Find the absolute pressure in the tank.

3. A vertical, frictionless piston-cylinder device contains a gas. The piston has a mass of 3 kg and a radius of 5 cm. A downward force of 75 N is applied to the piston. If the atmospheric pressure is 90 kPa, find the pressure inside the cylinder. (See Figure P6.3.)

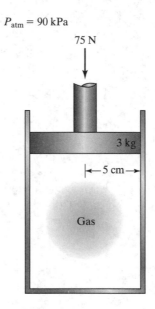

$P_{atm} = 90$ kPa

75 N

3 kg

5 cm

Gas

Figure P6.3.

4. A vacuum gauge connected to a tank reads 7.6 psi at a location where the atmospheric pressure is 13.8 psi. Find the absolute pressure in the tank.

5. A comfortable indoor air temperature is 70°F. What is this temperature in units of °R, °C, and K?

6. The average body temperature of a healthy adult is approximately 98.6°F. What is this temperature in units of °R, °C, and K?

7. Find the temperature at which the Fahrenheit and Celsius scales coincide.

8. Heat exchangers are devices that facilitate the transfer of thermal energy from one fluid to another across a solid wall. In a particular heat exchanger, glycerine enters the unit at a temperature of 30°C and exits the unit at a temperature of 47°C. What is the temperature change of the glycerine in units of °F, °R, and K?

9. As a 75-slug automobile travels along a horizontal, 1500-ft section of road, it changes speed from 5 mi/h to 60 mi/h. If the friction force acting on the automobile is 30 lb_f, what is the total work?

10. The pressure in a frictionless piston-cylinder device varies according to the function $P = CV^{-n}$, where C and n are constants and V is volume. Derive a relationship for the boundary work in terms of the initial and final volumes V_1 and V_2 and the constants C and n. What is the restriction on the constant n?

11.   A 180-lb$_m$ person climbs a stairway consisting of 150 stairs, each with a vertical rise of 8 in. How much work does this person do against gravity?

12.   A shaft connected to a motor does 400 kJ of work in 3 minutes. If the shaft rotates at 1750 rpm, what is the torque on the shaft?

13.   A crate is dragged across a rough floor by a force $F = 120$ N, as shown in Figure P6.13. A friction force of 40 N acts to retard the motion of the crate. If the crate is dragged 25 m across the floor, what is the work done by the 120-N force? By the friction force? What is the *net* work done?

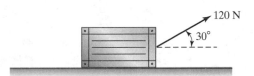

Figure P6.13.

For problems 14 through 39, use the general analysis procedure of (1) problem statement, (2) diagram, (3) assumptions, (4) governing equations, (5) calculations, (6) solution check, and (7) discussion.

14.   A 3-kg block is dropped from rest onto a linear elastic spring as shown in Figure P6.14. The spring is initially undeformed and has a spring constant of 1130 N/m. What is the deformation of the spring when the block momentarily stops? (*Hint:* Remember that the block travels 2 m *plus* a distance equal to the deformation of the spring.)

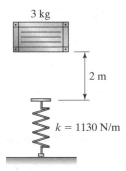

Figure P6.14.

15.   A piston-cylinder device containing a gas receives 25 kJ of heat. During the heating process, the gas expands, moving the piston outward, performing 10 kJ of boundary work. Also, an electric heating element imparts 6 kJ to the gas. If the heat loss from the device is 8 kJ, what is the change in internal energy of the gas during the process?

16. A machine shop is maintained at a constant temperature during the summer by small air-conditioning units with a rating of 8 kW. The rate of heat transfer from the surroundings to the machine shop is 24 kW. Five lathes and four mills dissipate a total of 4 kW, the lights in the shop dissipate 2.5 kW, and nine machinists dissipate a total of 3.5 kW. How many air-conditioning units are required?

17. The piston-cylinder device shown in the Figure P6.17 contains a fluid that can be stirred by a rotating shaft. The outer surface of the device is covered with a thick layer of insulation. The shaft imparts 50 kJ to the fluid during a process in which the pressure is held constant at 130 kPa as the piston moves outward. If the internal energy of the fluid increases by 20 kJ during the process, what is the axial displacement of the piston?

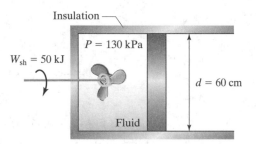

Insulation

$P = 130 \, kPa$

$W_{sh} = 50 \, kJ$

$d = 60 \, cm$

Fluid

Figure P6.17.

18. A closed tank containing hot air has an initial internal energy of 150 kJ. During the next 5 minutes, the tank loses heat to the surroundings at a rate of 1.2 kW, while an electrical element supplies 800 W to the air. Find the final internal energy of the air.

19. A small research facility in a remote polar region is maintained at a comfortable temperature by heaters that burn propane. The propane storage capacity of the facility is 5000 kg. If the rate of heat loss from the facility is 40 kW and the heat of combustion of propane is 46 MJ/kg, how long can the facility be continuously heated before depleting the propane? Assume that only 80 percent of the heat of combustion is utilized as useful energy.

20. A spherical hot-air balloon measuring 15 m in diameter flies at a constant altitude by periodically firing the burner system, maintaining the air within the canopy at a constant temperature. If the rate of heat loss per square meter through the canopy is 110 W/m^2, how much energy must the burner supply during a 1-hour period? If the burner system utilizes propane as fuel, how much propane is consumed during this time if the heat of combustion of propane is 46 MJ/kg? Assume that the entire heat of combustion is utilized to heat the air in the canopy.

21. A standard incandescent lightbulb operates on a voltage and current of 110 V and 0.91 A, respectively. What is the wattage of this lightbulb? What is the electrical work input to the lightbulb during a time interval of 2 min? In which of the three forms of heat transfer does thermal energy leave the lightbulb?

22. A cylindrical nuclear fuel rod with a length and diameter of 1.0 m and 2.0 cm, respectively, undergoes a fission process. During the fission process, the rod generates 3 GW/m^3 of energy uniformly within the rod. What is the total energy

generated by the rod during a time interval of 1 h? What is the rate of heat transfer from every square centimeter of rod surface, excluding the ends?

23. The change in internal energy, ΔU, for a closed system undergoing a thermo-dynamic process may be approximated by the relation

$$\Delta U = mc(T_2 - T_1)$$

where m is the mass of the substance within the system (kg), c is the average specific heat of the substance (J/kg $\cdot$ °C), and T_1 and T_2 are the initial and final temperatures of the substance (°C), respectively, for the process. A rigid tank contains 10 kg of steam at 250°C. During the next 5 min, the rate of heat transfer from the tank is 3 kW. What is the final temperature of the steam? For steam, let $c = 1.411$ kJ/kg $\cdot$ °C.

24. The change in internal energy, ΔU, for a closed system undergoing a thermo-dynamic process may be approximated by the relation

$$\Delta U = mc(T_2 - T_1)$$

where m is the mass of the substance within the system (kg), c is the average specific heat of the substance (J/kg $\cdot$ °C), and T_1 and T_2 are the initial and final temperatures of the substance (°C), respectively, for the process. A rigid tank contains 2 kg of air at 300°C. During the next 10 min, the rate of heat transfer from the tank is 1.3 kW, while during the same time, a rotating shaft does 500 kJ of work on the air. What is the final temperature of the air? For air, let $c = 0.718$ kJ/kg $\cdot$ °C.

25. A high-temperature source supplies a heat engine with 20 kJ of energy. The heat engine rejects 12 kJ of energy to a low-temperature sink. How much work does the heat engine produce?

26. During a time interval of 1h, a heat engine absorbs 360 MJ of energy from a high-temperature source while rejecting 40 kW to a low-temperature sink. How much power does the heat engine produce?

27. A heat engine produces 2 MW of power while rejecting 750 kW to the environment. What is the rate of heat transfer from the high-temperature source to the heat engine?

28. A heat engine produces 10 MW of power while absorbing 18 MW of power from a high-temperature source. What is the thermal efficiency of this heat engine? What is the rate of heat transfer to the low-temperature sink?

29. A heat engine rejects 2×10^6 Btu/h to a lake while absorbing 5×10^6 Btu/h from a furnace. What is the thermal efficiency of this heat engine? What is the power output?

30. The thermal efficiency of a heat engine is 70 percent. If the heat engine extracts 6 MJ of energy from a high-temperature source, how much energy is rejected to the low-temperature sink?

31. A heat engine absorbs 25 MW from a 400°C combustion chamber and rejects 15 MW to the atmosphere at 30°C. Find the actual and Carnot thermal efficiencies of this heat engine. How much power does the heat engine produce?

32.   A 2-GW steam power plant, which uses a nearby river as a low-temperature sink, has an actual thermal efficiency of 40 percent. The high-temperature source is a 400°C boiler, and the temperature of the river water is 10°C. Find the rate of heat transfer to the river and the ideal thermal efficiency of the power plant.

33.   An engineer proposes to design a heat engine that uses the atmosphere as the high-temperature source and a deep cavern as the low-temperature sink. If the temperature of the atmosphere and cavern are 25°C and 8°C, respectively, what is the maximum thermal efficiency that this heat engine can achieve? What is the maximum possible power output if the heat engine absorbs 300 kW from the atmosphere?

34.   A particular Carnot heat engine absorbs energy from a furnace and rejects energy to the atmosphere at 300 K. Graph the efficiency of this heat engine as a function of T_H, the temperature of the furnace. Use a range for T_H of 350 K to 2000 K. What can be concluded from this graph?

35.   An inventor submits a patent application for a heat engine that produces 1 kJ for every 2.2 kJ supplied to it. In the application, the inventor states that his heat engine absorbs energy from a 250°C source and rejects energy to a 40°C sink. Evaluate this patent.

36.   A 5-MW Carnot steam power plant operates between the temperature limits of 600°C and 20°C. Find the rates of heat transfer to and from the heat engine.

37.   A heat engine utilizes solar energy as its energy source. The heat engine incorporates a solar panel that intercepts a solar radiation flux of 900 W/m² of panel surface. Assuming that the solar panel absorbs 85 percent of the incident solar radiation, find the exposed surface area of the solar panel required to yield a thermal efficiency of 20 percent and a power output of 3.6 kW for the heat engine.

38.   What is the maximum possible power output of a heat engine operating between the temperature limits of 50°C and 800°C if 360 MJ of energy is supplied to the heat engine during a time period of 1 h? What is the actual power output if the heat engine rejects 216 MJ to the 50°C sink during the same time period?

39.   A coal-fired steam power plant is to be designed for the purpose of generating electrical power for a city with a population of 60,000 residents. Based on an order-of-magnitude analysis, it is estimated that each resident of the city will consume an average energy of 55 MJ per day. The coal-fired boiler supplies 70 MW to the steam while thermal energy is rejected to a nearby lake whose average temperature is 15°C. What is the minimum required temperature of the boiler to meet the power demands of the city?

7

Fluid Mechanics

7.1   INTRODUCTION

An important field of study in engineering is *fluid mechanics*. Many of the basic principles of fluid mechanics were developed in parallel with those of solid mechanics, and its historical roots can be traced to such great scientists and mathematicians as Archimedes (287–212 B.C.), Leonardo da Vinci (1425–1519), Isaac Newton (1642–1727), Evangelista Torricelli (1608–1647), Blaise Pascal (1623–1662), Leonhard Euler (1707–1783), Osborne Reynolds (1842–1912), and Ernst Mach (1838–1916). **Fluid mechanics** is the *study of fluids at rest and in motion*. As a subdiscipline of engineering mechanics, fluid mechanics is broadly divided into two categories, fluid statics and fluid dynamics. As the term implies, **fluid statics** is the branch of fluid mechanics that deals with the behavior of fluids at rest. **Fluid dynamics** is the branch of fluid mechanics that deals with the behavior of fluids in motion. In fluid statics, the fluid is at rest with respect to a frame of reference. This means that the fluid does not move with respect to a body or surface with which the fluid is in physical contact. In fluid dynamics, the fluid moves with respect to a body or surface, common examples being the flow of a fluid within a pipe or channel or around an immersed object such as a submarine or aircraft.

There are two primary physical states of matter—solid and fluid, the fluid state being subdivided into the liquid and gas states. A fourth state, referred to as the plasma state, refers to atoms and molecules that are ionized (electrically charged). Plasmas are categorized as special types of fluids that respond to electromagnetic fields. The analysis of plasmas is complex and will not be considered in this book. A fundamental question to be answered is "What is the difference between a solid and a fluid?" Casual observations tell us that solids are "hard" whereas fluids are "soft." Solids have a distinct size and shape and retain their basic dimensions even when large forces are

SECTIONS

OBJECTIVES

After reading this chapter, you will have learned

- The importance of fluid mechanics in engineering
- About density, specific weight, and specific gravity of fluids
- The concept of compressibility
- How viscosity affects shear forces in fluids
- To use the pressure-elevation relationship to find forces on submerged surfaces
- How to calculate volume flow rates and mass flow rates
- How to use the principle of continuity to analyze simple flow systems

applied to them. Fluids, however, do not really have a distinct size or shape unless they are confined in some manner by solid boundaries. When placed in a container, a fluid spreads throughout the container, taking on the shape of the container. Such phenomena occurs to one degree or another for liquids and gases. This behavior may be explained by examining the atomic and molecular structure of matter. In solids, the spacing of atoms or molecules is small, and there are large cohesive forces between these particles that enable solids to maintain their shape and size. In fluids, the atomic or molecular spacing is larger, and the cohesive forces are smaller, thereby permitting fluids more freedom of movement. At room temperature and atmospheric pressure, the average intermolecular spacing is approximately 10^{-10} m for liquids and 10^{-9} m for gases. The vast differences in cohesive forces in solids, liquids, and gases account for the rigidity of solids, the ability of liquids to fill containers from the bottom up, and the ability of gases to completely fill containers in which they are placed.

Although the differences between solids and fluids can be explained in terms of atomic or molecular structure, a more useful engineering explanation involves the response of solids and fluids to the application of external forces. Specifically, a **fluid** may be defined as *a substance that deforms continuously when acted upon by a shear stress of any magnitude*. Stress is a force that is applied over a specified area. A **shear stress** is produced when a force acts tangentially on a surface. When a solid material, such as metal, plastic, or wood, is subjected to a shear stress, the material deforms a small amount and maintains a deformed shape while the shear stress is applied. If the shear stress is not too great, the material even returns to its original shape when the force producing the stress is removed. When a fluid is subjected to a shear stress, however, the fluid continues to deform. Unlike a solid, a fluid cannot sustain a shear stress, so it continuously deforms (i.e., the fluid *flows* in response to the shear stress). Some substances, such as tar, toothpaste, putty, and other gunky and gooey materials, exhibit behavior that lies somewhere between solids and fluids. These types of substances will flow if the shear stress is high enough, but the analysis of these substances can be complex. We will therefore restrict our attention to common fluids such as water, oil, and air.

In most colleges and universities, one or more courses in fluid mechanics is required of mechanical, civil, and chemical engineering majors. Depending on the specific curricular policies of your school or department, other majors may also be required to take a course in fluid mechanics. Fluid mechanics is typically offered as part of a "thermofluid" sequence consisting of thermodynamics, fluid mechanics, and heat transfer, since these three disciplines are closely related to one another. Courses in statics, strength of materials, electrical circuits, and other analytically oriented courses round out the engineering science curriculum.

Engineers use principles of fluid mechanics to analyze and design a wide variety of devices and systems. Consider the plumbing fixtures in your home. The sink, bathtub or shower, toilet, dishwasher, and washing machine are supplied water by a system of pipes, pumps, and valves. When you turn on a faucet, the rate at which the water flows is determined by principles of fluid mechanics. The analysis and design of virtually every type of transportation system involves the use of fluid mechanics. Aircraft, surface ships, submarines, rockets, and automobiles require the application of fluid mechanics in their design. Mechanical engineers use fluid mechanics to design heating and air-conditioning systems, turbines, internal combustion engines, pumps, and air compressors. Aeronautical engineers use fluid mechanics to design aircraft, spacecraft, and missiles. Chemical engineers use fluid mechanics to design chemical processing equipment such as heat exchangers and cooling towers. Civil engineers use fluid mechanics to design water treatment plants, flood control systems, irrigation channels, and dams. Principles of fluid mechanics are even important in the design of ground-based structures. The collapse of the Tacoma Narrows Bridge in 1940 could have been prevented, had the designers paid attention to the possible effects of wind forces on suspension bridges. Principles of fluid mechanics are necessary for understanding winds and ocean currents. A proper understanding of fluid mechanics is also needed for studying blood flow in the human circulatory system. The list of fluid mechanics applications is long indeed. Figure 7.1, Figure 7.2, and Figure 7.3 show some engineering systems that involved the use of fluid mechanics in their design.

Figure 7.1. Aerial view of the Hoover Dam. Engineers used principles of fluid statics to determine the pressure forces acting on the structure.(Courtesy of the U.S. Department of the Interior Bureau of Reclamation, Lower Colorado Region.)

7.2   FLUID PROPERTIES

A *property* is a *physical characteristic or attribute of a substance*. Matter in either state, solid or fluid, may be characterized in terms of properties. For example, Young's modulus is a property of solids that relates stress to strain. Density is a property of solids and fluids that provides a measure of mass contained in a unit volume. In this section, we examine some of the more commonly used fluid properties. Specifically, we will discuss 1) density, specific weight, and specific gravity, 2) bulk modulus, and 3) viscosity.

Figure 7.2. Aerodynamics is a special discipline within fluid mechanics. Engineers used principles of aerodynamics to design the unique shape of the F-117 Nighthawk stealth fighter. (Courtesy of Lockheed Martin Corporation, Bethesda, MD.)

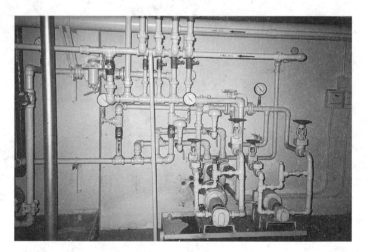

Figure 7.3. Principles of fluid dynamics are used to design and analyze complex piping systems.

7.2.1 Density, Specific Weight, and Specific Gravity

A fluid is a continuous medium; that is, a substance that is continuously distributed throughout a region in space. Because a fluid is a continuous medium, it would be rather awkward to analyze the fluid as a single entity with a total mass m, total weight W, or total volume V. It is more convenient to analyze the fluid in terms of the mass of fluid contained in a specified volume. **Density** is defined as *mass per unit volume*. Density is a property that applies to solids as well as fluids. The mathematical definition for density ρ is

$$\rho = \frac{m}{V} \tag{7.1}$$

The most commonly used units for density are kg/m^3 in the SI system and $slug/ft^3$ in the English system. Values for density can vary widely for different fluids. For example, the densities of water and air at 4°C and 1 atm pressure are about 1000 kg/m^3 (1.94 slug/ft^3) and 1.27 kg/m^3 $(0.00246 \text{ slug/ft}^3)$, respectively. Densities of liquids are higher than those of gases because the intermolecular spacing is smaller. Physical properties vary with temperature and pressure to some extent. For liquids, density does not vary significantly with changes in temperature and pressure, but the densities of gases are strongly influenced by changes in temperature and pressure.

A fluid property that is similar to density is specific weight. **Specific weight** is defined as *weight per unit volume*. The mathematical definition for specific weight γ is

$$\gamma = \frac{W}{V} \tag{7.2}$$

The most commonly used units for specific weight are N/m^3 in the SI system and lb_f/ft^3 in the English system. Note that the unit for specific weight in the English system is not lb_m/ft^3. The unit lb_m is a unit of mass, not a unit of weight. A quick inspection of Equation (7.1) and Equation (7.2) reveals that specific weight is essentially the same property as density with mass replaced by weight. A formula that relates density ρ and specific weight γ may be obtained by noting that the weight of a unit volume of fluid is $W = mg$, where g is the local gravitational acceleration. Substituting the relation for weight W into Equation (7.2) and combining the result with Equation (7.1), we obtain the relation

$$\gamma = \rho g \tag{7.3}$$

Using the standard value of gravitational acceleration, $g = 9.81 \text{ m/s}^2$, water at 4°C has a specific weight of

$$\gamma = \rho g$$
$$= (1000 \text{ kg/m}^3)(9.81 \text{ m/s}^2) = 9810 \text{ N/m}^3 = 9.81 \text{ kN/m}^3$$

Doing the same calculation in English units, noting that the standard value of gravitational acceleration is $g = 32.2 \text{ ft/s}^2$, water at 4°C (39.2°F) has a specific weight of

$$\gamma = \rho g$$
$$= (1.94 \text{ slug/ft}^3)(32.2 \text{ ft/s}^2) = 62.4 \text{ lb}_f/ft^3$$

An alternative form of Equation (7.3) is

$$\gamma = \frac{\rho g}{g_c} \tag{7.3a}$$

where g_c is a constant whose magnitude and units depend on the choice of units used for γ. For example, the specific weight of water in SI units may be calculated as

$$\gamma = \frac{\rho g}{g_c}$$

$$= \frac{(1000 \text{ kg/m}^3)(9.81 \text{ m/s}^2)}{1\dfrac{\text{kg} \cdot \text{m}}{\text{N} \cdot \text{s}^2}} = 9810 \text{ N/m}^3$$

Noting that 1 slug = 32.2 lb$_m$, the specific weight of water in English units may be calculated as

$$\gamma = \frac{\rho g}{g_c}$$

$$= \frac{(62.4 \text{ lb}_m/\text{ft}^3)(32.2 \text{ ft/s}^2)}{32.2 \dfrac{\text{lb}_m \cdot \text{ft}}{\text{lb}_f \cdot \text{s}^2}} = 62.4 \text{ lb}_f/\text{ft}^3$$

The density and specific weight of water, or any other substance for that matter, are numerically equivalent as long as the standard value of g is used. The rationale for finding the density and specific weight of water at 4°C in the foregoing discussion is that 4°C is a reference temperature on which specific gravity is based. **Specific gravity** is defined as the *ratio of the density of a fluid to the density of water at a reference temperature*. Typically, the reference temperature is taken as 4°C because the density of water is maximum (about 1000 kg/m^3) at this temperature. The mathematical definition for specific gravity sg is

$$sg = \frac{\rho}{\rho_{H_2O}@\ 4°C} \tag{7.4}$$

Because specific gravity is a ratio of two properties with the same units, it is a dimensionless quantity. Furthermore, the value of sg does not depend on the system of units used. For example, the density of mercury at 20°C is 13,550 kg/m^3 (26.29 slug/ft^3). Using SI units, the specific gravity of mercury is

$$sg = \frac{\rho}{\rho_{H_2O}@\ 4°C}$$

$$= \frac{13,550 \text{ kg/m}^3}{1000 \text{ kg/m}^3} = 13.55$$

Using English units, we obtain the same value.

$$sg = \frac{\rho}{\rho_{H_2O}@\ 4°C}$$

$$= \frac{26.29 \text{ slug/ft}^3}{1.94 \text{ slug/ft}^3} = 13.55$$

Specific gravity may also be defined as the *ratio of the specific weight of a fluid to the specific weight of water at a reference temperature*. This definition, which is derived by combining Equation (7.4) and Equation (7.3), is expressed as

$$sg = \frac{\gamma}{\gamma_{H_2O}@\ 4°C} \tag{7.5}$$

It does not matter whether Equation (7.4) or Equation (7.5) is used to find sg because both relations yield the same value. The definitions given by Equation (7.4) and Equation (7.5) apply regardless of the temperature at which the specific gravity is being determined. In other words, the reference temperature for water is always 4°C, but the density and specific weight of the fluid being considered are based on the temperature specified in the problem. Table 7.1 summarizes the reference values used in the definitions of specific gravity.

TABLE 7.1 Density and Specific Weight of Water at 4°C.

	ρ	γ
SI	1000 kg/m^3	9810 N/m^3
English	1.94 slug/ft^3	$62.4 \text{ lb}_f/\text{ft}^3$

7.2.2 Bulk Modulus

An important consideration in the analysis of fluids is the degree to which a given mass of fluid changes its volume (and therefore its density) when there is a change in pressure. Stated another way, how compressible is the fluid? **Compressibility** refers to the change in volume V of a fluid subjected to a change in pressure P. The property used to characterize compressibility is the **bulk modulus** K defined by the relation

$$K = \frac{-\Delta P}{\Delta V/V} \tag{7.6}$$

where ΔP is the change in pressure, ΔV is the change in volume, and V is the volume before the pressure change occurs. The negative sign is used in Equation (7.6) because an increase in pressure causes a decrease in volume, thereby assigning a negative sign to the quantity ΔV. The negative signs on ΔP and ΔV cancel, leaving a positive bulk modulus K, which is always a positive quantity. Because the ratio $\Delta V/V$ is dimensionless, the bulk modulus has units of pressure. Typical units used for K are MPa and psi in the SI and English systems, respectively. A large value of K means that the fluid is relatively incompressible (i.e., it takes a large change in pressure to produce a small change in volume). Equation (7.6) applies for liquids only. Compared with liquids, gases are considered compressible fluids, and the formula for bulk modulus depends on certain thermodynamic considerations. Only liquids will be considered here. Liquids are generally considered incompressible fluids because they compress very little when subjected to a large change in pressure. Hence, the value of K for liquids is typically large. For example, the bulk modulus for water at 20°C is $K = 2.24$ GPa. For mercury at 20°C, $K = 28.5$ GPa. A list of bulk modulus values for some common liquids is given in Table 7.2.

TABLE 7.2 Bulk Modulus for Common Liquids at 20°C.

LIQUID	K(GPa)	K(psi)
Benzene	1.48	2.15×10^5
Carbon tetrachloride	1.36	1.97×10^5
Castor oil	2.11	3.06×10^5
Glycerin	4.59	6.66×10^5
Heptane	0.886	1.29×10^5
Kerosene	1.43	2.07×10^5
Lubricating oil	1.44	2.09×10^5
Mercury	28.5	4.13×10^6
Octane	0.963	1.40×10^5
Seawater	2.42	3.51×10^5
Water	2.24	3.25×10^5

Compressibility is an important consideration in the analysis and design of hydraulic systems. Hydraulic systems are used to transmit and amplify forces by pressurizing a fluid in a cylinder. A tube or hose connects the fluid in the cylinder with a mechanical actuator. The hydraulic fluid completely fills the cylinder, connecting line, and actuator so that when a force is applied to the fluid in the cylinder, the fluid is pressurized with equal pressure everywhere in the system. A relatively low force applied to the fluid in the cylinder can produce a large actuator force because the cross-sectional area over which the pressure is applied is much larger in the actuator than in the cylinder. Thus, the force applied at the cylinder is amplified at the actuator. Hydraulic systems are used in a variety of applications, such as heavy construction equipment, manufacturing processes, and transportation systems. The brake system in your automobile is a hydraulic system. When you press the brake pedal, the brake fluid in the system is pressurized, causing the brake mechanism in the wheels to transmit friction forces to the wheels, thereby slowing the vehicle. Brake fluids must have high bulk modulus values for the brake system to function properly. If the value of the bulk modulus of the brake fluid is too low, a large change in pressure will produce a large change in volume that will cause the brake pedal to bottom out on the floor of the automobile, rather than activating the brake mechanism in the wheels. In principle, this is what happens when air becomes trapped inside the brake system. Brake fluid is incompressible, but air is compressible, so the brakes do not function. As an engineering student, you will understand the underlying engineering principles on which this hazardous situation is based. (See Figure 7.4.)

Figure 7.4. An engineering student explains a brake system failure. (Art by Kathryn Hagen.)

7.2.3 Viscosity

The fluid properties of density, specific weight, and specific gravity are measures of the "heaviness" of a fluid, but these properties do not completely characterize a fluid. Two different fluids, water and oil, for example, have similar densities, but exhibit distinctly different flow behavior. Water flows readily when poured from a container, whereas oil, which is a "thicker" fluid, flows more slowly. Clearly, an additional fluid property is required to adequately describe the flow behavior of fluids. **Viscosity** may be qualitatively defined as the *property of a fluid that signifies the ease with which the fluid flows under specified conditions.*

To investigate viscosity further, consider the hypothetical experiment depicted in Figure 7.5. Two parallel plates, one stationary and the other moving with a constant

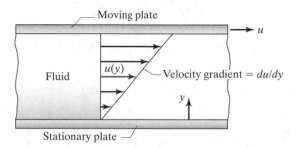

Figure 7.5.   A velocity gradient is established in a fluid between a stationary and a moving plate.

velocity u enclose a fluid. We observe in this experiment that the fluid in contact with both plates "sticks" to the plates. Hence, the fluid in contact with the bottom plate has a zero velocity, and the fluid in contact with the top plate has a velocity u. The velocity of the fluid changes linearly from zero at the bottom plate to u at the top plate, giving rise to a **velocity gradient** in the fluid. This velocity gradient is expressed as a derivative, du/dy, where y is the coordinate measured from the bottom plate. Because a velocity gradient exists in the fluid, adjacent parallel "layers" of fluid at slightly different y values have slightly different velocities, which means that adjacent layers of fluid slide over each other in the same direction as the velocity u. As adjacent layers of fluid slide across each other, they exert a shear stress τ in the fluid. Our experiment reveals that the shear stress τ is proportional to the velocity gradient du/dy, which is the slope of the function $u(y)$. Thus,

$$\tau \propto \frac{du}{dy} \tag{7.7}$$

The result indicates that for common fluids such as water, oil, and air, the proportionality in Equation (7.7) may be replaced by the equality

$$\tau = \mu \frac{du}{dy} \tag{7.8}$$

where the constant of proportionality μ is called the **dynamic viscosity**. Equation (7.8) is known as *Newton's law of viscosity*, and fluids that conform to this law are referred to as **Newtonian fluids**. Common liquids such as water, oil, glycerin, and gasoline are Newtonian fluids, as are common gases such as air, nitrogen, hydrogen, and argon. The value of the dynamic viscosity depends on the fluid. Liquids have higher viscosities than gases, and some liquids are more viscous than others. For example, oil, glycerin, and other gooey liquids have higher viscosities than water, gasoline, and alcohol. The viscosities of gases do not vary significantly from one gas to another, however.

Shear stress has the same units as pressure. In the SI system of units, shear stress is expressed in N/m^2, which is defined as a pascal (Pa). In the English system, shear stress is usually expressed in lb_f/ft^2 or lb_f/in^2 (psi). Velocity gradient has units of s^{-1}, so a quick inspection of Equation (7.8) shows that dynamic viscosity μ has units of $Pa \cdot s$ in the SI system. The units of $Pa \cdot s$ may be broken down into their base units of $kg/m \cdot s$. The units for μ are $lb_f \cdot s/ft^2$ or $slug/ft \cdot s$ in the English system.

Consider once again the configuration illustrated in Figure 7.5. As the fluid flows between the plates, shear forces caused by viscosity are resisted by inertia forces in the fluid. Inertia forces are forces that tend to maintain a state of rest or motion in all matter, as stated by Newton's first law. A second viscosity property that denotes the ratio of viscous forces to inertia forces in a fluid is kinematic viscosity. **Kinematic viscosity** ν is defined as *the ratio of dynamic viscosity to the density of the fluid*. Thus,

$$\nu = \frac{\mu}{\rho} \tag{7.9}$$

In the SI system of units, kinematic viscosity is expressed in m^2/s, and in the English system it is expressed in ft^2/s. Because the ratio of dynamic viscosity to density often appears in the analysis of fluid systems, kinematic viscosity may be the preferred viscosity property.

Viscosity, like all physical properties, is a function of temperature. For liquids, dynamic viscosity decreases dramatically with increasing temperature. For gases, however, dynamic viscosity increases, but only slightly, with increasing temperature. The kinematic viscosity of liquids behaves essentially the same as dynamic viscosity because liquid densities change little with temperature. However, because gas densities decrease sharply with increasing temperature, the kinematic viscosities of gases increase drastically with increasing temperature.

EXAMPLE 7.1

A graduated cylinder containing 100 mL of alcohol has a combined mass of 280 g. If the mass of the cylinder is 200 g, what is the density, specific weight, and specific gravity of the alcohol?

SOLUTION

The combined mass of the cylinder and alcohol is 280 g. By subtraction, the mass of the alcohol is

$$m = (0.280 - 0.200) \text{ kg} = 0.080 \text{ kg}$$

Converting 100 mL to m^3, we obtain

$$100 \text{ mL} \times \frac{1 \text{ L}}{1000 \text{ mL}} \times \frac{1 \text{ m}^3}{1000 \text{ L}} = 1 \times 10^{-4} \text{ m}^3$$

The density of the alcohol is

$$\rho = \frac{m}{V}$$

$$= \frac{0.080 \text{ kg}}{1 \times 10^{-4} \text{ m}^3}$$

$$= 800 \text{ kg/m}^3$$

The weight of the alcohol is

$$W = mg$$
$$= (0.080 \text{ kg})(9.81 \text{ m/s}^2)$$
$$= 0.7848 \text{ N}$$

so the specific weight is

$$\gamma = \frac{W}{V}$$
$$= \frac{(0.7848 \text{ N})}{1 \times 10^{-4} \text{ m}^3}$$
$$= 7848 \text{ N/m}^3$$

The specific gravity of the alcohol is

$$\text{sg} = \frac{\rho}{\rho_{H_2O}@ \, 4°C}$$
$$= \frac{800 \text{ kg/m}^3}{1000 \text{ kg/m}^3}$$
$$= 0.800$$

EXAMPLE 7.2

Find the change in pressure required to decrease the volume of water at 20° C by 1 percent.

SOLUTION

From Table 7.2, the bulk modulus of water at 20°C is $K = 2.24$ GPa. A 1 percent decrease in volume denotes that $\Delta V/V = -0.01$. Rearranging Equation (7.6) and solving for ΔP, we obtain

$$\Delta P = -K(\Delta V/V)$$
$$= -(2.24 \times 10^9 \text{ Pa})(-0.01)$$
$$= 22.4 \times 10^6 \text{ Pa} = 22.4 \text{ MPa}$$

EXAMPLE 7.3

Two parallel plates, spaced 3 mm apart, enclose a fluid. One plate is stationary, while the other plate moves parallel to the stationary plate with a constant velocity of 10 m/s. Both plates measure 60 cm × 80 cm. If a 12-N force is required to sustain the velocity of the moving plate, what is the dynamic viscosity of the fluid?

SOLUTION

The velocity varies from zero at the stationary plate to 10 m/s at the moving plate, and the spacing between the plates is 0.003 m. The velocity gradient in Newton's law of viscosity may be expressed in terms of differential quantities as

$$\Delta u/\Delta y = (10 \text{ m/s})/(0.003 \text{ m}) = 3333 \text{ s}^{-1}$$

The shear stress is found by dividing the force by the area of the plates. Thus,

$$\tau = \frac{F}{A}$$

$$= \frac{12\ \text{N}}{(0.6\ \text{m})(0.8\ \text{m})}$$

$$= 25\ \text{N/m}^2 = 25\ \text{Pa}$$

Rearranging Equation (7.8) and solving for dynamic viscosity μ we obtain

$$\mu = \frac{\tau}{\Delta u/\Delta y}$$

$$= \frac{25\ \text{Pa}}{3333\ \text{s}^{-1}}$$

$$= 7.50 \times 10^{-3}\ \text{Pa} \cdot \text{s}$$

PRACTICE!

1. A cylindrical container with a height and diameter of 16 cm and 10 cm, respectively, contains 1.1 kg of liquid. If the liquid fills the container, find the density, specific weight, and specific gravity of the liquid.
   Answer: 875 kg/m^3, 8585 N/m^3, 0.875

2. A swimming pool measuring 30 ft × 18 ft × 8 ft is to be filled by using a water truck with a capacity of 5500 gallons. How many trips does the water truck have to make to fill the pool? If the density of the water is 1.93 slug/ft^3, what is the mass and weight of the water in the pool after it has been filled?
   Answer: 6, 8381 slug, 2.70 × 10^5 lb$_f$

3. A cylinder containing benzene at 20°C has a piston that compresses the fluid from 0 to 37 MPa. Find the percent change in the volume of the benzene.
   Answer: −2.50%

4. Hydraulic fluid is compressed by a piston in a cylinder, producing a change in pressure of 40 MPa. Before the piston is activated, the hydraulic fluid fills a 20-cm length of the cylinder. If the axial displacement of the piston is 6.5 mm, what is the bulk modulus of the hydraulic fluid?
   Answer: 1.231 GPa

5. Glycerin at 20°C (ρ = 1260 kg/m^3, μ = 1.48 Pa·s) occupies a 1.6-mm space between two square parallel plates. One plate remains stationary while the other plate moves with a constant velocity of 8 m/s. If both plates measure 1 m on a side, what force must be exerted on the moving plate to sustain its motion? What is the kinematic viscosity of the glycerin?
   Answer: 7400 N, 1.175 × 10^{-3} m^2/s

7.3 FLUID STATICS

Fluid mechanics is broadly divided into two categories: *fluid statics* and *fluid dynamics*. Fluid statics, the subject of this section, is the branch of fluid mechanics that deals with the behavior of fluids at rest. In fluid statics, the fluid is at rest with respect to a frame of reference. This means that the fluid does not move with respect to a body or surface with which the fluid is in physical contact. Because the fluid is at rest, the fluid is in a state of equilibrium where the vector sum of the external forces acting on the fluid is zero. As a subject, fluid statics encompasses several areas for study, including forces on submerged surfaces, pressure measurement and manometry, buoyancy, stability, and fluid masses subjected to acceleration. Our treatment of fluid statics will focus on the most fundamental of these subjects: forces on submerged surfaces.

7.3.1 Pressure Elevation Relationship

Common experience tells us that the pressure increases with depth in a fluid. For example, a scuba diver experiences higher pressures as he descends below the water's surface. If we are to know how to analyze the effect of forces exerted on submerged surfaces, we must first understand how pressure changes with elevation (vertical distance) in a static fluid. To obtain a relationship between pressure and elevation in a static fluid, refer to the configuration shown in Figure 7.6. In Figure 7.6, we consider a static body of fluid with

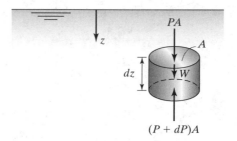

Figure 7.6. Differential fluid element used to derive the pressure elevation relation $\Delta P = \gamma h$.

density ρ. Because the entire body of fluid is in equilibrium, every particle of fluid must therefore be in equilibrium. Thus, we can isolate an infinitesimally small fluid element for analysis. We choose as our fluid element a cylinder of height dz whose top and bottom surface area is A. Treating the fluid element as a free body in equilibrium, we observe that there are three external forces acting on the element in the z direction. Two of the forces are pressure forces acting on the top and bottom surfaces of the element. The pressure force acting on the top surface is PA, the product of the pressure at a given z coordinate and the surface area. The pressure force acting on the bottom surface is $(P + dP)A$, the product of the pressure at $z + dz$ and the surface area. The pressure acting on the bottom surface is $(P + dP)$, because the pressure has increased a differential amount corresponding to an elevation change of dz. Note that both pressure forces are compressive forces. (There are also pressure forces acting around the perimeter of the cylinder on its curved surface, but these forces cancel one another and are not a function of elevation.) The third force acting on the fluid element is the weight of the fluid element W.

Writing a force balance on the fluid element in the z direction, we obtain

$$\Sigma F_z = 0 = PA - (P + dP)A - W \tag{7.10}$$

The weight of the fluid element is

$$W = mg = \rho V g = \rho g A dz \tag{7.11}$$

where the volume of the element is $V = Adz$. Substituting Equation (7.11) into Equation (7.10) and simplifying, we obtain

$$dP = \rho g dz \tag{7.12}$$

Equation (7.12) can now be integrated. Pressure is integrated from P_1 to P_2, and elevation is integrated from z_1 to z_2. Thus,

$$\int_1^2 dP = \rho g \int_1^2 dz \tag{7.13}$$

which yields

$$P_2 - P_1 = \rho g(z_2 - z_1) \tag{7.14}$$

In many instances, P_1 is taken as the pressure at the origin, $z = z_1 = 0$. The pressure P_2 then becomes the pressure at a depth, z_2, below the free surface of the fluid. We are usually not concerned with the force exerted by atmospheric pressure, so the pressure P_1 at the free surface of the fluid is zero (i.e., the *gauge* pressure at the free surface is zero, and P_2 is the gauge pressure at z_2). Equation (7.14) may be expressed in a simplified form by letting $\Delta P = P_2 - P_1$ and $h = z_2 - z_1$. Noting that $\gamma = \rho g$, Equation (7.14) reduces to

$$\Delta P = \gamma h \tag{7.15}$$

where γ is the specific weight of the fluid and h is the elevation change as referenced from the free surface. As h increases, pressure increases in accordance with our experience. We may draw some general conclusions from the relationship between pressure and elevation given by Equation (7.15):

1. Equation (7.15) is valid only for a homogenous static *liquid.* It does not apply to gases, because γ is not constant for compressible fluids.
2. The change in pressure is directly proportional to the specific weight of the liquid.
3. Pressure varies linearly with depth, the specific weight of the liquid being the slope of the linear function.
4. Pressure increases with increasing depth and vice versa.
5. Points on the same horizontal plane have the same pressure.

Another important conclusion that may be drawn from Equation (7.15) is that, for a given liquid, the pressure change is a function of elevation change h only. Pressure is independent of any other geometrical parameter. The containers illustrated in Figure 7.7

Figure 7.7.   For the same liquid, the pressures in these containers at a given depth h are equal, being independent of the shape or size of the container.

are filled to a depth h with the same liquid, so the pressure at the bottom of these containers is the same. Each container has a different size and shape, and therefore contains different amounts of liquid, but the pressure is a function of depth only.

7.3.2   Forces on Submerged Surfaces

Now that the relationship between pressure and elevation in static liquids has been established, let us apply the relationship to the analysis of forces on submerged surfaces. We will examine two fundamental cases. The first case involves forces exerted by static liquids on horizontal submerged surfaces. The second case involves forces exerted by static liquids on partially submerged vertical surfaces. In both cases, we will restrict our analysis to plane surfaces.

In the first case, we find that the force exerted by a static liquid on a horizontal submerged surface is determined by a direct application of Equation (7.15). Consider a container with a plane horizontal surface filled with a liquid to a depth h, as shown in Figure 7.8. The pressure at the bottom of the container is given by $P = \gamma h$. Because the

Figure 7.8.   The pressure is uniform on a horizontal submerged surface.

bottom surface is horizontal, the pressure is uniform across the surface. The force exerted on the bottom surface is simply the product of the pressure and the surface area. Thus, the force exerted on a horizontal submerged surface is

$$F = PA \qquad (7.16)$$

where $P = \gamma h$ and A is the surface area. Equation (7.16) is valid regardless of the shape of the horizontal surface. The force exerted on a horizontal submerged surface is equivalent to the weight W of the liquid above the surface. This fact is evident by writing Equation (7.16) as $F = \gamma(hA) = \gamma V = W$.

In the second case, we examine forces exerted on partially submerged vertical surfaces. One of the conclusions we gleaned from Equation (7.15) is that pressure varies linearly with depth in a static liquid. Consider the partially submerged vertical plane surface in Figure 7.9. The pressure (gauge pressure) is zero at the free surface of the liquid, and increases linearly with depth. At a depth h, below the free surface of the liquid, the gauge pressure is $P = \gamma h$. Because the pressure varies linearly from 0 to P over the range 0 to h, the average pressure, P_{avg}, is simply $P/2$. Thus

$$P_{\text{avg}} = \frac{P}{2} = \frac{\gamma h}{2} \qquad (7.17)$$

The average pressure is a constant pressure that, when applied across the entire surface, is equivalent to the actual linearly varying pressure. Like pressure, the force exerted by

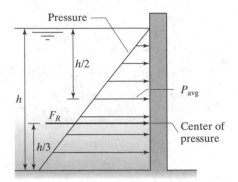

Figure 7.9. Pressure variation and resultant force on a partially submerged vertical surface.

the static liquid on the vertical surface increases linearly with depth. For purposes of structural design and analysis, we are generally interested in the *total force* or *resultant force* that acts on the vertical surface. The resultant force F_R is the product of the average pressure P_{avg} and the area A of the surface that is submerged. Hence,

$$F_R = P_{avg}A = \frac{\gamma hA}{2} \tag{7.18}$$

The resultant force is a concentrated force (a force applied at a point) that is equivalent to the linear force distribution on the vertical surface. In order to make use of the resultant force, the point of application of F_R must be known. From principles of statics, it can be shown that for a linearly varying force distribution, the point of application of the equivalent resultant force is two-thirds the distance from the end with the zero force. Consequently, as shown in Figure 7.9, the resultant force acts at a point $2h/3$ from the free surface of the liquid, or $h/3$ from the bottom of the vertical surface. The point at which the resultant force is applied is called the **center of pressure**. The resultant force, applied at the center of pressure, has the same structural effect on the surface as the actual linear force distribution. The reduction of a distributed force to a concentrated force simplifies the design and analysis of submerged surfaces such as dams, ship hulls, and storage tanks.

EXAMPLE 7.4

A small dam consists of a vertical plane wall with a height and width of 5 m and 30 m, respectively. The depth of the water ($\gamma = 9.81$ kN/m^3) is 4 m. Find the resultant force on the wall and the center of pressure.

SOLUTION

Using Equation (7.18) and noting that only 4 m of the dam wall is submerged, we find that the resultant force is

$$F_R = \frac{\gamma hA}{2}$$

$$= \frac{(9810 \text{ N/m}^3)(4 \text{ m})(4 \times 30) \text{ m}^2}{2}$$

$$= 2.35 \times 10^6 \text{ N} = 2.35 \text{ MN}$$

The center of pressure is located two-thirds from the free surface of the water. Thus, the center of pressure, which we denote by z_{cp}, is

$$z_{cp} = \frac{2h}{3}$$

$$= \frac{2(4 \text{ m})}{3} = 2.67 \text{ m (from the free surface)}$$

PRACTICE!

1. A barrel of motor oil ($\gamma = 8.61 \text{ kN/m}^3$) is filled to a depth of 1.15 m. Neglecting atmospheric pressure, what is the pressure on the bottom of the barrel? If the radius of the barrel's bottom is 20 cm, what is the force exerted by the motor oil on the bottom?

   Answer: 9.902 kPa, 78.8 kN

2. The bottom portion of the hull of a barge is submerged 12 ft in seawater ($\gamma = 64.2 \text{ lb}_f/\text{ft}^3$). The hull is horizontal and measures 30 ft × 70 ft. Find the total force exerted by the seawater on the hull.

   Answer: $1.618 \times 10^6 \text{ lb}_f$

3. The gauge pressure at the bottom of a tank containing ethyl alcohol ($\gamma = 7.87 \text{ kN/m}^3$) is 11 kPa. What is the depth of the alcohol?

   Answer: 1.398 m

4. A vertical gate in an irrigation canal holds back 2.2 m of water. Find the total force on the gate if its width is 3.6 m.

   Answer: 85.5 kN

5. A simple dam is constructed by erecting a vertical concrete wall whose base is secured firmly to the ground. The width of the wall is 16 m, and 5 m of the wall is submerged in water. Find the moment of force about the base of the wall. (*Hint*: The moment of force is the product of the resultant force and the perpendicular distance from the center of pressure to the base of the wall.)

   Answer: 3.27 MN · m

7.4   FLOW RATES

The concept of flow rate is fundamental to the understanding of elementary fluid dynamics. In general terms, flow rate refers to the time it takes a quantity of fluid to pass a specified location. Virtually all engineering systems that incorporate moving fluids for their operation involve the principle of flow rate. For example, the pipes in your home carry water at certain flow rates to various fixtures and appliances such as sinks, bathtubs, and washing machines. Heating and air-conditioning systems supply air at specified flow rates to the rooms in a building to achieve the desired heating or cooling effects. Minimum flow rates are required to produce the lifting forces that sustain the flight of aircraft. The design of turbines, pumps, compressors, heat exchangers, and other fluid-based devices involves the use of flow rates.

In fluid dynamics, there are primarily two types of flow rates: volume flow rate and mass flow rate. **Volume flow rate** is the *rate at which a volume of fluid passes a location per unit time*. **Mass flow rate** is the *rate at which a mass of fluid passes a location per unit time*. These are general definitions that apply to all fluid dynamic situations, but our application of these definitions will be limited to the flow of fluids in conduits such as pipes, ducts, and channels. Volume flow rate Q is calculated by using the relation

$$Q = AV \qquad (7.19)$$

where A is the inside cross-sectional area of the conduit and V is the *average* velocity of the fluid. The word "average" is emphasized here because the velocity of the fluid in a conduit is not constant. Effects of viscosity produce a velocity gradient or profile in the fluid across the breadth of the conduit. (Be careful not to confuse fluid velocity with volume because the symbol V is used for both quantities. Similarly, do not confuse volume flow rate with heat, which both use the symbol Q.) Common units for volume flow rate are m^3/s in the SI system and ft^3/s in the English system. Other units frequently used for volume flow rate are L/min and gal/min or L/h and gal/h. Equation (7.19) applies to any conduit, regardless of its cross-sectional shape. For example, if the conduit is a circular pipe or tube, then $A = \pi R^2$, where R is the inside radius, whereas if the conduit is a duct with a square cross section, then $A = L^2$, where L is the inside dimension of the duct. Mass flow rate m is calculated by using the relation

$$\dot{m} = \rho Q \qquad (7.20)$$

where ρ is the density of the fluid and Q is the volume flow rate given by Equation (7.19). The 'dot' over the m denotes a time derivative or a rate quantity. However, by convention, the dot is not used for Q. Common units for mass flow rate are kg/s in the SI system and slug/s or lb_m/s in the English system. Equations (7.19) and (7.20) apply to liquids and gases.

EXAMPLE 7.5

A pipe with an inside diameter of 5 cm carries water at an average velocity of 3 m/s. Find the volume flow rate and mass flow rate.

SOLUTION

The cross-sectional area of the pipe is

$$A = \frac{\pi D^2}{4}$$
$$= \frac{\pi (0.05 \text{ m})^2}{4} = 1.963 \times 10^{-3} \text{ m}^2$$

The volume flow rate is

$$Q = AV$$
$$= (1.936 \times 10^{-3} \text{ m}^2)(3 \text{ m/s})$$
$$= 5.89 \times 10^{-3} \text{ m}^3/s$$

Taking the density of water to be $\rho = 1000 \text{ kg/m}^3$, the mass flow rate is

$$\dot{m} = \rho Q$$
$$= (1000 \text{ kg/m}^3)(5.89 \times 10^{-3} \text{ m}^3/s)$$
$$= 5.89 \text{ kg/s}$$

PROFESSIONAL SUCCESS: THINGS TO CONSIDER AT THE "HUMP"

A traditional engineering bachelor's degree takes four years to complete. The times that mark the conclusion of the freshman, sophomore, junior, and senior years of a college career are sometimes facetiously referred to as the bump, hump, slump, and dump, respectively. (The fact that you are reading this book suggests that you have not bumped yet.) By the time you hump, you should begin thinking about what you want to do after graduation. Should you accept an engineering position immediately after graduation or go to graduate school? What about going to work as an engineer while working on a graduate degree part time? Should you work for a few years and then go back to school for a graduate degree? Should you obtain your professional engineering license? Should you pursue a nontechnical graduate degree to complement your engineering background? These are some of the questions that you should be asking yourself about midway through your undergraduate engineering program.

A bachelor's degree in engineering paves the road to a rewarding career with a very respectable salary, so many four-year engineering graduates do not pursue graduate studies. However, many engineering companies have a need for engineers with in-depth expertise in specific technical disciplines, so engineers with graduate degrees are in high demand. In general, engineers with graduate degrees have higher salaries than their coworkers with only a bachelor's degree and are frequently well positioned for supervisory and managerial roles. If a graduate degree is in the future for you, is it better to enter graduate school immediately after you graduate with your four-year degree, or should you accrue some engineering experience first and then pursue a graduate degree while you are working? That depends on your personal circumstances, the nature of the graduate school you wish to attend and the policies of your employer. Many people feel that their lives are busy enough with a full-time job, family, and other responsibilities without adding graduate school to the list. Some graduate programs may not look favorably upon part-time graduate students who, because of work commitments, cannot devote their whole body and soul to their graduate studies.

However, many schools are quite willing to work with (and even welcome) part-time graduate students in their engineering programs. Most engineering companies offer educational assistance to their engineers who wish to pursue graduate studies. This assistance most often comes in the form of tuition reimbursement and flexible working schedules so their employees can take graduate courses at a nearby university. As for pursuing a graduate degree in engineering or a nontechnical field such as business or management, you should examine your personal educational and career goals. Do you want to advance technically in a specific discipline, or do you want to climb the management ladder?

Should you become professionally licensed? Regardless of your answer to this question, you should seriously consider taking the Fundamentals of Engineering (FE) examination in the junior or senior year of your program. This test is a state-sponsored exam offered twice a year and administered at your own school or at a school in your area. The FE exam, or the EIT (Engineer in Training) exam, as it is sometimes called, is an 8-hour exam that covers the fundamental principles of engineering that are covered in a typical undergraduate engineering curriculum. Some engineering schools require their students to pass the FE exam to graduate. After you have passed the exam and have worked a few years as a practicing engineer, you can take the PE (Professional Engineer) exam that is specific to your discipline. If you pass that exam and if you have satisfied the other licensing requirements specified by your state, you can write the initials PE after your name on official letters, drawings, and other documents. A professional engineer is an engineer who is officially recognized by the state as having demonstrated proficiencies in a specific engineering discipline. Most engineering companies do not require their engineers to be professionally licensed, but some firms, particularly state and municipal governments, have strict rules about employing engineers who are professionally licensed. Thus, your decision to become professionally licensed or not may be largely based on the requirements or recommendations of your employer.

PRACTICE!

1. A steel tube carries gasoline ($\rho = 751$ kg/m^3) at an average velocity of 0.85 m/s. If the inside diameter of the tube is 7 mm, find the volume flow rate and mass flow rate.

   Answer: 3.271×10^{-5} m^3/s, 0.0246 kg/s

2. Water flows through a plastic pipe at a volume flow rate of 160 gal/min. What is the inside radius of the pipe if the average velocity of the water is 8 ft/s? Express your answer in inches and centimeters.

   Answer: 1.43 in, 3.63 cm

3. A blower forces air ($\gamma = 11.7$ N/m^3) through a rectangular duct with a 50 cm $\times$ 80 cm inside cross section. If the average velocity of the air is 7 m/s, find the volume flow rate and mass flow rate. Express the volume flow rate in m^3/s and ft^3/min (CFM) and the mass flow rate in kg/s and slug/h.

   Answer: 2.80 m^3/s, 5933 ft^3/min, 3.339 kg/s, 824 slug/h

4. A pump removes water from a 1200-gallon storage tank at a rate of 0.05 m^3/s. How long will it take the pump to empty the tank? If a pipe with an inside diameter of 6 cm connects the pump to the tank, what is the average velocity of the water in the pipe?

   Answer: 17.7 m/s

5. Air flows through a duct with a rectangular cross section at an average velocity of 20 ft/s and a volume flow rate of 3000 CFM. If the inside dimension of one side of the duct measures 18 in, what is the dimension of the other side?

   Answer: 1.667 ft

7.5 CONSERVATION OF MASS

Some of the most important fundamental principles used to analyze engineering systems are the conservation laws. A conservation law is an immutable law of nature declaring that certain physical quantities are conserved. Defined another way, a conservation law states that the total amount of a particular physical quantity is constant during a process. A familiar conservation law is the first law of thermodynamics, which states that energy is conserved. According to the first law of thermodynamics, energy may be converted from one form to another, but the total energy is constant. Another conservation law is Kirchhoff's current law, which states that the algebraic sum of the currents entering a circuit node is zero. Kirchhoff's current law is a statement of the law of conservation of electric charge. Other quantities that are conserved are linear and angular momentum.

In this section, we examine the principal conservation law used in fluid mechanics, the law of conservation of mass. Like the first law of thermodynamics, the law of conservation of mass is an intuitive concept. To introduce the conservation of mass principle, consider the system shown in Figure 7.10. The system may represent any region in space chosen for analysis. The boundary of the system is the surface that separates the system from the surroundings. We may construct a mathematical representation of the conservation of mass principle by applying a simple physical argument. If an amount of mass m_{in} is supplied *to* the system, that mass can either *leave* the system or *accumulate* within the system, or both. The mass that leaves the system is m_{out}, and the change in mass within the system is Δm. Thus, the mass that enters the system equals the mass that leaves the system plus the change in mass within the system. The conservation of mass principle may therefore be expressed mathematically as

$$m_{in} = m_{out} + \Delta m \tag{7.21}$$

We see that the conservation of mass law is nothing more than a simple accounting principle that maintains the system's "mass ledger" in balance. In fact, this conservation law is often referred to as a *mass balance* because that is precisely what it is. Equation (7.21)

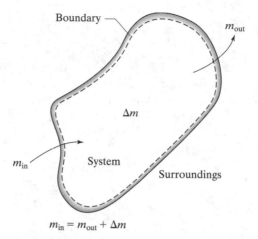

Boundary

m_{out}

Δm

m_{in}

System

Surroundings

$m_{in} = m_{out} + \Delta m$

Figure 7.10. The law of conservation of mass.

is more useful when expressed as a rate equation. Dividing each term by a time interval Δt, we obtain

$$\dot{m}_{in} = \dot{m}_{out} + \Delta m/\Delta t \qquad (7.22)$$

where $\dot{m}_{in}$ and $\dot{m}_{out}$ are the inlet and outlet mass flow rates, respectively, and $\Delta m/\Delta t$ is the rate at which mass accumulates within the system. The law of conservation of mass is called the **continuity principle**, and Equation (7.22), or a similar relation, is referred to as the *continuity equation*.

We now examine a special case of the configuration given in Figure 7.10. Consider the converging pipe shown in Figure 7.11. The dashed line outlines the boundary of

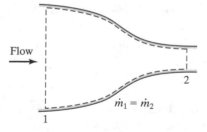

Flow

2

$\dot{m}_1 = \dot{m}_2$

1

Figure 7.11. Continuity principle for a converging pipe.

the flow system defined by the region inside the pipe wall and between sections 1 and 2. A fluid flows at a constant rate from section 1 to section 2. Because fluid does not accumulate between sections 1 and 2, $\Delta m/\Delta t = 0$, and Equation (7.22) becomes

$$\dot{m}_1 = \dot{m}_2 \qquad (7.23)$$

where the subscripts 1 and 2 denote the input and output, respectively. Thus, the mass of fluid flowing past section 1 per unit time is the same as the mass of fluid flowing past section 2 per unit time. Because $\dot{m} = \rho Q$, Equation (7.23) may also be expressed as

$$\rho_1 Q_1 = \rho_2 Q_2 \tag{7.24}$$

where ρ and Q denote density and volume flow rate, respectively. Equation (7.23), and its alternative form, Equation (7.24), are valid for liquids and gases. Hence, these relations apply to compressible and incompressible fluids. If the fluid is incompressible, the fluid density is constant, so $\rho_1 = \rho_2 = \rho$. Dividing Equation (7.24) by density ρ, yields

$$Q_1 = Q_2 \tag{7.25}$$

which may be written as

$$A_1 V_1 = A_2 V_2 \tag{7.26}$$

where A and V refer to cross-sectional area and average velocity, respectively. Equations (7.25) and (7.26) apply strictly to liquids, but these relations may also be used for gases with little error if the velocities are below approximately 100 m/s.

The continuity principle can also be used to analyze more complex flow configurations, such as a flow branch. A flow branch is a junction where three or more conduits are connected. Consider the pipe branch shown in Figure 7.12. A fluid enters a junction from a supply pipe, where it splits into two pipe branches. The flow rates in the branching pipes depend on the size of the pipes and other characteristics of the system, but

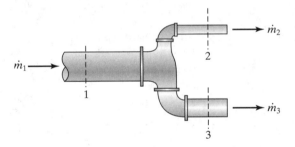

Figure 7.12. A pipe branch.

from the continuity principle it is clear that the mass flow rate in the supply pipe must equal the sum of the mass flow rates in the two pipe branches. Thus, we have

$$\dot{m}_1 = \dot{m}_2 + \dot{m}_3 \tag{7.27}$$

Junctions in flow branches are analogous to nodes in electrical circuits. Kirchhoff's current law, which is a statement of the law of conservation of electric charge, states that the algebraic sum of the currents entering a node is zero. For a flow branch, the continuity principle states that *the algebraic sum of the mass flow rates entering a junction is zero*. The mathematical expression for this principle is similar to Kirchhoff's current law and is written as

$$\Sigma \dot{m}_{in} = 0 \tag{7.28}$$

The continuity relation given by Equation (7.27) for the specific case illustrated in Figure 7.12 is equivalent to the general form of the relation given by Equation (7.28). We have

$$\begin{aligned} \Sigma \dot{m}_{in} &= 0 \\ &= \dot{m}_1 - \dot{m}_2 - \dot{m}_3 \end{aligned} \tag{7.29}$$

where minus signs are used for mass flow rates $\dot{m}_2$ and $\dot{m}_3$ because the fluid in each pipe branch is leaving the junction. The mass flow rate $\dot{m}_1$ is positive because the fluid in the supply pipe is entering the junction.

In the next example, we analyze a basic flow system by using the general analysis procedure of (1) problem statement, (2) diagram, (3) assumptions, (4) governing equations, (5) calculations, (6) solution check, and (7) discussion.

EXAMPLE 7.6

Problem statement

A converging duct carries oxygen ($\rho = 1.320 \text{ kg/m}^3$) at a mass flow rate of 110 kg/s. The duct converges from a cross-sectional area of 2 m² to a cross-sectional area of 1.25 m². Find the volume flow rate and the average velocities in both duct sections.

Diagram

The diagram for this problem is shown in Figure 7.13.

Assumptions

1. The flow is steady.
2. The fluid is incompressible.
3. There are no leaks in the duct.

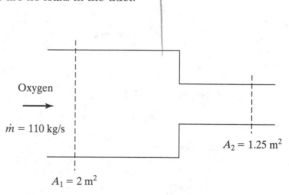

Oxygen

$\dot{m} = 110 \text{ kg/s}$

$A_2 = 1.25 \text{ m}^2$

$A_1 = 2 \text{ m}^2$

Figure 7.13. Converging duct for Example 7.6.

Governing equations

Two equations are needed to solve this problem—the relation for mass flow rate and the continuity relation:

$$\dot{m} = \rho Q$$

$$Q = A_1 V_1 = A_2 V_2$$

Calculations

By continuity, the volume flow rate and mass flow rate are equal at sections 1 and 2. The volume flow rate is

$$Q = \frac{\dot{m}}{\rho}$$

$$= \frac{110 \text{ kg/s}}{1.320 \text{ kg/m}^3}$$

$$= \underline{83.33 \text{ m}^3/\text{s}}$$

The average velocity in the large section is

$$V_1 = \frac{Q}{A_1}$$

$$= \frac{83.33 \text{ m}^3/\text{s}}{2 \text{ m}^2}$$

$$= \underline{41.7 \text{ m/s}}$$

and the average velocity in the small section is

$$V_2 = \frac{Q}{A_2}$$

$$= \frac{83.33 \text{ m}^3/\text{s}}{1.25 \text{ m}^2}$$

$$= \underline{66.7 \text{ m/s}}$$

Solution check

After a careful review of our solution, no errors are found.

Discussion

Note that velocity and cross-sectional area are inversely related. The velocity is low in the large portion of the duct and high in the small portion of the duct. The maximum velocity in the duct is below 100 m/s, so the oxygen may be considered an incompressible fluid with little error. Our assumption that the fluid is incompressible is therefore valid.

APPLICATION: ANALYZING A PIPE BRANCH

Pipe branches are used frequently in piping systems to split a stream into two or more flows. Consider a pipe branch similar to the one shown in Figure 7.12. Water enters the pipe junction at a volume flow rate of 350 gal/min, and the flow splits into two branches. One pipe branch has an inside diameter of 7 cm, and the other pipe branch has an inside diameter of 4 cm. If the average velocity of the water in the 7-cm branch is 3 m/s, find the mass flow rate and volume flow rate in each branch and the average velocity in the 4-cm branch. A flow schematic with the pertinent information is shown in Figure 7.14. First, we convert the volume flow rate in the supply pipe to m³/s.

$$Q_1 = 350 \frac{\text{gal}}{\text{min}} \times \frac{1 \text{ m}^3}{264.17 \text{ gal}} \times \frac{1 \text{ min}}{60 \text{ s}} = 0.02208 \text{ m}^3/\text{s}$$

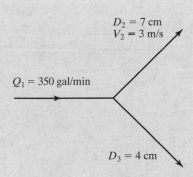

Figure 7.14. Flow schematic for a pipe branch.

The cross-sectional areas of the pipes branches are

$$A_2 = \frac{\pi D_2^2}{4}$$
$$= \frac{\pi (0.07 \text{ m})^2}{4} = 3.848 \times 10^{-3} \text{ m}^2$$
$$A_3 = \frac{\pi D_3^2}{4}$$
$$= \frac{\pi (0.04 \text{ m})^2}{4} = 1.257 \times 10^{-3} \text{ m}^2$$

The volume flow rate in the 7-cm pipe branch is

$$Q_2 = A_2 V_2$$
$$= (3.848 \times 10^{-3} \text{ m}^2)(3 \text{ m/s})$$
$$= 0.01154 \text{ m}^3/\text{s}$$

and the mass flow rate is

$$\dot{m}_2 = \rho Q_2$$
$$= (1000 \text{ kg/m}^3)(0.01154 \text{ m}^3/\text{s})$$
$$= 11.54 \text{ kg/s}$$

In order to find the flow rates in the other pipe branch, the continuity principle is required:

$$Q_1 = Q_2 + Q_3$$

Solving for Q_3, we obtain

$$Q_3 = Q_1 - Q_2$$
$$= (0.02208 - 0.01154) \text{ m}^3/\text{s}$$
$$= 0.01054 \text{ m}^3/\text{s}$$

and the corresponding mass flow rate is

$$\dot{m}_3 = \rho Q_3$$
$$= (1000 \text{ kg/m}^3)(0.01054 \text{ m}^3/\text{s})$$
$$= 10.54 \text{ kg/s}$$

Finally, the average velocity in the 4-cm branch is

$$V_3 = \frac{Q_3}{A_3}$$
$$= \frac{0.01054 \text{ m}^3/\text{s}}{1.257 \times 10^{-3} \text{ m}^2}$$
$$= 8.38 \text{ m/s}$$

As a way of checking for errors, we use the continuity principle to assure that our flow rates are correct. The volume flow rate into the junction must equal the sum of the volume flow rates out of the junction. Thus,

$$Q_1 = Q_2 + Q_3$$
$$0.02208 \text{ m}^3/\text{s} = 0.01154 \text{ m}^3/\text{s} + 0.01054 \text{ m}^3/\text{s}$$

The flow rates balance, so our answers are correct. Notice that the flow rates in the pipe branches are nearly equal (11.54 kg/s and 10.54 kg/s), but the velocities are quite different (3 m/s and 8.38 m/s). This is due to the difference in diameters of the pipes. The velocity is nearly three times higher in the 4-cm pipe than in the 7-cm pipe.

PRACTICE!

1. Water flows through a converging pipe at a mass flow rate of 25 kg/s. If the inside diameters of the pipes sections are 7 cm and 5 cm, find the volume flow rate and the average velocity in each pipe section.
   Answer: 0.025 m³/s, 6.50 m/s, 12.7 m/s

2. A fluid flows through a pipe whose diameter decreases by a factor of three from section 1 to section 2 in the direction of the flow. If the average velocity at section 1 is 10 ft/s, what is the average velocity at section 2?
   Answer: 90 ft/s

3. Air enters a junction in a duct at a volume flow rate of 2000 CFM. Two square duct branches, one measuring 12 in × 12 in and the other measuring 16 in × 16 in, carry the air from the junction. If the average velocity in the small branch is 20 ft/s, find the volume flow rates in each branch and the average velocity in the large branch.
   Answer: 20 ft³/s, 13.3 ft³/s, 9.98 ft/s

4. Two water streams, a cold stream and a hot stream, enter a mixing chamber where both streams combine and exit through a single tube. The mass flow rate of the hot stream is 5 kg/s, and the inside diameter of the tube carrying the cold stream is 3 cm. Find the mass flow rate of the cold stream required to produce an exit velocity of 8 m/s in a tube with an inside diameter of 4.5 cm.
   Answer: 7.72 kg/s

KEY TERMS

bulk modulus	fluid dynamics	specific gravity
center of pressure	fluid mechanics	specific weight
compressibility	fluid statics	velocity gradient
continuity principle	kinematic viscosity	viscosity
density	mass flow rate	volume flow rate
dynamic viscosity	Newtonian fluid	
fluid	shear stress	

REFERENCES

Fox, R.W., A.T. McDonald, and P.J. Pritchard, *Introduction to Fluid Mechanics*, 6th ed. NY: John Wiley & Sons, 2004.

Munson, B.R., D.F. Young, and T.H. Okiishi, *Fundamentals of Fluid Mechanics*, 4th ed., NY: John Wiley & Sons, 2002.

Douglas, D.F., J.M. Gasiorek, and J.A. Swaffield, *Fluid Mechanics*, 4th ed., Upper Saddle River, NJ: Prentice Hall, 2001.

Street, R.L., G.Z. Watters, and J.K. Vennard, *Elementary Fluid Mechanics*, 7th ed., NY: John Wiley & Sons, 1996.

White, W.M., *Fluid Mechanics*, 5th ed., NY: McGraw-Hill, 2002.

Problems

1. The specific gravity of a liquid is 0.920. Find the density and specific weight of the liquid in SI and English units.

2. A 12-cm diameter cylindrical can is filled to a depth of 10 cm with motor oil ($\rho = 878$ kg/m³). Find the mass and weight of the motor oil.

3. The fuel tank of a truck has a capacity of 0.12 m³. If the tank is full of gasoline (sg = 0.751), what is the mass and weight of the gasoline?

4. A 5-m diameter spherical balloon contains hydrogen. If the density of the hydrogen is $\rho = 0.0830$ kg/m³, what is the mass and weight of the hydrogen in the balloon?

5. Find the volume of mercury (sg = 13.55) that weighs the same as 0.04 m³ of ammonia (sg = 0.600).

6. Find the pressure change required to produce a 1.5 percent decrease in the volume of carbon tetrachloride at 20°C.

7. Hydraulic fluid is compressed by a piston in a cylinder producing a change in pressure of 120 MPa. Before the piston is activated, the hydraulic fluid fills a 16-cm length of the cylinder. If the axial displacement of the piston is 8 mm, what is the bulk modulus of the hydraulic fluid?

8. The pressure change in a hydraulic cylinder is 180 MPa for an axial displacement of 15 mm of the piston. If the bulk modulus of the hydraulic fluid is 4500 MPa, what is the minimum length of cylinder required?

9. What is the percent change in the volume of a liquid whose bulk modulus value is 100 times the change in pressure?

10. The velocity gradient $u(y)$ near the surface of a single plate over which a fluid flows is given by the function

$$u(y) = ay + by^2 + cy^3$$

where y is the distance from the plate's surface and a, b, and c are constants with the values $a = 10.0 \text{ s}^{-1}$, $b = 0.02 \text{ m}^{-1}\text{s}^{-1}$, and $c = 0.005 \text{ m}^{-2}\text{s}^{-1}$. If the fluid is water at 20°C ($\mu = 1.0 \times 10^{-3}$ Pa·s), find the shear force at the surface of the plate (at $y = 0$).

For problems 11 through 31, use the general analysis procedure of (1) problem statement, (2) diagram, (3) assumptions, (4) governing equations, (5) calculations, (6) solution check, and (7) discussion.

11. Two square parallel plates enclose 20°C glycerin ($\mu = 1.48$ Pa·s) as illustrated in Figure P7.11. The bottom plate is fixed, and the top plate is attached to a hanging mass by a cord that passes over a frictionless pulley. What mass m is required to sustain a constant velocity of 1.5 m/s for the top plate?

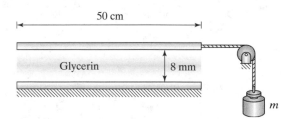

50 cm

Glycerin   8 mm

m

Figure P7.11.

12. The deepest known point in the oceans of the earth is the Mariana Trench, east of the Philippines, with a depth of approximately 10.9 km. Taking the specific gravity of seawater as sg = 1.030, what is the pressure at the bottom of the Mariana Trench? Express your answer in kPa and atmospheres.

13. The average depth of the world's oceans is 5000 m, and the oceans cover 71 percent of the earth's surface. What is the approximate total force exerted by the oceans on the earth's surface? The earth is nearly spherical with an average diameter of approximately 12.7×10^6 m, and the specific weight of seawater is $\gamma = 10.1$ kN/m³.

14. A storage tank containing heavy fuel oil ($\rho = 906 \text{ kg/m}^3$) is filled to a depth of 7 m. Find the gauge pressure at the bottom of the tank.

15. A container holds three immiscible liquids as shown in Figure P7.15. Find the gauge pressure at the bottom of the container.

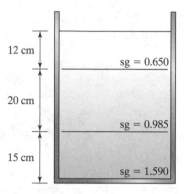

12 cm

sg = 0.650

20 cm

sg = 0.985

15 cm

sg = 1.590

Figure P7.15.

16. The side of a barge is submerged 6 m below the surface of the ocean ($\gamma = 10.1 \text{ kN/m}^3$). The length of the barge is 40 m. Treating the side of the barge as a plane vertical surface, what is the total force exerted by the ocean on the side of the barge?

17. To what depth would a container of glycerin (sg = 1.26) have to be filled to yield the same pressure at the bottom of a container with 4.5 in of mercury (sg = 13.55)?

18. Calculate the gauge pressure at the bottom of an open 2-liter container full of soft drink.

19. Calculate the force required to remove a 5-cm diameter plug from a hot-tub drain when the hot tub is filled with water to a depth of 60 cm. Neglect friction.

20. A cast iron pipe carries waste water at an average velocity of 5 m/s. If the inside diameter of the pipe is 10 cm, find the volume flow rate and the mass flow rate.

21. An open canal with the cross section shown in Figure P7.21 carries irrigation water at an average velocity of 2 m/s. Find the volume flow rate and mass flow rate.

22. An intravenous device for administering a sucrose solution to a hospital patient deposits a drop of solution into the mouth of a delivery tube every two seconds. The drops are spherical in shape with a diameter of 3.5 mm. If the inside diameter of the delivery tube is 2.0 mm, what is the mass flow rate and the average velocity of sucrose solution in the tube? If the plastic vessel containing the sucrose solution holds 500 mL, how long will it take to empty? The sucrose solution has a specific weight of $\gamma = 10.8 \text{ kN/m}^3$.

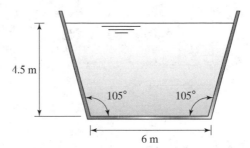

Figure P7.21.

23. A furnace requires 1500 lb_m/h of cold air for efficient combustion. If the air has a specific weight of 0.064 lb_f/ft^3, find the required volume flow rate.

24. A ventilation duct supplies fresh filtered air to a clean room where semiconductor devices are manufactured. The cross section of the filter medium is 1.2 m × 1.6 m. If the volume flow rate of air to the clean room is 3 m^3/s, find the average velocity of the air as it passes through the filter. If $\rho = 1.194$ kg/m^3 for the air, find the mass flow rate.

25. A converging rectangular duct carries nitrogen ($\rho = 1.155$ kg/m^3) at a mass flow rate of 4 kg/s. The small section of the duct measures 30 cm × 40 cm, and the large section measures 50 cm × 60 cm. Find the volume flow rate and the average velocities in each section.

26. A nozzle is a device that accelerates the flow of a fluid. A circular nozzle that converges from an inside diameter of 10 cm to 6 cm carries a gas at a volume flow rate of 0.25 m^3/s. Find the change in average velocity of the gas.

27. A diffuser is a device that decelerates the flow of air in order to recover a pressure loss. For the diffuser shown in Figure P7.27, find the mass flow rate and the average velocity of the air at the exit of the diffuser. For air, let $\rho = 1.194$ kg/m^3.

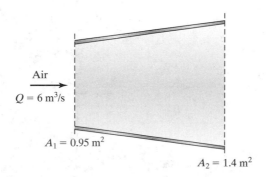

Figure P7.27.

28.   A pipe branch is illustrated in schematic form in Figure P7.28. Find the mass flow rate $\dot{m}_4$. Does the fluid in branch 4 enter or leave the junction?

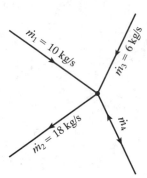

Figure P7.28.

29.   The average velocity of the water in a 0.75-in diameter pipe connected to a shower head is 12 ft/s. The shower drain, which is partially clogged with hair, allows 0.03 slug/s to flow into the drain. If the bottom of the shower measures 2.5 ft square and 6 inch deep, how much time is required for the shower to overflow?

30.   A duct carrying conditioned air from a refrigeration unit splits into two separate ducts that supply cool air to different parts of a building. The supply duct has a 1.8 m × 2.2 m inside cross section, and the two duct branches have inside cross sections of 0.9 m × 1.2 m and 0.65 m × 0.8 m. The average velocity of the air in the supply duct is 7 m/s, and the average velocity of the air in the smaller duct branch is 18 m/s. Find the volume flow rates, and mass flow rates, in each branch. For air, use $\rho = 1.20$ kg/m^3.

31.   The mixing chamber shown in Figure P7.31 facilitates the blending of three liquids. Find the volume flow rate and mass flow rate of the mixture at the exit of the chamber.

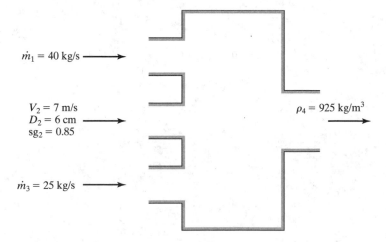

Figure P7.31.

8

Data Analysis: Graphing

8.1 INTRODUCTION

A friend of mine is an avid outdoors man and photographer who spends much of his leisure time hiking through the national parks in the western part of the United States. His long treks take him into pristine rugged backcountry where natures's scenery is spectacular. Along the way, he takes hundreds of photographs of mountains, rivers and streams, lakes and waterfalls, sunsets and sunrises, tress and flowers, and animals. While he shared some of these images with me during a recent slide presentation, I marveled at how effectively the visual information conveyed the beauty and grandeur of nature's handiwork. I almost felt as if I was actually there. I wondered, "Could his presentation impart the same sense of majesty without the photographs?" I concluded that simply telling me about the national parks without the accompanying images would be less than satisfying. Short of actually visiting them, how could I appreciate the national parks without seeing pictures of them? The old maxim "one picture is worth a thousand words" certainly holds true. A 1000-word essay cannot fully capture the essence of the place shown in Figure 8.1.

Describing technical data without graphs is somewhat like describing national parks without photographs. Only a certain amount of information can be effectively conveyed verbally or by using the written word; pictures must be used to communicate the whole message. A **graph** is a *special visual representation of the relationship between two or more physical quantities*. For example, Figure 8.2 is a graph of national SAT scores from 1967 through 2001. Two quantities—verbal score and mathematics score—are plotted as a function of the quantity time, measured in years. The graph readily shows a drop in both verbal and mathematics scores from 1967 through 1981, followed by a general rise in both scores, and a crossover in 1991, when the mathematics score

OBJECTIVES

After reading this chapter, you will have learned

- How to collect and record experimental data
- How to properly construct graphs
- How to fit data to common mathematical functions
- How to do interpolation and extrapolation

Figure 8.1. Wizard Island, Crater Lake National Park. (Courtesy of Art Lee, AllParks.com, © 2002.)

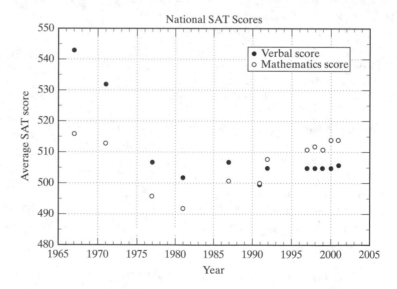

Figure 8.2. A graph of national SAT scores. (Courtesy of the U.S. Department of Education.)

rose above the verbal score. Mathematics scores increased from 500 in 1981 to 514 in 2001 (a 2.8 percent increase), which may be attributed to the emphasis placed on mathematics and science education in the United States during this time period.

Another example of a graph is shown in Figure 8.3. In this graph, the axial stress in a specimen of mild steel is plotted as a function of axial strain. Note that the data points nearly lie in a straight line, so they have been fitted with a "best-fit" straight line in order to demonstrate a linear relationship between the two quantities stress and

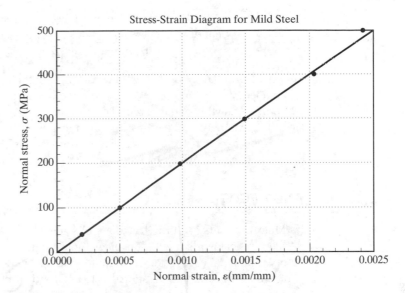

Figure 8.3. A graph representing the stress-strain diagram in the elastic region for mild steel.

strain. A linear relationship between stress and strain indicates that, for a limited range of stresses, the steel behaves elastically, which means that the specimen returns to its original length after the force that causes the axial stress has been removed. This kind of graph is useful for investigating certain structural properties of materials.

Engineers are designers, analysts, researchers, consultants, and managers. Regardless of the engineering role they assume, engineers are communicators, and graphs are effective ways to communicate technical information. Engineers work in a world of technical data, which generally consist of measurements of various physical quantities such as voltage, stress, temperature, velocity, flow rate, viscosity, frequency, and many others.

There are five main categories having to do with engineering measurements:

1. **Performance evaluation** involves making measurements to ascertain that a system is functioning properly. For example, a pressure sensor can indicate whether the pressure in a boiler is sufficient to deliver enough steam to a building's heating system.

2. **Process control** involves a feedback operation in which measurements are used to maintain processes within specified operating conditions. By continually monitoring indoor air temperature, for example, the thermostats in our homes signal heating and cooling equipments to cycle on and off, thereby maintaining comfortable living conditions.

3. **Accounting** involves keeping a record of the use or flow of a specific quantity, such as water flow from a reservoir.

4. **Research** involves investigating fundamental scientific phenomena. In engineering research, experiments are developed and measurements taken to support or confirm theoretical notions. For example, miniature flow sensors may be used to measure the flow of blood in arteries to enable biomedical engineers to develop flow models for the human heart.

5. **Design** involves testing new products and processes in order to verify their functionality. For example, if a materials engineer designs a new type of

insulation for controlling noise in a commercial aircraft, he or she would conduct some acoustics tests to ascertain that the new material functions properly for the intended application.

Testing is almost always the "last word" in the world of engineering design. Rarely do engineers design a product or process without testing it prior to manufacturing and marketing. Analytical and theoretical considerations alone are almost never sufficient to establish the viability of a new design. Carefully performed tests validate analyses and theories, but poorly conducted tests validate neither. In this chapter, we present the fundamentals of data analysis, which includes the gathering and graphing of data from measurements.

8.2   COLLECTING AND RECORDING DATA

Measurements form the backbone of science and engineering, because descriptions of the physical world are impossible without them. Imagine attempting to characterize the operation of a hard disk in a computer without measurements of data retrieval rate, voltage and current, and rotational velocity. Before we can construct a graph of data, we must take measurements of the quantities we wish to investigate. Engineering **measurement** is *the act of using instruments to determine the numerical value of a physical quantity*. For example, we use a scale (the instrument) to determine the weight (the quantity) of a person, which may be 160 lb$_f$ (the numerical value). A thermometer (the instrument) is used to determine the temperature (the quantity) of the air in a building, which may be 70°F (the numerical value). An ohmmeter (the instrument) is used to determine the electrical resistance (the quantity) of a resistor, which may be 10 kΩ (the numerical value). Many other examples could be cited.

An extensive discussion of engineering measurement is beyond the scope of this book, but a few fundamental concepts are worth covering. An engineer must be able to identify the kinds of data desired and how to associate various quantities. He or she must also understand that no measurement can be taken with ultimate accuracy or precision and that errors are an integral part of engineering measurement.

8.2.1   Data Identification and Association

To help us understand how to properly identify and associate data, let's use a simple and familiar example. Suppose that we wanted to measure the performance of a long-distance runner. First, we have to decide what kinds of data are required. To characterize the performance of the runner, we obviously want to know how fast he runs. We are not directly concerned with his body temperature, the electrical resistance of his limbs, the viscosity of his sweat, or his blood pressure. We want to know his speed, defined as distance divided by time. Thus, we have *identified* the data to be measured (distance and time) and *associated* these two quantities via the quantity (speed). Distance can be determined with a measuring tape or some other instrument, and time can be measured by using a stop watch or other suitable timing device. A graph of distance as a function of time, shown in Figure 8.4, reveals a *meaningful* relationship between these two quantities, and the quotient of distance and time gives the runner's average speed at various times throughout the race. Furthermore, the graph shows that the runner's speed decreases with time, indicating that he may lack the physical conditioning for a long-distance race or that he did not pace himself properly.

In the example just given, the data were properly identified and associated, which resulted in a meaningful graph. Now consider a situation in which the data are neither properly identified nor associated. In Figure 8.5, the electrical resistance of geometrically identical wire specimens made of different metal alloys is plotted as a function of melting point of the

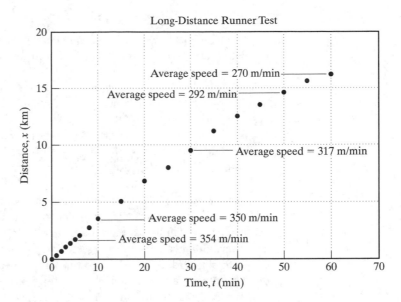

Figure 8.4. A graph of the quantities distance and time reveals a meaningful relationship for a long-distance runner.

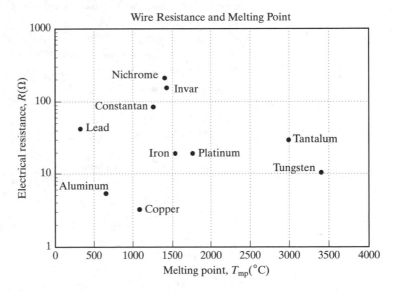

Figure 8.5. This graph shows that there is no relationship between electrical resistance and melting point of geometrically identical wires. Electrical resistance of all wires is based on a temperature of 20°C.

alloys. Because the data points are scattered randomly on the graph, there does not seem to be any meaningful relationship between electrical resistance and melting point of wires. We might as well have graphed electrical resistance as a function of the national deficit. Clearly, electrical resistance does not *depend* on melting point. Stated another way, melting point does not *affect* electrical resistance, so a graph of these two quantities is not useful. This does not mean, however, that a graph of seemingly unrelated data should never be constructed. In some engineering work, particularly research, we may not know in advance whether certain data are related or not. By graphing such data, physical relationships between the quantities may be manifested that otherwise would have gone unnoticed had a graph not been done.

8.2.2 Accuracy, Precision, and Error

At one time or another, virtually all engineers participate in taking measurements. The nature of engineering measurements encountered by engineers depend to a large extent on the type of product or process being developed or investigated. For example, a mechanical engineer who deals with thermal management of electronics would want to ascertain that the microprocessor in a computer is not going to fail thermally during operation. What does "fail thermally" mean? It means that the microprocessor will not function properly because its temperature falls outside the temperature limits specified by the manufacturer of that device. Thus, the engineer identifies temperature as the quantity to be measured. In order to measure the temperature of the microprocessor, the engineer must decide what type of instrument to use. Obviously, an instrument for measuring temperature must be used, but there are numerous types of thermometers available, such as liquid in glass, bimetal, thermocouple, and thermistor. The engineer must also decide where and how to attach the thermometer to the microprocessor, as well as how many thermometers to use at a given time. Is one thermometer sufficient, or are five needed to simultaneously measure the temperature at various locations on the microprocessor? In order to obtain meaningful measurements, the engineer must address these kinds of questions and many others. This is part of what makes taking measurements so challenging.

Issues common to all types of measurements are accuracy, precision, and error. **Accuracy** refers to *how close a measured value is to the true or correct value*. **Precision** refers to *the repeatability of a measurement* (i.e., how close successive measurements are to each other). **Error** is *the deviation of a measured value from the true or correct value*. From these definitions, it is clear that accuracy and error are closely related. Because error is the deviation of a measured value from the true or correct value, the magnitude of that deviation is indicative of the accuracy of the measurement. The difference between accuracy and precision is illustrated in Figure 8.6. Suppose that four shooters fire bullets

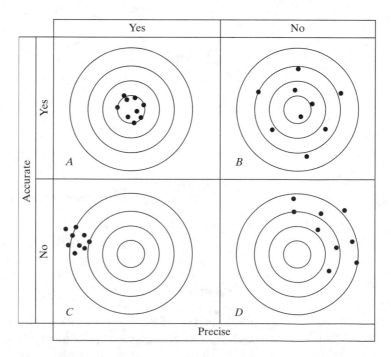

Figure 8.6. An illustration of accuracy and precision.

at different, but identical targets. Shooter *A* is accurate because all his bullets are close to the bull's-eye, and he is precise because his bullets are clustered closely together. Shooter *B* is accurate because her bullets are evenly distributed around the bull's-eye, but she is not precise because the bullets are scattered widely. Shooter *C* is precise because her bullets are clustered closely together, but she is not accurate because the cluster is far away from the bull's-eye. Shooter *D* is not accurate because the spread of bullets is not evenly distributed around the bull's-eye, and he is not precise because the bullets are not clustered closely together. Specific reasons for poor accuracy and precision may relate to the shooter's breathing, body position, vision, and other factors. Furthermore, the gun sights may need adjustment, the inside of the barrel may be dirty, or the wind may be blowing steadily or gusting randomly. The list of possible causes for the bullet spreads of shooters *B*, *C*, and *D* may be long, which leads us into a discussion of error.

No measurement is without error, so it is important that engineers recognize potential sources of errors and how to minimize them. Errors may be generally categorized as gross, systematic, and random. *Gross errors* are errors that virtually invalidate the measurement. These errors are caused by misuse of instruments, use of inappropriate or unsuitable instruments, incorrect recording of data, and failure to follow proper measurement procedures. For example, in Section 8.2.2, we discussed a mechanical engineer who wanted to ascertain that the micro processor in a computer would not fail thermally during operation. A gross error on the engineer's part would be his or her failure to wait for the microprocessor to achieve a steady thermal condition before recording the data. Electronic devices require time to heat up to their operating temperatures, so if measurements of temperature are recorded too early after the device is powered on, the data will be useless. Gross errors should be unequivocally eliminated from engineering practice.

Systematic errors are errors that exhibit regular or orderly behavior. Systematic errors may be caused by the measuring instrument, the environment, or the observer conducting the measurement. For example, consider viscosity measurements of a lubricating oil with the use of a simple falling-sphere viscometer instrument. Over the course of the measurements, we notice that the time for the spheres to fall a given distance in the oil gradually decreases. The tests started in the early morning and concluded around noon. We discover that our measurements of time reflect a systematic error caused by a gradual rise in the temperature of the sample of oil caused by poor heating equipment in the building and the morning sun shining through a nearby laboratory window. This particular systematic error, like all systematic errors, can be corrected. Here, the error could be corrected by simultaneously monitoring the temperature of the sample of oil so that the fall-time data reflect the temperature of the sample and, therefore, its viscosity. A different way to correct the error would be to regulate the temperature of the sample.

Another example of a systematic error is *parallax*. Parallax is an observational error that can occur when reading a dial gauge. To demonstrate how parallax works, hold the tip of your pencil about 1 cm above the second *a* in the word *parallax* in this sentence. Holding the pencil steady, move your head from side to side while observing the letter directly behind the pencil tip. If you move your head far enough to the right, the letter *r* lines up with the pencil tip, whereas if you move your head far enough to the left, the letter *l* lines up with the pencil tip. In this simple demonstration, the pencil tip represents a needle or dial, and the letters on the page represent a numerical scale. An observer who reads a dial gauge from one side or the other will introduce a systematic error in the data.

A type of systematic error caused by the instrument is called *hysteresis*. An instrument exhibits hysteresis when there is a difference in readings depending on whether the value of the measured quantity is approached from above or below. Hysteresis may be caused by mechanical friction, magnetic fields, elastic deformation, or thermal effects within the instrument.

Random errors are errors that are caused by chance-related phenomena. Let's consider the previously discussed scenario of the four shooters, where wind gust is a an example of a random error. Wind gusts occur at unpredictable times and have unpredictable speeds, which can randomly divert the path of a bullet from the bull's-eye. Yet another example of a random error is dirt or other foreign matter in the gun barrel. Some barrels may be cleaner than others, and the amount of contaminants in a gun barrel is basically a random variable.

All the different types of systematic and random errors are too numerous to cover in this book. Shown in Figure 8.7 are the primary sources of gross, systematic, and random errors, categorized in an organized fashion. Note that some causes of error, such as friction and vibration, may be systematic or random. For more in-depth coverage of error sources, the references at the end of this chapter should be consulted.

8.2.3   Recording Data

While measurements are being taken, it is important to record the data in a systematic and organized manner when preparing for graphing. For the data to be meaningful, thorough recording procedures must be followed. Standard engineering practice dictates the use of laboratory notebooks with data sheets, similar to the one shown in Figure 8.8, for recording data and documenting all aspects of the measurements. Recording data on loose papers is strongly discouraged, because loose pages can easily be lost, damaged, or improperly sorted with the other pages. The following information should be recorded in a *bound* notebook:

- Title of test
- Name of person(s) conducting the test
- Date, time of day, and location of test
- Page numbers
- List of instruments and equipment used
- Sketches of test setup
- Data

    - Neatly written
    - Arranged in a tabular format
    - Clearly identified and labeled with units
    - Recorded with the correct number of significant figures

- Short explanatory notes of data, as needed

The information contained in a laboratory notebook is regarded as the "raw data" for a test and constitutes the foundation of all subsequent graphs, analyses, and evaluations of the test. For this reason, the information recorded in the laboratory notebook is not to be tampered with in any way. Under no circumstances should the data be discarded, erased, or altered. To do so constitutes a breach of professional engineering ethics. If all or part of the data turns out to be bad for some reason, corrective steps should be taken to flush out problems that may exist with the instruments or the experimental procedure, and the test should be conducted again. Generating meaningful experimental data can be very time consuming, but the value of that data makes the effort worth it.

While laboratory notebooks are used for recording data manually, data may also be collected electronically and stored by using chart recorders and data loggers (which interface with the gauges or sensors that measure the desired physical quantities). Chart recorders utilize an automated mechanical marker that produces a graphical record of the measurements. Data loggers convert analog electrical signals from the sensors into a digital form that can be stored on a computer for later processing. Electronic systems used to collect and store data are sometimes referred to as data acquisition systems.

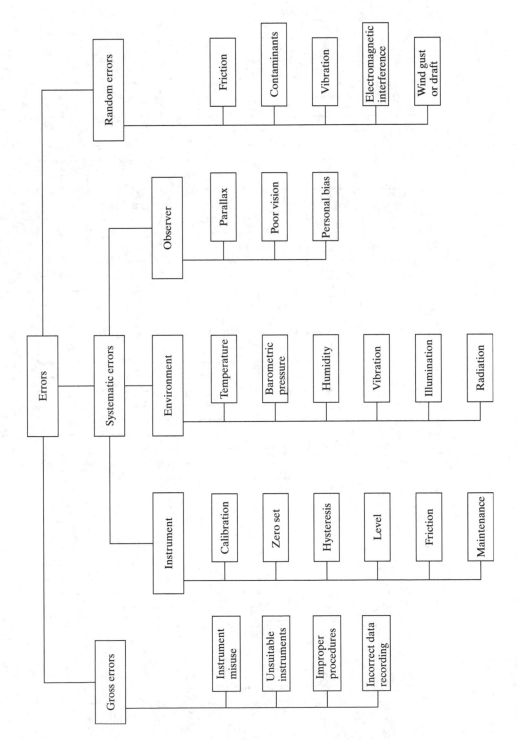

Figure 8.7. Categorization of gross, systematic, and random errors.

Engineering Laboratory, General Engineering Corporation

Title of test _____ Page _____
Test conducted by _____
Date _____ Time _____ Location _____
Equipment List _____
_____
_____
_____

	1	2	3	4	5	6	7	8	9	10	11	12	13	14	15

Figure 8.8. Laboratory notebook data sheet.

PRACTICE!

1. An environmental engineer wants to evaluate the supply of water for a town located at the base of a mountain range. The water supply for the town is based solely on runoff from the snowpack that accumulates in the nearby mountains during the winter. Describe the kinds of data that the engineer should collect and how graphs of the data could be used to assess the town's water supply.

2. In a plant that manufacturers plasterboard, an industrial engineer wants to optimize the production of plasterboard sheets by investigating the effects of

heating rates for curing the gypsum as the boards travel along a conveyor. Describe the kinds of data that the engineer should collect and how graphs of the data could be used to optimize the production of plasterboard.

3. In a drilling operation, a petroleum engineer wants to determine, based on oil production at various depths, when drilling at a particular site should be terminated. Describe the kinds of data that the engineer should collect and how graphs of the data could be used by the engineer to make the decision to terminate drilling.

4. Classify the following errors as gross (G), systematic (S), or random (R) (in some cases, more than one classification may apply):

Error	Classification (G, S, or R)
a. Not using a micrometer properly	_____
b. Dust and grease in a mass-balance linkage	_____
c. Reading a pressure gauge dial at a 60° angle	_____
d. Voltmeter incorrectly zeroed	_____
e. Ventilation drafts in a laboratory	_____

Answer: a. G b. R c. S d. S e. R

8.3   GENERAL GRAPHING PROCEDURE

In this section, a general procedure for graphing engineering data is presented. The procedure applies to all engineering disciplines and functions and, when applied consistently and correctly, leads to meaningful graphs that enable engineers to assess the performance of engineered systems, control processes, keep an accounting of physical quantities, conduct engineering research, and design products and processes. Practicing engineers in all disciplines have been using this general graphing procedure in one form or another for a long time with great success. It is vitally important that students learn proper graphing procedures and apply them in their course work. Establishing good graphing habits while still in school will make it that much easier for students to make a successful transition into engineering practice.

After data has been collected and recorded, graphs of these data can be constructed. As shown in Figure 8.9, there are numerous types of graphs that may be used to show relationships between various kinds of data. The most widely used graphs for engineering applications are the scatter graph and line graph. A *scatter graph* is a graph consisting of data points only, without lines drawn through them. (See Figures 8.4 and 8.5.) A *line graph* is a graph consisting of one or more lines without data points. Line graphs are typically used to show relationships between continuous quantities generated by mathematical equations. The other types of graphs shown in Figure 8.9 are used less frequently for engineering work. A *bar graph* is typically used to show distributions of quantities for purposes of statistical analysis. A *pie graph* is commonly used to show percentages or fractions of a whole in financial and business applications. A *polar graph* is used to show how quantities vary with angle. A *contour graph* shows how a quantity varies on a two-dimensional surface. A *3-D surface graph* shows how a quantity varies in three-dimensional space. However, since the scatter graph is predominantly used in engineering work, we will devote our entire attention to this type of graph.

The general procedure for constructing a graph of experimental data can be outlined in a step-by-step fashion. Each step in the procedure will be explained and illustrated in the sections that follow. The procedure applies to the manual construction of a graph, as well as to the construction of a graph with the use of a computer software

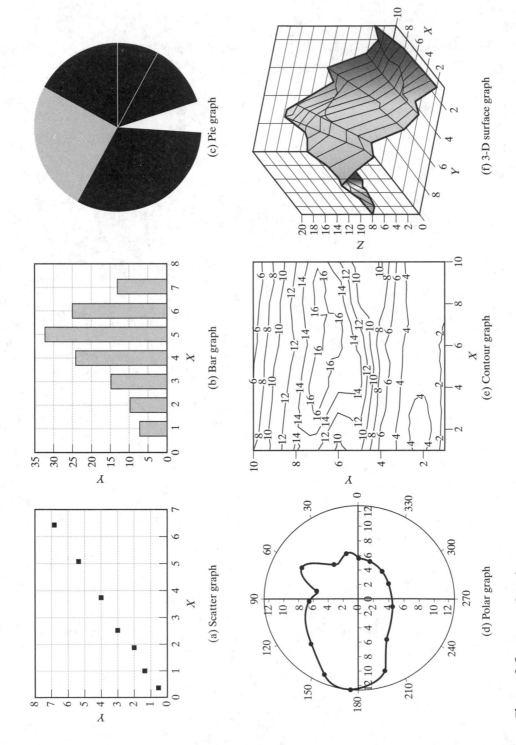

Figure 8.9. Types of graphs.

(a) Scatter graph

(b) Bar graph

(c) Pie graph

(d) Polar graph

(e) Contour graph

(f) 3-D surface graph

package. It is recommended that the student learn the graphing procedure by applying it manually before attempting to construct graphs via computer software. Once the techniques of graphing have been mastered with pencil and paper, the student will find that, after learning how to use the software, computer-aided graphing is straightforward.

General Graphing Procedure

The general graphing procedure is as follows:

1. Determine what data are to be graphed (i.e., the *dependent* variable and the *independent* variable). These data are obtained from the laboratory notebook.
2. Determine the *range* for the dependent and independent variables.
3. Select the *graph paper* based on the type of scale desired: linear, semi log, or full log. If semi log or full log is chosen, determine how many cycles (powers of 10) are required.
4. Based on the ranges of the variables, chose the *location* on the graph paper for the horizontal and vertical *axes*.
5. *Calibrate* and *graduate* the axes.
6. Label the axes.
7. *Plot* the data points by using appropriate symbols.
8. If a curve fit is desired, draw a *curve* or *curves* through the data points.
9. Identify multiple curves with a *legend*, and add a *title* to the graph.

In the sections that follow, each of these steps is discussed in detail.

8.3.1 Dependent and Independent Variables

Perhaps the most crucial step in graphing is the proper identification of the dependent and independent variables. These variables were identified and associated when they were measured in the laboratory, so they should be recorded in the laboratory notebook, or, if a data acquisition system is used, they should be stored electronically. A **dependent variable** is *a quantity that depends on another quantity*. Stated another way, a dependent variable is a quantity that changes in response to changes of another variable. The dependent variable depends on the independent variable, which is autonomous as far as the measurements are concerned. The **independent variable** is *the variable that can be controlled by the experimenter*. There is a cause–effect relationship between the dependent and independent variables. The independent variable (the cause) influences in some way the dependent variable (the effect). In mathematical terms, we say that the dependent variable is a *function* of the independent variable. For example, a biomedical engineer may want to investigate the factors that influence the length of a person's stride so that a prosthetic may be designed. Stride length depends on variables such as leg length, hip strength, and knee flexibility. Hence, stride length is the dependent variable, and leg length, hip strength, and knee flexibility are the independent variables (i.e., stride length is a function of these three variables).

A common mistake that beginning students make is to juxtapose the dependent and independent variables. This mistake is usually due to an incomplete understanding of the physical nature of the quantities involved or a lack of critical reasoning skills. Let's suppose that we wanted to study the relationship between altitude and air density by constructing a graph. Which of these quantities is the dependent variable and which is the independent variable? Does density depend on altitude or does altitude depend on density? We know that as we ascend a mountain, for example, the air gets "thinner," which means that the density of the air decreases as altitude increases. This would suggest that density depends on altitude, but can we say that altitude depends on density? Air density can be changed in a variety of ways, such as filling a flat tire with air at the gas station. The pressure and, therefore, the density of the air in the tire increases as we put more air into it. As the density of the air in the tire increases, does the altitude of the air in the tire change? Obviously not! Altitude can be controlled by the experimenter, but it does not depend on density. Thus, air density is the dependent variable and altitude is the independent variable. Students who struggle to properly identify dependent and independent variables are encouraged to carefully think about the nature of the physical quantities involved and apply some basic scientific and practical reasoning to the problem.

8.3.2 Variable Ranges

After the dependent and independent variables have been identified, the range for both variables should be determined. *Range* refers to the extent of numerical values over which the variable is to be graphed. For example, a civil engineer may wish to graph the flow rate of water in a natural irrigation system as a function of the time of year. She may have recorded flow rates of water from 2 to 30 m^3/s in a notebook, but she is only interested in graphing flow rates from 5 to 20 m^3/s. Thus, the range of flow rates is 5 to 20 m^3/s. In order to properly graph the data, there must be some flow-rate data between the low value of 5 m^3/s and the high value of 20 m^3/s. Of course, for each flow rate over this range there is a corresponding value of time.

8.3.3 Graph Paper

Graph paper is commercially available in most college bookstores and office supply stores and may be downloaded and printed from various Internet websites. **Graph paper** has preprinted horizontal and vertical grid lines with special spacing. The type of graph paper to be used for a particular graph depends on the nature of the data being graphed and the relationship between the dependent and independent variables. In general, graph paper comes in three types, each type being distinguished by the grid spacing or *scale*: linear, semilog, and full log. The most common size for graph paper is the standard letter size ($8\frac{1}{2} \times 11$ inch), but other sizes, such as $8\frac{1}{2} \times 14$ inch and 11×17 inch, may also be available.

On *linear* graph paper, the horizontal and vertical grid lines are equally spaced, as shown in Figure 8.10(a). The spacing is typically 5, 10, or 20 divisions per inch, but other spacings are available. The grid spacing in Figure 8.10(a) is 10 divisions per inch.

On *semilog* graph paper, the grid lines in one direction are equally spaced, but the grid lines in the other direction follow a logarithmic relationship. Typically, as shown in Figure 8.10(b), the grid spacing in the vertical direction is logarithmic and the grid spacing in the horizontal direction is linear. The vertical scale shown in Figure 8.10(b) has one *cycle*, which means that the maximum data range is one power of 10. That range can be 1 through 10, 10 through 100, 100 through 1000, 0.01 through 0.1, 0.1 through 1, or any other similar range, as long as it spans one power of 10. The grid on a semilog graph paper with two cycles spans a maximum range of two powers of 10, such as 0.1 through 10, 10 through 1000, or 10^5 through 10^7.

On *full-log* paper, the grid lines in both directions follow a logarithmic relationship, as illustrated in Figures 8.10(c) and 8.10(d). The graph paper shown in Figure 8.10(c) is designated as 1×1 cycles, because the maximum data range in both directions is one power of 10. The maximum data ranges do not have to be identical, however. For example, the ranges in one direction could be 10 through 100, while the range in the other direction could be 0.1 through 1. The graph paper in Figure 8.10(d) is designated as 2×2 cycles, because the maximum data range in both directions is two powers of 10. Once, again the maximum data ranges do not have to be identical.

8.3.4 Location of Axes

The *axes* of a graph consist of two intersecting straight lines, usually at or near the lower left corner of the graph paper. The horizontal axis is the *abscissa* (the x-axis), and the

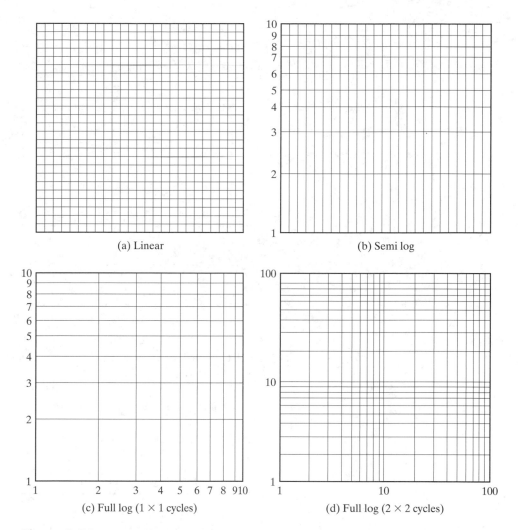

Figure 8.10. Types of graph paper.

vertical axis is the *ordinate* (the *y*-axis). The point of intersection of the two axes is the *origin* of the graph. Note that

> *It is standard graphing practice to associate the independent variable with the abscissa and the dependent variable with the ordinate.*

This standard is illustrated in Figure 8.11. Although most graphs of engineering data follow the standard, there can be some exceptions.

In most engineering applications, the dependent and independent variables are limited to positive values, so the origin of the graph is located near the lower left corner of the graph paper. If the variables contain both positive and negative values, the axes (and therefore the origin) of the graph must be shifted in order to accommodate the negative values, resulting in a graph consisting of four quadrants, as shown in Figure 8.12.

Regardless of where the axes of a graph are located, care should be taken to use as much of the graph paper as possible to make the graph more readable. This is made possible by properly marking out the calibrations, which is explained in the next section.

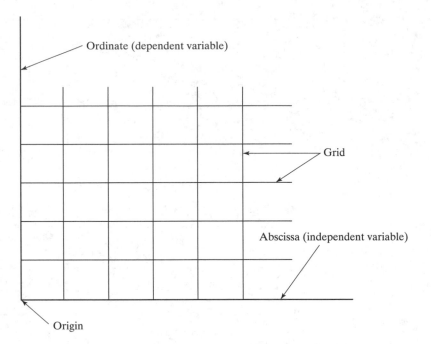

Figure 8.11. Parts of a graph, including the abscissa (x-axis) and the ordinate (y-axis).

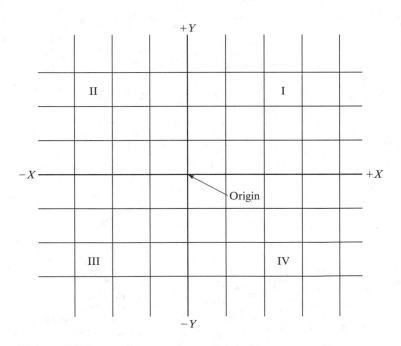

Figure 8.12. Axis locations for positive and negative variables.

8.3.5 Graduation and Calibration of Axes

Prior to plotting any data points, the axes must be graduated and calibrated. *Graduations* are a series of marks on the axis that define the type of scale used. As discussed in Section 8.3.3, the three common types of scales on graph paper are linear, semilog, and full log. On a linear scale, the marks are equally spaced, whereas on a logarithmic scale, the marks are not equally spaced, but follow a logarithmic function. *Calibrations* are the numerical values assigned to the graduations. After defining the locations for the axes, the next step is to calibrate both axes based on the variable ranges defined in step 2 of the graphing procedure. The axes should be calibrated by using as much of the graph paper as possible. As an example, suppose our variable ranges are

> independent variable: 0 to 300
> dependent variable: 0 to 40

Assuming we are using graph paper with a linear scale, it is preferable to calibrate the axes as illustrated in Figure 8.13(a), which uses most of the available space on the graph paper. It is not advisable to calibrate the axes in the manner shown in Figure 8.13(b), because the actual graph area is too small, making it difficult to read the graph's detailed features. If the graph is constructed with the use of computer software, the software will probably do the calibration automatically, based on the variable ranges supplied by the user. Alternatively, the calibration can be done manually. Whether a graph is constructed manually or with the use of computer software, the graph should fill as much of the page as possible to enhance readability. As shown in Figure 8.13, if the graph paper does not have much space between the borders of the graph and the edge of the paper, the axes should be drawn slightly inside the borders of the graph, thus providing space for the axis calibrations and axis labels, which are discussed in the next section.

Graduations, sometimes called *ticks* or *tick marks*, are referred to as either *major* or *minor*. As shown in Figure 8.14, major graduations are usually calibrated and are drawn slightly longer than minor graduations. Because graph paper typically consists of horizontal and vertical grid lines that extend the entire width and height of the graph, major graduations should be preselected to coincide with the major divisions of the graph paper. Major graduations can be emphasized by drawing tick marks. [See Figure 8.13(a)]. As illustrated in Figure 8.15, the tick marks may be drawn inward (away from the calibrations), outward (toward the calibrations), or both.

Minor graduations are located between major graduations, and should follow the **1, 2, 5 rule**, as illustrated in Figure 8.16. This rule states that *the smallest division for the minor graduations is 1, 2, or 5*. The 1, 2, 5 rule readily enables interpolation of data by subdividing the interval between major graduations, using commonly used integers. Graduations that do not follow the 1, 2, 5 rule are undesirable, producing noninteger subdivisions that render point plotting awkward. Exceptions to the 1, 2, 5 rule are graduations that represent units of time, such as years, months, or days.

Although the number of graduations to use should follow the 1, 2, 5 rule, the number of calibrations to use is discretionary. The most common mistake is the overuse of calibrations. Too many calibrations makes the axis look crowded, so only the minimum number of calibrations required to read the graph should be used. The axis shown in Figure 8.17(a) is easy to read, but the axis shown in Figure 8.17(b), even though it follows the 1, 2, 5 rule, has too many calibrations.

8.3.6 Axis Labels

Graduations and calibrations are meaningless unless the axes are labeled. An *axis label* is a designated name of the variable and its corresponding unit. Referring back to Figure 8.3, the designated name for the independent variable is "Normal strain, ϵ", and the corresponding unit (enclosed in parentheses) is "mm/mm". The designated name for the dependent variable is "Normal stress, σ" and the corresponding unit is "MPa," enclosed in parentheses. Note that the designated name of both variables in Figure 8.3 consists of a word name plus an algebraic symbol, separated by a comma. An alternative label might be the word name without the algebraic symbol, so the label for the independent variable would be "Normal strain (mm/mm)"; the label for the dependent variable would be "Normal stress (MPa)." Unsuitable labels for the independent and dependent variables,

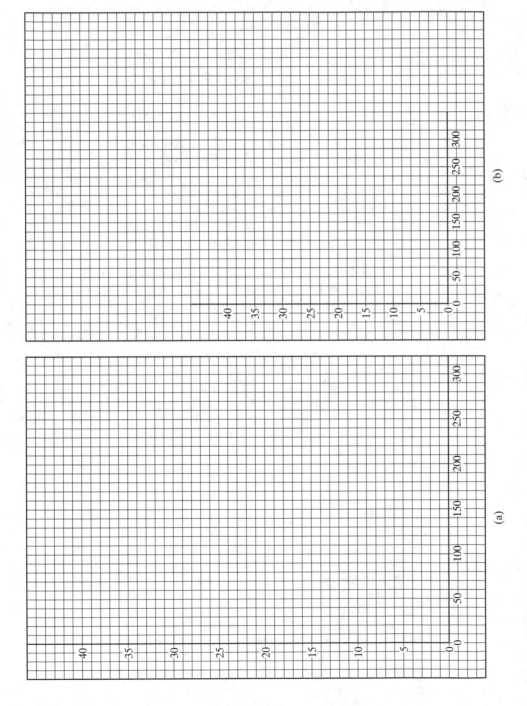

(a)

(b)

Figure 8.13. Calibration of axes done (a) correctly and (b) incorrectly.

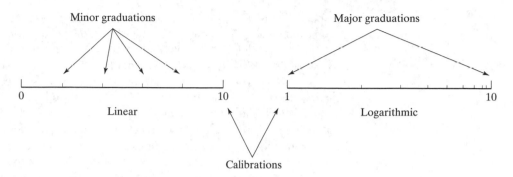

Figure 8.14. Major and minor graduations.

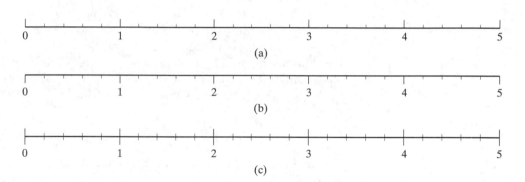

Figure 8.15. Graduations may be drawn (a) inward, (b) outward, or (c) both.

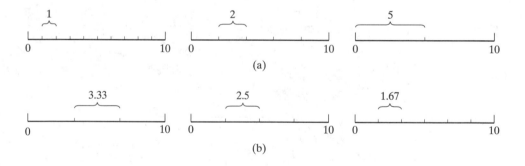

Figure 8.16. (a) Proper calibrations follow the 1, 2, 5 rule. (b) Poor calibrations do not.

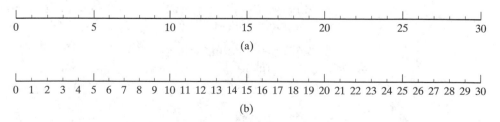

Figure 8.17. (a) Calibrations that are easy to read. (b) Calibrations that are too crowded.

respectively, are "ϵ (mm/mm)" and "σ (MPa)" because the reader may not know the physical meanings of "ϵ" and "σ" without referencing another document or the supporting text for the graph. The axis labels must be specific enough for the graph to stand on its own merits without forcing the reader to reference other sources.

Axes should never be labeled simply as x and y just because you did it that way in math courses or because the abscissa is the "x-axis" and the ordinate is the "y-axis." These generic labels fail to tell the reader the true identity of the independent and dependent variables. Tell the reader of the graph precisely what the variables are by using specific labels.

Engineering data sometimes consist of very small or very large numbers. If a quantity is expressed in SI units, prefixes should be used to simplify the calibrations, as illustrated in Figure 8.18. Note carefully the correct reading of the graduations shown in Figure 8.18. The energy unit used in the first example is MJ (megajoule), which means 10^6 J. The linear scale is calibrated in tenths of MJ, so the interval between two adjacent major graduations is 0.1×10^6 J $= 10^5$ J. The pressure unit used in the second example is kPa (kilopascal), which means 10^3 Pa. The linear scale is calibrated in whole numbers of kPa, so the interval between two adjacent major graduations is 1×10^3 Pa $= 1000$ Pa. The current unit used in the third example is mA (milliampere), which means 10^{-3} A. The logarithmic scale is calibrated in whole numbers of mA, so the interval between two adjacent major graduations is 1×10^{-3} A $= 0.001$ A.

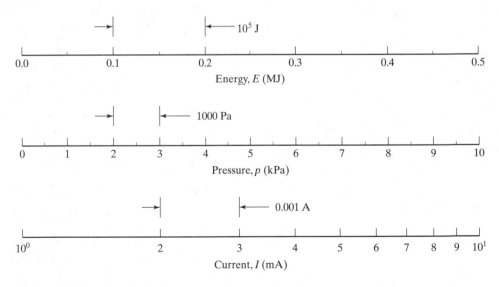

Figure 8.18.   Examples of axis labels, using prefixes of SI units.

If English units are used in the axis label, it is customary to use powers of 10 instead of prefixes, since most English units do not have prefixes. One exception is the unit, ksi, defined as 10^3 lb$_f$/in^2.

8.3.7   Data Point Plotting

Recall that the laboratory notebook contains a tabular record of data that represents independent and dependent variables. *Plotting* a data point means to place on the graph a mark that represents a data pair of corresponding independent and dependent variables. Consider the graph shown in Figure 8.19, which was constructed from the data in Table 8.1. The independent variable is current and the dependent variable is voltage. The data come from current and voltage measurements for a resistor in a power circuit. There are six data pairs in the table, so there are six data points on the graph, one for each data pair.

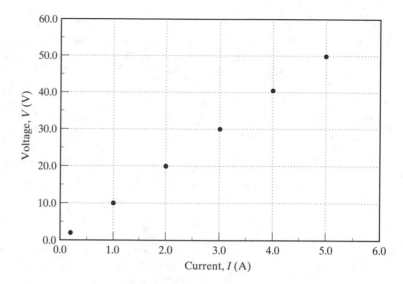

Figure 8.19. Symbols represent data pairs of the independent and dependent variables. (Refer to Table 8.1.)

TABLE 8.1 Current and Voltage Data for Constructing the Graph Shown in Figure 8.19.

Current, *I* (A)	Voltage, *V* (V)
0.2	2.1
1.0	10.1
2.0	19.8
3.0	30.3
4.0	40.5
5.0	49.5

The marks representing experimental data points on the graph are made with *symbols*. The most commonly used symbols are the circle, square, triangle, and diamond, and, as shown in Figure 8.20, they may be filled or unfilled. The symbols shown in Figure 8.19 are filled circles. When plotting data points with these symbols, there are some basic guidelines to follow. First, the centers of the symbols should coincide with the numerical values of the data points. Second, the symbols should be large enough to be easily identified, but they should not be so large that they run into each other on the graph. Third, if two or more sets of data are plotted on the same graph, as illustrated in Figure 8.2, different symbols must be used for each data set in order to distinguish them. By using the symbols shown in Figure 8.20, up to eight different data sets may be plotted on the same graph, which is sufficient for the majority of graphs. If more than eight data sets are plotted, additional unique symbols are required. If the graph is constructed manually, drawing templates may be used to create the symbols. If the graph is constructed via computer software, the software should have a ready-to-use list of symbols, including the ones shown in Figure 8.20.

Figure 8.20. Common data-point symbols.

8.3.8 Curves

A *curve* is a line drawn through data points on a graph. The manner in which the curve is drawn depends on the type of data shown. Data points on a graph are generally categorized as *observed*, *empirical*, or *theoretical*. Observed data are presented on a scatter graph without attempting to correlate the data or fit the data to a mathematical function. Graphs of observed data are shown in Figures 8.2 and 8.4. Empirical data are presented with a smooth curve drawn through the symbols for the purpose of showing a physical correlation or interpretation. The curve may be straight or not, and the symbols may not necessarily lie on the curve. A graph of empirical data is shown in Figure 8.3. Theoretical data are generated by mathematical functions and are represented by continuous smooth curves without symbols. Graphs of theoretical data do not show symbols because the curve is not derived from measurements of discrete physical quantities. Every "point" on the curve is a calculated, not a measured, value. A graph of theoretical data is shown in Figure 8.21.

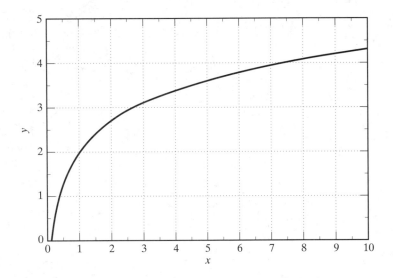

Figure 8.21. A graph of the equation $y = 2 + \ln(x)$.

As with data point symbols, there are some standard line types that are commonly used for drawing curves, as illustrated in Figure 8.22. These line types can be drawn by using manual drafting tools. (Computer software typically allows the user to select from a ready-to-use list of line types, including those shown here.) It is recommended that lines not be drawn through unfilled symbols because the symbols may be mistakenly interpreted as filled symbols.

8.3.9 Legends and Titles

The last step in the general graphing procedure is to place a legend and a title on the graph. A *legend* is a key that differentiates two or more data sets on the same graph by labeling the symbols or line types, or both. The graph shown in Figure 8.2 has a legend that differentiates the verbal SAT score (filled circles) from the mathematics SAT score (unfilled circles). The best location for the legend is inside the boundaries of the graph, but if space is limited, the legend may be placed just outside the boundaries of the graph. However, in order to help the reader to locate the legend, it should be enclosed in a border.

A *title* is a caption that concisely describes the graph. The titles for the graphs shown in Figures 8.2, 8.3, and 8.4 are "National SAT Scores," "Stress-Strain Diagram for

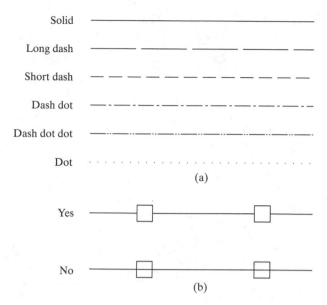

Figure 8.22.   (a) Common line types. (b) Proper use of lines with symbols.

Mild Steel," and "Long-Distance Runner Test," respectively. In printed matter, such as books and journals, the title of a graph is typically the figure caption and is usually located below the graph. For a graph prepared manually, the title may be placed below, above, or even inside the graph to suit one's personal preference. For a graph prepared with the use of computer software, the location of the title may be fixed by the software, or the user may have control over where to place the title.

8.3.10 Graphing with Computer Software

To the engineer, the computer is an indispensable tool. Engineers use computers to write technical reports, prepare drawings, conduct analyses, collect data, and, of course, construct graphs. The primary advantages of using a computer for graphing are speed and appearance. Assuming the engineer is familiar with the mechanics of the graphing software, he or she can probably construct a graph faster by using a computer software package than by using a pencil and graph paper. Moreover, a graph prepared with the use of a computer will have a more professional appearance than a graph that was prepared manually. Moreover, a graph constructed via computer software can also be electronically imported into technical documents (e.g., memos and reports).

Several software packages for graphing are commercially available. Perhaps the most commonly used software for graphing is the *spreadsheet*. Spreadsheets are widely available and relatively inexpensive software packages that were initially developed for business and accounting applications, but have also been extensively used for scientific and engineering work. Spreadsheets consist of an array of rows and columns, making them ideally suited for tables of data from which graphs can be made. When a spreadsheet for graphing is used, the procedure to follow is fundamentally the same as that outlined in the previous sections. However, spreadsheets do not have ultimate graphing capability and flexibility. For example, spreadsheets may restrict the type and number of symbols and curves or may not allow the user to define the length of the major and minor tick marks. Because these restrictions and limitations are usually minor, spreadsheets are still a popular software tool for graphing in most engineering applications. The graph shown in Figure 8.23(a) was built by using Microsoft® Excel.

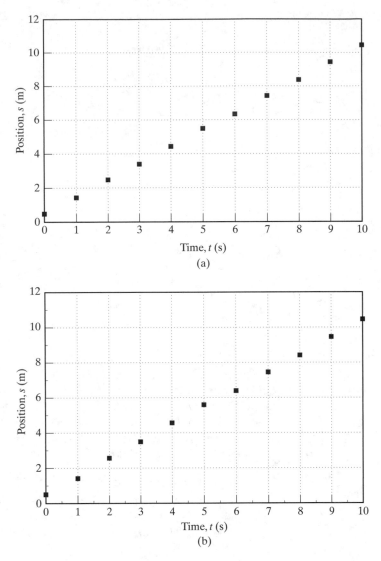

Figure 8.23. Graph of position as a function of time by using (a) EXCEL and (b) SigmaPlot. EXCEL is a registered trademark of Microsoft Corp., and SigmaPlot is a registered trademark of SPSS, Inc.

Furthermore, if the engineer desires more powerful graphing features and capabilities, he or she may utilize a software that is designed specifically for graphing. These high-end graphing packages are more sophisticated than spreadsheets, allowing the engineer to investigate data by using more advanced types of graphs and to fully manipulate all the features of the graph. Furthermore, specialized graphing packages typically include advanced mathematical and statistical routines that spreadsheets may not have. Figure 8.23(b) shows a graph that was prepared by using Sigma Plot, which is a specialized graphing package. However, please note that for the simple application illustrated in Figure 8.23, both software packages are capable of producing nearly identical graphs.

APPLICATION: GRAPHING WIND DATA TO SELECT A SITE FOR A WIND TURBINE

In order to determine a suitable site for a wind turbine near the mouth of a canyon in the Rocky Mountains, an engineer takes some wind-speed measurements. At this particular mountain location, the wind blows in a westerly direction through the canyon toward a wide valley. To measure wind speed, the engineer uses the cup anemometers, (typically used by meteorologists). He places one instrument at the mouth of the canyon and a second instrument one mile directly downstream from the mouth of the canyon where it opens into the valley. Both anemometers are mounted on 30-ft-high towers. Data-acquisition equipment is set to take the measurements at 10-second intervals and to calculate and record hourly wind-speed averages. The data shown in Table 8.2 are the recorded average hourly wind speeds for a 24-hour period on January 30, 2003. The time of day is indicated in military time to facilitate easy graphing, and the wind speed is measured in miles per hour.

TABLE 8.2 Average Hourly Wind Speed for Mountain Canyon.

	Wind speed (mi/h)	
Time (h)	Mouth	1 mi downstream from mouth
0100	7.8	6.7
0200	7.5	6.4
0300	8.5	8.0
0400	8.4	6.5
0500	11.1	6.9
0600	14.5	11.5
0700	22.6	15.6
0800	34.9	20.0
0900	30.0	18.5
1000	29.5	18.0
1100	21.3	14.5
1200	20.5	13.4
1300	18.4	13.0
1400	15.6	8.9
1500	14.0	8.5
1600	13.9	8.8
1700	13.0	7.4
1800	11.8	6.3
1900	12.4	6.8
2000	13.4	5.4
2100	5.6	4.7
2200	6.7	3.6
2300	4.2	3.2
2400	5.9	3.8

A scatter graph of the data in Table 8.2 is shown in Figure 8.24. Both data sets are plotted on the same graph to compare wind speeds at both locations. The graph clearly shows that wind speed achieves a maximum value at 8:00 A.M. at both locations, and, that, for a given time of day, the average wind speed is always higher at the mouth of the canyon than at the location 1 mile directly downstream from the mouth of the canyon. This phenomenon is consistent with a meteorological version of the conservation of mass principle in fluid mechanics. One would expect a higher wind speed at the mouth of the canyon than 1 mile downstream, because as the air exits the canyon into the valley, the air is allowed to spread over a cross-sectional area that is

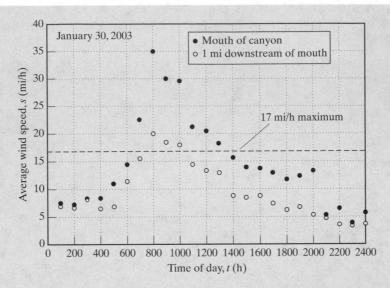

Figure 8.24. A graph of wind speed that could be used to determine a suitable site for a wind turbine near the mouth of a mountain canyon.

much larger than the canyon mouth itself, thereby undergoing a reduction in velocity.

For this particular wind turbine application, the maximum wind speed must not exceed 17 mi/h for an appreciable length of time to minimize the potential for rotor damage. At the mouth of the canyon, the average wind speed exceeds 17 mi/h from about 6:00 A.M. to about 2:00 P.M., a time interval of nearly 8 hours, whereas at 1 mile downstream from the mouth of the canyon, the average wind speed exceeds 17 mi/h for about 3 hours. Using the graph as a decision-making tool, we conclude that the wind turbine should not be located at the mouth of the canyon, but at a more suitable location downstream, where the wind speeds are more moderate. Of course, the graph shown in Figure 8.24 applies only for one winter day, but a similar graphing process could be used to ascertain a suitable location for the wind turbine by accounting for wind-speed measurements during an entire month or year.

PRACTICE!

In order to characterize the fuel economy of three different automobiles, consider the following set of data:

	Fuel Economy (mi/gal)		
Speed (mi/h)	Vehicle A	Vehicle B	Vehicle C
5.0	29.4	26.7	24.6
10.0	30.3	27.5	26.5
20.0	31.0	28.3	27.0
30.0	32.4	30.7	28.9
40.0	34.5	31.2	28.1
50.0	33.8	31.9	27.4
60.0	32.6	29.5	26.2
70.0	28.1	26.8	25.0

Using the general graphing procedure outlined in this section, construct a suitable graph of the data. Based on the graph, what conclusions can be drawn about the fuel economy of each vehicle?

8.4   CURVE FITTING

Frequently, the data on a graph indicate a specific physical interpretation. For example, the graph shown in Figure 8.19 suggests a linear relationship between the electrical current in a resistor and the voltage across it. This relationship is experimental evidence for a physical law known as Ohm's law, which states that the voltage V is directly proportional to the current I:

$$V \propto I \qquad (8.1)$$

By introducing a constant of proportionality R, Equation (8.1) may be written as an equality

$$V = RI \qquad (8.2)$$

where R is resistance. A simple rearrangement of Equation (8.2) shows that resistance R is voltage divided by current. If a best-fit straight line is drawn through the data points, as illustrated in Figure 8.25, the value of R may be obtained. The slope (rise over run) of

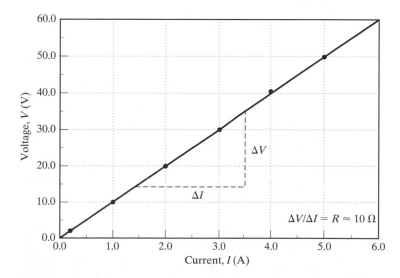

Figure 8.25.   Resistance is the slope of a best-fit straight line drawn through current–voltage data.

the line is the resistance R. A quick inspection of the graph indicates that the resistance is approximately $10\ \Omega$.

The line drawn through the data points in Figure 8.25 is an example of curve fitting. **Curve fitting** means *to draw a smooth line through the data points for the purpose of approximating a mathematical relationship or function.* In the example just cited, the mathematical relationship is Ohm's law, and the line is straight. In general, a line can be straight or curved, and the line does not have to pass directly through all the data points. In fact, the line rarely passes through all the data points, because errors produce some *scatter* about the expected trend, yielding roughly equal numbers of data points on both sides of the line.

Before covering specific curve-fitting methods, a brief discussion of some common mathematical functions is in order.

8.4.1 Common Mathematical Functions

The physical world exhibits remarkable order, which enables scientists and engineers to utilize mathematics as a modeling tool. Many physical systems and processes follow simple mathematical relationships. We just saw that voltage and current in a resistor have a linear relationship: $V = RI$. From fundamental physics, we know that the distance an object falls (neglecting air friction) varies quadratically with time according to the relation $y = \frac{1}{2}gt^2$, where y is distance, g is gravitational acceleration, and t is time. Radioactive decay follows the exponential relationship $N = N_0\,e^{-\lambda t}$, where N is the number of nuclei present at time t, N_0 is the number of nuclei present at $t = 0$, and λ is the decay constant.

The foregoing three examples typify the kinds of physical phenomena that can be described by using common mathematical functions. These functions are denoted as linear, power, and exponential. A **linear function** is expressed in the familiar form $y = mx + b$, where m is the slope of the line and b is the y-intercept. Ohm's law is a specific version of the linear function with $b = 0$. A **power function** has the form $y = bx^m$, where b and m are constants. An **exponential function** has the form $y = be^{mx}$, where again, b and m are constants.

A linear function $y = mx + b$ when plotted on a linear graph will appear as a straight line. However, the power and exponential functions do not represent linear relationships between the x and y variables, so something must be done in order to curve-fit data that are described by these functions. The power and exponential functions can be transformed into linear functions by doing a little algebra. For the power function, the steps are as follows:

power function
$$y = bx^m$$
$$\log(y) = \log(bx^m)$$
$$\log(y) = \log(x^m) + \log(b)$$
$$\log(y) = m\log(x) + \log(b) \tag{8.3}$$

Compare Equation (8.3) with the standard linear function $y = mx + b$. Equation (8.3) is a linear function in $\log(y)$ and $\log(x)$, where m is the slope of the line, and $\log(b)$ is the y-intercept. Note that the natural logarithm ln could also be used. *Data that follow a power function appear as a straight line when y is plotted as a function of x on a full-log graph.* In addition, data that follow a power function appear as a straight line when $\log(y)$ is plotted as a function of $\log(x)$ on a linear graph. The linearization of the exponential function is similar to that of the power function, but it is more expedient to use the natural logarithm:

exponential function
$$y = be^{mx}$$
$$\ln(y) = \ln(be^{mx})$$

$$\ln(y) = \ln(e^{mx}) + \ln(b)$$
$$\ln(y) = mx + \ln(b) \qquad (8.4)$$

Equation (8.4) is a linear function in $\ln(y)$ and x, where m is the slope of the line, and $\ln(b)$ is the y-intercept. *Data that follow an exponential function appear as a straight line when y is plotted as a function of x on a semilog graph (where the x-axis scale is linear and the y-axis scale is logarithmic).* Moreover, data that follow an exponential function appear as a straight line when $\ln(y)$ is plotted as a function of x on a linear graph.

In summary, curve fitting for our purposes means to draw a smooth line through data points on a graph for the intent of approximating the data with a linear, power, or exponential function. If the function is a power or exponential function, the data must be linearized before implementing the two curve-fitting methods discussed here: (1) the method of selected points and (2) least squares linear regression.

8.4.2   Method of Selected Points

The **method of selected points** is based on a *visual* best fit of a straight line to the data on a graph. In the following procedure, y refers to the dependent variable, and x refers to the independent variable.

Procedure: Method of Selected Points

A procedure for the method of selected points is as follows:

1. Plot y as a function of x on a graph with a *linear* scale. If the data points suggest a straight line, we have a linear function. Proceed to step 4.
2. Plot y as a function of x on a graph with a *full-log* scale. If the data points suggest a straight line, we have a power function. Proceed to step 4.
3. Plot y as a function of x on a graph with a *semi-log* scale. If the data points suggest a straight line, we have an exponential function. Proceed to step 4.
4. Using a transparent straightedge, draw a line through the data points such that the line is as close as possible to all the data points, with roughly the same number of data points on each side of the line. (A transparent straightedge makes this task a little easier because all the symbols can be seen at once.)
5. Select two points *on the line* that are widely separated. (These two points should not be data points.) Record the values of these two points on a separate paper, and label them points A and B.
6. If the function is *linear*, substitute the values of points A and B into the two equations

$$A_y = mA_x + b$$
$$B_y = mB_x + b$$

Solve these equations simultaneously for the slope m and the y-intercept b. A linear function $y = mx + b$ that fits the data has now been determined.
7. If the function is a *power* function, substitute the values of points A and B into the two equations

$$\log(A_y) = m \log(A_x) + \log(b)$$
$$\log(B_y) = m \log(B_x) + \log(b)$$

Solve these equations simultaneously for the slope m and the y-intercept b. A power function $y = bx^m$ that fits the data has now been determined.

8. If the function is an *exponential* function, substitute the values of points A and B into the two equations

$$\ln(A_y) = mA_x + \ln(b)$$
$$\ln(B_y) = mB_x + \ln(b)$$

Solve these equations simultaneously for the slope m and the y-intercept b. An exponential function $y = be^{mx}$ that fits the data has now been determined.

The method of selected points is illustrated for the linear, power, and exponential functions in the examples that follow.

EXAMPLE 8.1

The position of a linear actuator in a machine is measured at specified times, as shown in the table that follows. Using the method of selected points, determine a mathematical function that fits the data.

Time, t (s)	Position, s (cm)
0.0	0.40
1.0	2.49
2.0	4.37
3.0	5.66
4.0	7.92
5.0	8.47
6.0	11.8
7.0	12.4

SOLUTION

After plotting the data on a graph with a linear scale, we see that, as shown in Figure 8.26, the data suggest a linear function. Using a transparent straightedge, we draw a best-fit straight line through the data. Note that there are equal numbers of data points on

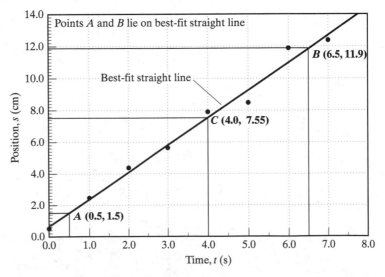

Figure 8.26. Method of selected points for Example 8.1.

each side of the line. We then select two points on the line that are widely separated. For the two points, we arbitrarily choose 0.5 and 6.5 as the x-coordinates (time coordinates), which yields 1.5 and 11.9, respectively, for the y-coordinates (position coordinates). Thus,

$$A_x = 0.5 \qquad A_y = 1.5$$
$$B_x = 6.5 \qquad B_y = 11.9$$

Points A and B and the other illustrative elements in Figure 8.26 are shown on the graph for instructional purposes only and should not be shown on the actual graph. Following step 6, we set up two simultaneous equations,

$$1.5 = m(0.5) + b$$
$$11.9 = m(6.5) + b$$

Solving for the slope m and the y-intercept b, we obtain

$$m = 1.73 \text{ cm/s} \qquad b = 0.63 \text{ cm}$$

Hence, the equation for the position of the linear actuator as a function of time is

$$s = 1.73\, t + 0.63 \quad \text{(cm)}$$

Now that we have an equation that fits the data, we can determine the position of the linear actuator for other time values. As a check of our solution, we substitute $t = 4.0$ s into the equation,

$$s = (1.73 \text{ cm/s})(4.0 \text{ s}) + 0.63 \text{ cm}$$
$$= 7.55 \text{ cm}$$

which agrees with the coordinates of point C on the graph.

EXAMPLE 8.2

As the accompanying table shows, the power dissipated by a large transformer with a resistance of 5 Ω is measured for several values of current that pass through the transformer's windings. Using the method of selected points, determine a mathematical function that fits the data.

Current, I (A)	Power, P (W)
1.05	5.63
1.25	7.58
1.75	16.9
2.50	32.1
3.0	48.0
4.0	78.2
5.0	126.0
6.0	188.0
8.0	315.1
10.0	490.3

SOLUTION

After plotting the data on a graph with a linear scale, shown in Figure 8.27, we see that the data do not suggest a linear function. Plotting the data on a full-log scale, shown in Figure 8.28, we see that the data points suggest a straight line, which means that we have a power function. Using a transparent straightedge, we draw a best-fit straight line

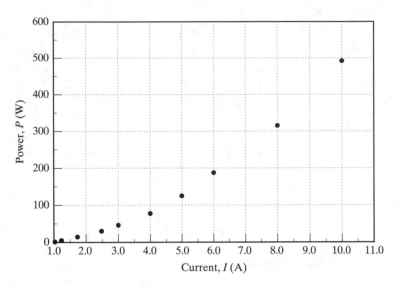

Figure 8.27. Graph of power dissipation for Example 8.2. The data points do not suggest a linear function.

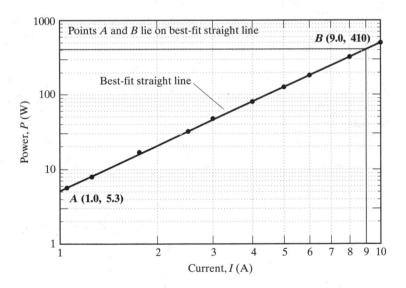

Figure 8.28. Method of selected points for Example 8.2.

through the data. For the two points, we arbitrarily choose 1.0 and 9.0 as the x-coordinates (current coordinates), which yields 5.3 and 410, respectively, for the y coordinates (power coordinates). Thus,

$$A_x = 1.0 \qquad A_y = 5.3$$

and

$$B_x = 9.0 \qquad B_y = 410$$

Once again, *points* A *and* B *and the other illustrative elements on the graph are shown for instructional purposes only and should not be shown on the actual graph.* Following step 7, we set up two simultaneous equations:

$$\log(5.3) = m \log(1.0) + \log(b)$$

and

$$\log(410) = m \log(9.0) + \log(b)$$

Because of our selection for point A, a simultaneous solution is not necessary to solve for b, because $\log(1.0) = 0$, yielding a value of $b = 5.30 \; \Omega$ directly from the first equation. Solving for the slope, we obtain $m = 1.98$. The quantity m is an exponent and has no units. Hence, the equation for the power dissipated by the transformer as a function of current is

$$P = 5.30 \; I^{1.98} \quad \text{(W)}$$

The result is consistent with the fundamental equation from electrical circuit theory,

$$P = I^2 R$$

where R is resistance. Note the similarity between this equation and the equation resulting from the curve fit. The value of b is approximately equal to the resistance of the transformer 5 Ω and the exponent on current I is 1.98, which is very close to the theoretical value of 2.

EXAMPLE 8.3

Density of atmospheric air is measured for various altitudes, as indicated in the table that follows. Using the method of selected points, determine a mathematical function that fits the data.

Altitude, z (m)	Density, ρ (kg/m^3)
0	1.225
400	1.179
1000	1.112
2000	1.007
3000	0.909
4000	0.819
5000	0.736
7000	0.590
10,000	0.413
14,000	0.227
18,000	0.121
20,000	0.088
25,000	0.0395
30,000	0.018

SOLUTION

After plotting the data on a graph with a linear scale and a full-log scale, we see that the data do not suggest a linear or power function, as shown in Figures 8.29 and 8.30, respectively. Plotting the data on a semi-log scale, shown in Figure 8.31, we see that the data points suggest a straight line, which means that we have an exponential

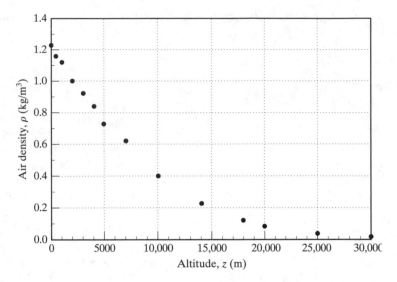

Figure 8.29. Graph of air density for Example 8.3. The data points do not suggest a linear function.

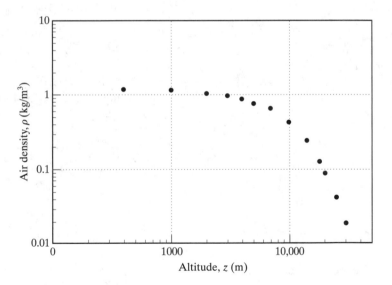

Figure 8.30. Graph of air density for Example 8.3. The data points do not suggest a power function.

function. Using a transparent straightedge, we draw a best-fit straight line through the data. For the two points, we arbitrarily choose 2500 and 29,000 as the x-coordinates (altitude coordinates), which yields 1.00 and 0.024, respectively, for the y-coordinates (density coordinates). Hence,

$$A_x = 2500 \qquad A_y = 1.00$$

and

$$B_x = 29,000 \qquad B_y = 0.024$$

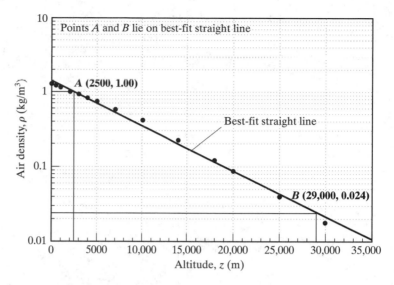

Figure 8.31. Method of selected points for Example 8.3.

Once again, *points* A *and* B *and the other illustrative elements on the graph are shown for instructional purposes only and should not be shown on the actual graph.* Following step 8, we set up two simultaneous equations:

$$\ln(1.00) = m(2500) + \ln(b)$$
$$\ln(0.024) = m(29000) + \ln(b)$$

Solving for the slope m and the y-intercept b we obtain

$$m = -1.41 \times 10^{-4}\, \text{m}^{-1} \qquad b = 1.42\, \text{kg/m}^3$$

The exponential function for density can be simplified by expressing the altitude in units of km instead of m, which has the effect of changing the value of the slope to

$$m = -0.141\, \text{km}^{-1} \quad (z\ \text{in km})$$

Hence, the equation for the density of atmospheric air as a function of altitude is

$$\rho = 1.42\, e^{-0.141\, z} \quad (\text{kg/m}^3)$$

The accuracy of this function can be verified by substituting various values of altitude z in units of km and comparing the calculated values of density with those obtained visually from the graph.

8.4.3   Least Squares Linear Regression

The main drawback of the method of selected points is that it relies on the judgment of the person doing the curve fitting. This is particularly troublesome if the data shows considerable scatter. If 10 different people were to use the method of selected points to fit widely scattered data to a straight line, we would probably obtain 10 different slopes and 10 different y-intercepts. Least squares linear regression is superior to the method of selected points because it employs a precise mathematical technique for finding the best-fit straight line for the data. The basic idea underlying **least squares linear regression** is to *find a straight line such that the difference between a data point and the corresponding* *point predicted by the line is minimized for all data points on the graph.* Referring to Figure 8.32, the objective is to minimize the differences or residuals d_i, which results in a straight line that is as close as possible to all the data points. The residual d_i is defined as the difference between a data point and the corresponding point on the line

$$d_i = y_i - (mx_i - b) \tag{8.5}$$

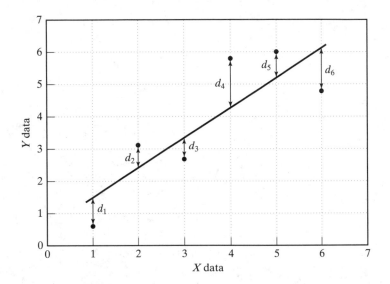

Figure 8.32. Least squares linear regression is based on minimizing the residuals, d_i.

where the subscript i is an index that refers to the data point number, 1, 2, 3, etc. As shown in Figure 8.32, the residual has both positive and negative values, depending on whether the data point is above or below the line. The residuals may be limited to positive values by squaring them, so the best-fit straight line would be obtained by minimizing the sum S of the squares of all residuals, which is written as

$$S = \sum d_i^2 = d_1^2 + d_2^2 + \cdots + d_n^2 = \sum [y_i - (mx_i + b)]^2 \tag{8.6}$$

where the symbol $\sum$ denotes a sum and n is the number of data points. The minimization of Equation (8.6) involves calculus of partial derivatives, which is beyond the scope of this book. After performing the minimization and solving for the slope m and the y-intercept b, we obtain

$$m = \frac{n(\sum x_i y_i) - (\sum x_i)(\sum y_i)}{n(\sum x_i^2) - (\sum x_i)^2} \tag{8.7}$$

$$b = \frac{\sum y_i - m(\sum x_i)}{n} \tag{8.8}$$

Upon finding the slope and y-intercept by using least squares linear regression, the question still remaining is "how well does the line fit the data?" Clearly, if all the data points fall precisely on the line described by the equation $y = mx + b$, then the fit would be "perfect." This rarely, if ever, happens, however, so the degree to which the line "correlates" with the data is an important consideration in curve fitting. A statistical parameter called the **coefficient of determination** (r^2) is used to ascertain the "goodness of fit" of a straight line to the data. The coefficient of determination is given by the equation

$$r^2 = \frac{n(\sum x_i y_i) - (\sum x_i)(\sum y_i)}{\sqrt{n(\sum x_i^2) - (\sum x_i)^2}\;\sqrt{n(\sum y_i^2) - (\sum y_i)^2}} \qquad (8.9)$$

The range of values for r^2 is 0 to 1. Within this range, high values of r^2 indicate a good fit, whereas low values of r^2 indicate a poor fit. Note that many of the terms appearing in Equation (8.9) also appear in Equations (8.7) and (8.8).

EXAMPLE 8.4

Using least squares linear regression, rework Example 8.1. Also, find the coefficient of determination r^2.

SOLUTION

The table containing time and position data for the linear actuator is repeated here:

Time, t (s)	Position, s (cm)
0.0	0.40
1.0	2.49
2.0	4.37
3.0	5.66
4.0	7.92
5.0	8.47
6.0	11.8
7.0	12.4

In order to use least squares linear regression, it is beneficial to construct a special table that enables us to readily calculate the terms in Equations (8.7), (8.8), and (8.9). (See Table 8.3.)

TABLE 8.3 Data for Example 8.4.

Data Point i	Time, t (s) x_i	Position, s (cm) y_i	$x_i y_i$	x_i^2	y_i^2
1	0.0	0.40	0.00	0.00	0.16
2	1.0	2.49	2.49	1.00	6.20
3	2.0	4.37	8.74	4.00	19.10
4	3.0	5.66	16.98	9.00	32.04
5	4.0	7.92	31.68	16.00	62.73
6	5.0	8.47	42.35	25.00	71.74
7	6.0	11.8	70.80	36.00	139.24
8	7.0	12.4	86.80	49.00	153.76
$n = 8$	$\sum x_i = 28.0$	$\sum y_i = 53.51$	$\sum x_i y_i = 259.84$	$\sum x_i^2 = 140.00$	$\sum y_i^2 = 484.97$

From Equation (8.7), the slope of the line is

$$m = \frac{n(\sum x_i y_i) - (\sum x_i)(\sum y_i)}{n(\sum x_i^2) - (\sum x_i)^2} = \frac{8(259.84) - (28.0)(53.51)}{8(140.00) - (28.0)^2}$$

$$= 1.728 \text{ cm/s}$$

and from Equation (8.8), the y-intercept is

$$b = \frac{\sum y_i - m(\sum x_i)}{n} = \frac{53.51 - 1.728(28.0)}{8}$$
$$= 0.641 \text{ cm}$$

Thus, the equation for the position of the actuator as a function of time is

$$s = 1.728\, t + 0.641 \quad (\text{cm})$$

Using the method of selected points, we found that the values of m and b were 1.73 and 0.63, respectively, indicating that, for this problem, at least, the method of selected points yielded an excellent curve fit to the data. From Equation (8.9), the coefficient of determination is

$$r^2 = \frac{n(\sum x_i y_i) - (\sum x_i)(\sum y_i)}{\sqrt{n(\sum x_i^2) - (\sum x_i)^2}\,\sqrt{n(\sum y_i^2) - (\sum y_i)^2}}$$
$$= \frac{8(259.84) - (28.0)(53.51)}{\sqrt{8(140.00) - (28.0)^2}\,\sqrt{8(484.97) - (53.51)^2}}$$
$$= 0.993$$

which indicates an excellent fit.

EXAMPLE 8.5

Using least squares linear regression, rework Example 8.2. Also, find the coefficient of determination.

SOLUTION

The table containing current and power data for the transformer is repeated here:

Current, I (A)	Power, P (W)
1.05	5.63
1.25	7.58
1.75	16.9
2.50	32.1
3.0	48.0
4.0	78.2
5.0	126.0
6.0	188.0
8.0	315.1
10.0	490.3

Recall that the data in this example follows a power function. This means that, in order to use least squares linear regression, the data must be manipulated before we can use Equations (8.7), (8.8), and (8.9). As suggested by Equation (8.3), in place of the independent variable x_i, we use $\log(x_i)$, in place of the dependent variable y_i, we use $\log(y_i)$, and in place of the y-intercept b, we use $\log(b)$. Again, we construct a special table that enables us to calculate the terms in the equations. (See Table 8.4.)
From Equation (8.7), the slope of the line is

$$m = \frac{n\sum(\log x_i)(\log y_i) - (\sum \log x_i)(\sum \log y_i)}{n\sum(\log x_i)^2 - (\sum \log x_i)^2}$$

TABLE 8.4 Data for Example 8.5.

Data Point i	Current, I (A) $\log(x_i)$	Power, P (W) $\log(y_i)$	$(\log x_i)(\log y_i)$	$(\log x_i)^2$	$(\log y_i)^2$
1	0.0212	0.7505	0.0159	4.49×10^{-4}	0.5633
2	0.0969	0.8797	0.0852	9.39×10^{-3}	0.7739
3	0.2430	1.2279	0.2984	0.0590	1.5077
4	0.3979	1.5065	0.5995	0.1583	2.2695
5	0.4771	1.6812	0.8022	0.2276	2.8264
6	0.6021	1.8932	1.1398	0.3625	3.5842
7	0.6990	2.1004	1.4681	0.4886	4.4117
8	0.7782	2.2742	1.7696	0.6056	5.1720
9	0.9031	2.4984	2.2563	0.8156	6.2420
10	1.0000	2.6905	2.6905	1.0000	7.2388
$n = 10$	$\sum \log x_i =$ 5.2185	$\sum \log y_i =$ 17.5025	$\sum (\log x_i)(\log y_i) =$ 11.1255	$\sum (\log x_i)^2 =$ 3.7270	$\sum (\log y_i)^2 =$ 34.5895

$$= \frac{10(11.1255) - (5.2185)(17.5025)}{10(3.7270) - (5.2185)^2}$$

$$= 1.984$$

and from Equation (8.8), we have

$$\log(b) = \frac{\sum \log y_i - m(\sum \log x_i)}{n} = \frac{17.5025 - 1.984(5.2185)}{10}$$

$$= 0.715$$

Hence, the y-intercept is

$$b = 10^{0.715} = 5.19$$

Therefore, the equation for the power dissipated by the transformer as a function of current is

$$P = 5.19 \, I^{1.984} \quad (\text{W})$$

Using the method of selected points, we found that the values of m and b were 1.98 and 5.30, respectively, in good agreement with the values calculated here by using least squares linear regression. From Equation (8.9), the coefficient of determination is

$$r^2 = \frac{n \sum (\log x_i)(\log y_i) - (\sum \log x_i)(\sum \log y_i)}{\sqrt{n \sum (\log x_i)^2 - (\sum \log x_i)^2} \, \sqrt{n \sum (\log y_i)^2 - (\sum \log y_i)^2}}$$

$$= \frac{10(11.1255) - (5.2185)(17.5025)}{\sqrt{10(3.7270) - (5.2185)^2} \, \sqrt{10(34.5895) - (17.5025)^2}}$$

$$= 0.999$$

which indicates an excellent fit.

Least squares linear regression can be manually applied by using Equations (8.7), (8.8), and (8.9), but this can be tedious when dealing with a large number of data points. The most efficient way to apply the method is to use a software package that has a built-in least squares linear regression routine. Virtually all graphing and spreadsheet packages consist of a least squares linear regression routine that readily calculates the slope and *y*-intercept of the best-fit line, as well as the coefficient of determination. Some packages, particularly graphing packages, have more advanced curve-fitting routines that also enable the user to fit data to various nonlinear functions.

| PRACTICE! | Using least squares linear regression, rework Example 8.3. Also, find the coefficient of determination. |

8.5   INTERPOLATION AND EXTRAPOLATION

Sometime during the data analysis process, it may be necessary to determine data points that are not part of the original data set used to construct the graph. **Interpolation** is *a process used to find data points between known data points*, whereas **extrapolation** is *a process used to find data points beyond known data points*. Consider the graph shown in Figure 8.33. The *y* value of the extrapolated data point at *x* = 8 is estimated based on the shape of the curve and its slope at *x* = 5, the last known data point. Extrapolation can be a risky process, because the behavior of the data beyond the last measured data point may be unpredictable, so extrapolation is generally not recommended. If reliable data beyond the current data set are desired, additional measurements should be taken. Interpolation, on the other hand, is generally reliable, because the known data points on both sides of the unknown data point serve as lower and upper limits, thereby bracketing the value of the unknown data point within a known range.

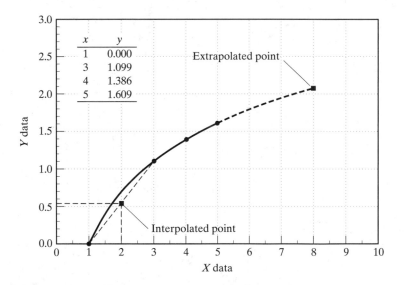

Figure 8.33.   Interpolation and extrapolation.

Referring again to Figure 8.33, suppose that we wanted to *calculate* the corresponding *y* value for *x* = 2, which is not in the original data set and therefore not plotted on the graph. In the absence of an equation for the smooth curve drawn through the data points, we cannot calculate this value (although we can estimate it graphically). To calculate the *y* value for *x* = 2, we approximate the curve between the adjacent known data points as a straight line, which is known as *linear interpolation*. To illustrate how linear interpolation works, let's examine the lower left portion of the graph, which is shown in

Figure 8.34. In this figure, x_1 and y_1 are the coordinates of the known data point on the left of the interpolated point, and x_2 and y_2 are the coordinates of the known data point on the right of the interpolated point. The coordinates of the interpolated point are x

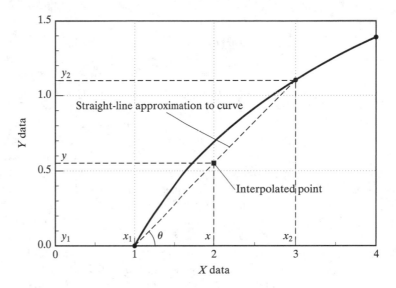

Figure 8.34. Linear interpolation.

and y. A straightforward way to derive a formula for the y value of the unknown data point is to use a familiar concept from geometry: similar triangles. We have two similar triangles having a common angle θ, with the straight line approximation serving as the hypotenuse. For similar triangles, the ratio of the opposite sides equals the ratio of the adjacent sides. Noting the coordinates of the corners of the triangles, the equality is written as

$$\frac{y_2 - y_1}{x_2 - x_1} = \frac{y - y_1}{x - x_1} \tag{8.10}$$

Solving for the unknown y, we obtain

$$y = y_1 + \frac{y_2 - y_1}{x_2 - x_1}(x - x_1) \tag{8.11}$$

Using linear interpolation, Equation (8.11) is a general formula for the y value of a data point.

Applying Equation (8.11) to the given data, we get

$$y = y_1 + \frac{y_2 - y_1}{x_2 - x_1}(x - x_1) = 0 + \frac{1.099 - 0}{3 - 1}(2 - 1) = 0.550$$

Hence, using linear interpolation, the coordinates of the interpolated data point are $x = 2, y = 0.550$. The data points in Figure 8.33 were deliberately selected to fit the function $y = \ln x$. The actual y value for $x = 2$ is $y = \ln(2) = 0.693$. For the first two data points, linear interpolation does a rather poor job of approximating the y value for $x = 2$. However, as shown in Figure 8.33, the function $y = \ln x$ begins to resemble a straight line with increasing values of x, so linear interpolation should be more accurate for the other data points. Let's use linear interpolation to calculate the y value for $x = 4.5$:

$$y = y_1 + \frac{y_2 - y_1}{x_2 - x_1}(x - x_1) = 1.386 + \frac{1.609 - 1.386}{5 - 4}(4.5 - 4) = 1.498$$

The actual y value for $x = 4.5$ is $y = \ln(4.5) = 1.504$, a marked improvement compared with the previous interpolation. Obviously, linear interpolation is most accurate if the data points describe a linear relationship. If the data points describe a nonlinear relationship, as illustrated in Figure 8.34, linear interpolation yields an approximate value whose accuracy depends on the type of function described by the data points, the region of the data where the interpolation is made, and how close the two known data points are to each other.

Linear interpolation can be performed on tabulated data without regard to a graph. Consider the values of saturation temperatures and pressures of water in Table 8.5. A fundamental concept from thermodynamics is that saturation temperature and saturation pressure are dependent properties (i.e., for every value of saturation temperature, there is a unique value of saturation pressure). Suppose that we want to determine the saturation pressure of water at a saturation temperature of 77.0°C. This temperature is not listed in Table 8.5, so we cannot simply read from the table a corresponding value of pressure. We can, however, estimate a saturation pressure by using linear interpolation. A saturation temperature of 77.0°C lies between two known temperatures, 75.0°C and 80.0°C. The corresponding saturation pressures for 75.0°C and 80.0°C are 38.58 kPa and 47.39 kPa, respectively.

TABLE 8.5 Saturation Temperature and Pressure of Water.

Temperature, T (°C)	Pressure, P (kPa)
65.0	25.03
70.0	31.19
75.0	38.58
80.0	47.39
85.0	57.83

Focusing on the portion of Table 8.5 that is of interest, we construct an interpolation table. (See Table 8.6.) Denoting saturation temperatures as the x values and saturation pressures as the y values, we want to calculate the corresponding y value for $x = 77.0$. Using linear interpolation, Equation (8.11) is the general formula for the y value of a data point. Hence, we have

$$y = y_1 + \frac{y_2 - y_1}{x_2 - x_1}(x - x_1) = 38.58 + \frac{47.39 - 38.58}{80.0 - 75.0}(77.0 - 75.0)$$

$$= 42.10$$

TABLE 8.6 Interpolation Table for Data in Table 8.5.

Temperature, T (°C)	Pressure, P (kPa)
x_1	y_1
75.0	38.58
x ——— 77.0	y
80.0	47.39
x_2	y_2

Thus, the corresponding saturation pressure for a saturation temperature of 77.0°C is 42.10 kPa.

It is readily apparent that linear interpolation may be performed by recognizing that the same fractional amount must exist between the known x value and either x_1 or x_2 and the unknown y value and either y_1 or y_2. Using Table 8.6 as an example, we find that a temperature of 77.0°C must be the same fractional amount between 75.0°C and 80.0°C as the interpolated pressure is between 38.58 kPa and 47.39 kPa. The mathematical version of this statement is Equation (8.10), which was derived from geometrical considerations.

As you progress through your engineering coursework, you will undoubtedly encounter numerous situations where linear interpolation of tabulated data is required. Equation (8.11) may be used for any tabulated data regardless of whether the x and y numerical values are ascending or descending. In addition, to help you in your calculations, you might want to write a simple linear interpolation program on your scientific calculator. Over time, however, you will probably become so proficient in using linear interpolation that you'll be able do the calculations without referring to Equation (8.11).

PRACTICE!

1. For the data in Table 8.5, use linear interpolation to find the pressure for a temperature of 66.5°C.
   Answer: 26.88 kPa

2. For the data in Table 8.5, use linear interpolation to find the temperature for a pressure of 50.0 kPa.
   Answer: 81.3°C

3. For the data in Table 8.5, use extrapolation to estimate the pressure for a temperature of 90.0°C.
   Answer: 69 kPa

KEY TERMS

1, 2, 5 rule
accuracy
coefficient of
   determination
curve fitting
dependent variable
error
exponential function

extrapolation
graph
graph paper
independent variable
interpolation
least squares linear
   regression
linear function

measurement
method of selected
   points
power function
precision

REFERENCES

Tufte E.R., *The Visual Display of Quantitative Information*, 2nd ed., Cheshire, CT: Graphics Press, 2001.

Tufte, E.R., *Visual Explanations: Images and Quantities, Evidence and Narrative*, Cheshire, CT: Graphics Press, 1997.

Taylor, J.R., *An Introduction to Error Analysis*, 2nd ed., Herndon, VA: University Science Books, 1997.

Henry, G.T., *Graphing Data: Techniques for Display and Analysis*, Thousand Oaks, CA: SAGE Publications, 1995.

Harris, R.L., *Information Graphics: A Comprehensive Illustrated Reference*, Oxford, NY: Oxford University Press, 1999.

Holman, J.P., *Experimental Methods for Engineers*, 7th ed., NY: McGraw-Hill, 2000.

Problems

1.   A chemical engineer wants to analyze the effects of sodium chloride in a new medicine being developed. On a mass basis, the medicine consists of 85 percent water, a maximum of 10 percent of other chemicals, and a maximum of 5 percent sodium chloride. Describe the kinds of data that the engineer should collect and how graphs of the data could be used to evaluate the new medicine.

2.   In a production facility where pistons are machined on numerically controlled lathes, a manufacturing engineer wants to study the effect of feed rate on the production and surface finish of the parts. To maximize the production rate, a large feed rate is desired, but if the feed rate is too high, a poor surface finish results. Describe the kinds of data that the engineer should collect and how graphs of the data could be used to determine the appropriate feed rate.

3.   An electrical engineer wants to assess the effects of temperature, humidity, and barometric pressure on the electrical resistance of a large wire-wound ceramic resistor. The expected ranges of these environmental variables are

temperature: 10°C to 80°C
humidity: 10 percent to 90 percent relative humidity
barometric pressure: 0.7 atm to 1.1 atm

Estimate how many unique measurements should be taken to adequately characterize the resistor. Which of these variables do you think would have the most pronounced effect on resistance? How could the engineer use graphs of the data to evaluate the effects of these variables on the resistance?

4.   Classify the following errors as gross (G), systematic (S), or random (R) (in some cases, more than one classification may apply):

Error	Classification (G, S or R)
a. Dropping a pressure gauge on the floor	_____
b. Air conditioning running from 3 pm to 7 pm in the lab	_____
c. Mass balance not zeroed	_____
d. Surface plate not leveled	_____
e. Ohmmeter set to wrong scale	_____
f. Flow meter calibrated 5 years ago	_____
g. Sensitive electromagnetic tests conducted near a large transformer	_____
h. Using a carpenter's tape to measure distances to a ±0.02 inch accuracy	_____

5.   The graph shown in Figure P8.5 is incorrectly drawn. Referring to the general graphing procedure, identify the problems.

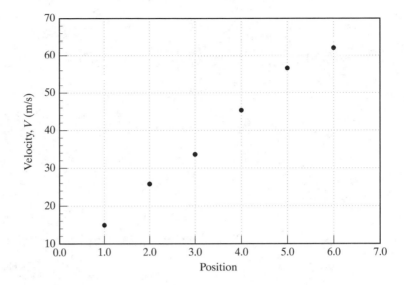

Figure P8.5.

6.   The graph shown in Figure P8.6 is incorrectly drawn. Referring to the general graphing procedure, identify the problems.

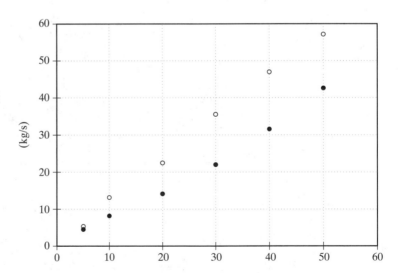

Figure P8.6.

7.   The graph shown in Figure P8.7 is incorrectly drawn, and the method of selected points is used to fit the data to a straight line. Identify the problems.

8.   Table P8.8 lists the annual salary of an engineer from 1976 through 2002. Using the general graphing procedure, construct a graph of the data.

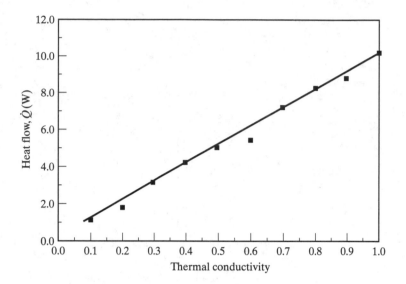

Figure P8.7.

TABLE P8.8

Year	Salary, $	Year	Salary, $
1976	15,450	1989	36,300
1977	16,120	1990	38,100
1978	17,840	1991	40,560
1979	18,900	1992	42,850
1980	20,680	1993	44,995
1981	21,975	1994	47,540
1982	23,050	1995	50,125
1983	24,800	1996	52,980
1984	26,100	1997	55,050
1985	27,960	1998	57,160
1986	29,200	1999	59,850
1987	32,450	2000	62,100
1988	34,250	2001	64,740
		2002	67,250

9.   The solar intensity, measured in W/m^2, for a horizontal surface, as shown in Table P8.9, is measured at one-hour intervals during a partly cloudy day. Using the general graphing procedure, construct a graph of the data.

10. Listed in Table P8.10 are measured rotational speeds in rpm of a pump and the corresponding pump output in gallons per minute.

   a.   Identify the independent and dependent variables.

   b.   Use the method of selected points to obtain an equation for the curve.

   c.   Using the equation found in part (b), find the pump output for a speeds of 175, 350, and 475 rpm.

TABLE P8.9

Time, t (h)	Solar intensity, I (W/m^2)
0500	9.7
0600	35.9
0700	43.0
0800	228
0900	357
1000	518
1100	624
1200	739
1300	701
1400	612
1500	456
1600	178
1700	41.5
1800	13.2
1900	3.5

TABLE P8.10.

Pump rotational speed, S (rpm)	Pump output, Q (gal/min)
0	0
100	0.52
230	1.45
325	2.80
400	4.30
500	6.10

11.   Using least squares linear regression, work Problem 10 and find the coefficient of determination.

12.   A 10-V battery is connected across a variable resistor whose resistance is varied from 2 kΩ to 10 kΩ. An ammeter is used to measure the current. Table P8.12 lists the values of resistance and current.

TABLE P8.12

Resistance, R (kΩ)	Current, I (mA)
2.0	5.66
3.0	3.37
4.0	2.42
5.0	2.01
6.0	1.62
8.0	1.18
10.0	0.995

    a.   Identify the independent and dependent variables.

    b.   Use the method of selected points to obtain an equation for the curve.

    c.   Using the equation found in part (b), find the current for resistance values of 2.5, 4.7, and 5.6 kΩ.

13.   Using least squares linear regression, work Problem 12 and find the coefficient of determination.

14.   Given in Table P8.14 is the air velocity and drag-force data for a wind-tunnel test of a new airfoil.

    a.   Identify the independent and dependent variables.

    b.   Use the method of selected points to obtain an equation for the curve.

    c.   Using the equation found in part (b), find the drag force for velocities of 12, 45, and 60 m/s.

TABLE P8.14

Velocity, v(m/s)	Drag force, F (N)
2	3.5
5	22
15	176
20	330
30	728
50	1970
75	4560

15.   Using least squares linear regression, work Problem 14 and find the coefficient of determination.

16.   The measured-temperature history of a small metal forging after removal from a heat-treating oven is shown in Table P8.16.

    a.   Identify the independent and dependent variables.

    b.   Use the method of selected points to obtain an equation for the curve.

    c.   Using the equation found in part (b), find the temperature for times of 7 and 18 s.

TABLE P8.16

Time, t(s)	Temperature, T(°C)
0	200
1	180
2	166
3	151
4	132
5	125
10	72
15	46
20	28

17. Using least squares linear regression, work Problem 16 and find the coefficient of determination.

18. The voltage, measured in mV, produced by a type-K thermocouple for various junction temperatures is shown in Table P8.18.

TABLE P8.18

Temperature, T (°C)	Voltage, V (mV)
50	1.98
100	4.35
200	7.76
300	12.51
400	16.70
500	19.62
700	29.43
1000	42.02

a. Identify the independent and dependent variables.
b. Use the method of selected points to obtain an equation for the curve.
c. Using the equation found in part (b), find the voltage for temperatures of 150, 575, and 850 °C.

19. Using least squares linear regression, work Problem 18 and find the coefficient of determination.

20. The variation with temperature of the solubility, measured in kg, of calcium bicarbonate, $Ca(HCO_3)_2$, in 100 kg of water is shown in Table P8.20.
a. Identify the independent and dependent variables.
b. Use the method of selected points to obtain an equation for the curve.
c. Using the equation found in part (b), find the solubility for temperatures of 302 and 330 K.

TABLE P8.20

Temperature, T (K)	Solubility, S (kg)
273	16.15
280	16.30
290	16.53
300	16.75
310	16.98
320	17.20
350	17.88
373	18.40

21. Using least squares linear regression, work Problem 20 and find the coefficient of determination.

22. In a drilling operation, the material removal rate (MRR) is measured for a range of drill diameters, as shown in Table P8.22.

a. Identify the independent and dependent variables.
b. Use the method of selected points to obtain an equation for the curve.

c. Using the equation found in part (b), find the material removal rate for drill diameters of 0.875 and 1.25 in.

TABLE P8.22

Diameter, d (in)	MRR, M (in^3/min)
0.375	1.41
0.500	2.36
0.625	4.06
0.750	5.43
1.000	10.8
1.500	21.3

23. Using least squares linear regression, work Problem 22 and find the coefficient of determination.

24. The power required for an automobile to overcome aerodynamic drag at various speeds is shown in Table P8.24.

   a. Identify the independent and dependent variables.
   b. Use the method of selected points to obtain an equation for the curve.
   c. Using the equation found in part (b), find the power for a speed of 55 mi/h.

TABLE P8.24

Speed, s (mi/h)	Power, P (hp)
10	0.060
20	0.478
30	1.61
40	3.83
50	7.47
60	12.9

25. Using least squares linear regression, work Problem 24 and find the coefficient of determination.

26. Table P8.26 shows the variation of the specific heat of liquid water with temperature. Using linear interpolation, calculate specific heat for temperatures of 75, 120, and 180°C.

TABLE P8.26

Temperature, T (°C)	Specific heat, c (kJ/kg · °C)
0	4.217
10	4.193
20	4.182
30	4.179
50	4.181
100	4.216
150	4.310
200	4.497

27. Table P8.27 shows the variation of electrical resistance of a 1000-m length of copper wire with a range of wire gauge numbers. Using extrapolation, estimate the resistance of a 1000-m length of copper wire for a wire gauge number of 34.

TABLE P8.27

Wire gage number	Resistance (Ω)
20	33.31
22	52.96
24	84.21
26	133.9
28	212.9
30	338.6

9

Data Analysis: Statistics

9.1 INTRODUCTION

Statistics is *a branch of applied mathematics dealing with the collection, presentation, analysis, and interpretation of data.* Statistics is used to study phenomena in which randomness or uncertainty play a role. For example, the simple act of flipping a coin is a random process that can be described with the tools of statistics. Results of political elections can be projected, weather conditions can be forecasted, and the outcomes of sporting events can be predicted by using statistical methods. Because randomness and uncertainty are integral parts of these and other phenomena, statistics can only provide information that is imperfect and incomplete. The information is imperfect due to unavoidable random variation in the measurements, and it is incomplete because we seldom know or can measure all the influential variables that affect the phenomena. Hence, statistics does not provide the absolute "truth," but an approximation of it. Statistics, properly used, helps us move toward the truth; however, it cannot guarantee that we will reach it, nor can it tell us whether we have done so. Statistics enables us to make scientifically honest assessments about the *likelihood* of certain phenomena.

When misused or misinterpreted, statistics leads to conclusions that are, at best, misleading and, at worst, completely wrong. A social statistic quoted in a doctoral dissertation stated, "Every year since 1950, the number of American children gunned down has doubled." Let's examine this statement. Assuming there was only one child gunned down in 1950, there would have been two children gunned down in 1951, four in 1952, eight in 1953, and so on. By 1995, the year of publication, there would have been about 2×10^{13} children gunned down—more than one thousand times the world's population. Where did this erroneous statistic come from? The author obtained it from the Children's Defense Fund, whose 1994 yearbook states, "The number of American children killed by guns

OBJECTIVES

After reading this chapter, you will have learned

- How to group data into classes
- How to find measures of central tendency
- How to find measures of variation
- How to use the normal distribution

has doubled since 1950." Note the difference in wording. The original claim was that the number of children killed by guns doubled over the time period from 1950 through 1995; however, the doctoral student misinterpreted the statistic to claim that the number of children gunned down from 1950 to 1995 doubled *each year*, resulting in a completely different meaning. The notion that we need to beware of bad statistics is not new. You have no doubt heard the adage "You can prove anything with statistics." The famous aphorism by the British statesman Benjamin Disraeli (1804–1881) is "There are three kinds of lies: lies, damned lies, and statistics."

While there is a certain amount of substance to these tongue-in-cheek maxims, we must not diminish the importance of statistics in everyday life and in engineering. Statistics is an indispensable decision-making and design tool. For example, transportation engineers use statistics to determine the anticipated lifetime of roads, highways, and bridges. Chemical engineers and medical researchers use statistics to identify effective drugs and medicines. Manufacturing and industrial engineers use statistics to assure the quality of products and processes. Nuclear engineers use statistics to evaluate the reliability of safety systems in nuclear power plants. Materials engineers and scientists use statistics to optimize the properties of new metal alloys and composites for aerospace and medical applications. Electrical engineers use statistics to reduce noise from transmitted signals in communications systems. These are but a few engineering applications of statistics.

Traditionally, engineers have been taught to approach an analytical problem in terms of a *deterministic model* without regard for variability of the quantities. A deterministic model is strictly and accurately described by a governing equation derived from a conservation law or some other fundamental physical principle. A typical example is Ohm's law, which states that voltage V equals the product of current I and resistance R:

$$V = IR \qquad (9.1)$$

According to Ohm's law, if the current and resistance are known, then voltage is exactly determined, and hence there is no need to measure it. But if we set up a simple laboratory experiment where 25 students measure the voltage across the same resistor through which the same current flowed, we would obtain 25 different voltages. This does not mean that Ohm's law is invalid. All 25 voltage measurements would be close to the value obtained using Equation (9.1), and the small differences would be due to inherent measurement deviations. A *statistical* model of Ohm's law accounts for deviations from the values obtained from the *deterministic* model given by Equation (9.1), and would be expressed as

$$V = IR + \epsilon \qquad (9.2)$$

where ϵ represents deviations from the expected voltage. In a similar manner, statistical models of other deterministic models in science and engineering could be written.

Statistics may be generally categorized as either *descriptive* or *inferential*. The aim of descriptive statistics is to describe the principal characteristics of a set of data without inferring conclusions that go beyond the data. The aim of statistical inference is to make general inferences based on a limited set of data. To illustrate the difference between these two categories, suppose that we wish to determine the average height of the residents in Anytown, USA, whose total citizenry is 5000. In statistics, **population** is defined as the total number of measurements or observations. In this example, the population is 5000. Because it is impractical to measure the height of every resident, we randomly select every 50th resident, making a total of 100 measurements. The total, 100, is a subset representative of the population and is defined as a **sample**. If the average height for the sample is 5.43 ft, we could simply state that 5.43 ft describes the average height of 100 randomly selected residents of Anytown, without attempting to draw any general conclusions about the average height of all 5000 residents. Alternatively, we could infer that a height of 5.43 ft for the sample should relate in a specific way to the average height for the population. Basic techniques for relating the characteristics of a sample to the population are discussed in this chapter.

9.2 DATA CLASSIFICATION AND FREQUENCY DISTRIBUTION

Unprocessed data recorded in a laboratory notebook are referred to as *raw data*. A complete statistical analysis requires that the raw data be processed in some meaningful way. One way of processing the raw data is to *sort* it into a list of ascending or descending values. After the data has been sorted, it can be grouped into *classes*. To illustrate how this works, let's return to our hypothetical height study of the residents of Anytown, USA. Once again, we refer to the subset (100) representative of the total population, and we record them in a table. (See Table 9.1.) Because our measurements are taken for 100

TABLE 9.1 Heights of 100 Residents of Anytown, USA (ft).

4.98	5.23	5.46	4.36	4.94
5.01	5.92	5.98	4.23	4.79
5.42	4.76	5.38	5.85	5.10
4.75	5.02	5.88	5.65	5.43
6.05	5.67	5.32	4.97	5.55
5.87	6.12	5.68	5.39	5.99
4.93	5.27	5.59	6.20	4.96
5.03	5.26	5.29	5.40	6.31
4.65	5.19	5.38	5.78	5.99
4.82	6.22	5.45	5.21	5.87
5.07	5.68	5.34	5.34	5.06
5.33	5.89	5.01	6.10	6.29
4.67	5.20	5.31	5.78	5.92
4.81	6.19	5.47	5.01	5.87
5.09	5.69	5.37	5.56	5.93
5.05	5.63	5.35	4.43	5.59
5.84	6.10	5.77	5.33	5.01
4.91	6.02	5.56	6.25	4.99
4.56	5.60	5.23	5.25	5.89
5.24	5.87	5.43	5.98	6.03

residents selected at random, the numbers in the table are not ordered in any particular way. The values are listed in the order that the measurements are taken.

Data Classification Guidelines

The height data in Table 9.1 can be classified by following some simple guidelines:

1. Select classes (ranges) for the data. A rule of thumb is to subdivide the data into $\sqrt{n}$ classes, where n is the number of data points. No fewer than six classes should be used.
2. Select classes that encompass the entire data range.
3. Select classes such that no data point falls into more than one class.
4. Make the class intervals of equal size.

There are 100 data points, so we subdivide the data into 10 classes. An inspection of the data in Table 9.1 reveals that the minimum and maximum heights are 4.23 ft and 6.31 ft, respectively. If we define the data range as 4.00 ft to 6.50 ft, we obtain 10 classes with a size of 0.25 ft each. The classification of the height data is given in Table 9.2. *Frequency*

TABLE 9.2 Classification of Height Data in Table 9.1.

Class	Height range (ft)	Tally	Frequency
1	4.00–4.25	|	1
2	4.26–4.50	||	2
3	4.51–4.75	||||	4
4	4.76–5.00	ⲯⲯ ⲯⲯ |	11
5	5.01–5.25	ⲯⲯ ⲯⲯ ⲯⲯ |||	18
6	5.26–5.50	ⲯⲯ ⲯⲯ ⲯⲯ ⲯⲯ |	21
7	5.51–5.75	ⲯⲯ ⲯⲯ ||	12
8	5.76–6.00	ⲯⲯ ⲯⲯ ⲯⲯ ||||	19
9	6.01–6.25	ⲯⲯ ⲯⲯ	10
10	6.26–6.50	||	2
		Total	100

denotes the number of data points that falls into each class. The sum of the frequencies equals the total number of data points, which is 100 for this example. The data in Table 9.2 can be conveniently displayed as a bar graph in which frequency is plotted as a function of height, as shown in Figure 9.1. This type of graph is called a **histogram**. Each bar

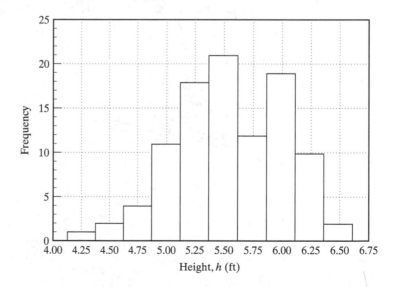

Figure 9.1. Histogram for Anytown, USA.

in the histogram represents a data class (i.e., a range of heights). The calibrations along the horizontal axis denote the upper limits on the range of heights for each class listed in Table 9.2. The first bar, which indicates a frequency of 1, represents heights from 4.00 ft to 4.25 ft. The second bar, which indicates a frequency of 2, represents heights from 4.26 ft to 4.50 ft. The third bar, which indicates a frequency of 4, represents heights from 4.51 ft to 4.75 ft, and so on.

Histograms are valuable statistical tools for showing the **frequency distribution** of data. From the frequency distribution, we can usually draw certain conclusions about the data. The information about location, spread, and shape that is portrayed on the histogram can provide clues concerning the function of a physical process that generated the data. It can also suggest the nature of and potential improvements for the physical mechanisms at work in the process. For example, consider the diameters of machined metal rods purchased from a supplier. The customer specifications require that the diameter of the rods is 5.000 ± 0.002 cm, which means that the desirable diameter is 5.000 in, but the acceptable range of diameters is 4.998 to 5.002 cm. Let's assume that 100 rods are measured. If the rod diameters follow a *normal* or *bell-shaped* frequency distribution, the histogram would look like Figure 9.2(a), suggesting that the majority of the rods have diameters very close to

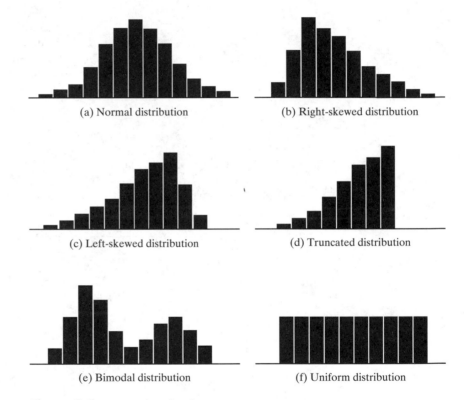

(a) Normal distribution   (b) Right-skewed distribution

(c) Left-skewed distribution   (d) Truncated distribution

(e) Bimodal distribution   (f) Uniform distribution

Figure 9.2.   Frequency distributions.

5.000 cm and that the number of rods with lower and higher diameters symmetrically "tails off" on both sides of the "hump." If the rod diameters are *skewed* toward 4.998 cm or 5.002 cm, the histogram would look like Figures 9.2(b) and 9.2(c), respectively. Note that the terms "right skew" and "left skew" refer to the location of the tail of the frequency distribution, and not the hump. A skewed frequency distribution may indicate a systematic error of some kind in the machining equipment. If the histogram resembles either the left or right half of a normal histogram, the frequency distribution is called *truncated*. A truncated distribution, shown in Figure 9.2(d), may suggest that the machine operator deliberately produced parts with diameters of less than 5.000 cm or greater than 5.000 cm, or that an inspection resulted in a removal of rods with high or low diameters. If two humps appear in the histogram, as shown in Figure 9.2(e), the frequency distribution is called *bimodal*. A bimodal distribution suggests that the machining was performed on more than one machine, or by more than one operator, or at more than one time. A *uniform* frequency distribution, as shown in Figure 9.2(f), indicates that there is no variation in the data. For our sample, a uniform distribution suggests that the number of rods with a diameter of 5.000 cm is equal to the number of rods with any other diameter.

PRACTICE!

1. A manufacturer of toy cars orders a supply of plastic wheels from an injection-molding supplier. The wheels are to have a diameter of 0.750 ± 0.010 in. A quality engineer randomly selects 30 wheels as a sample and measures the following diameters:

0.741	0.750	0.759	0.755	0.754
0.750	0.747	0.743	0.746	0.752
0.751	0.745	0.748	0.757	0.755
0.748	0.752	0.749	0.753	0.752
0.750	0.747	0.758	0.754	0.751
0.749	0.750	0.754	0.752	0.749

    Using the data classification guidelines, subdivide the data into classes and construct a histogram. What kind of frequency distribution is suggested by the histogram?

9.3   MEASURES OF CENTRAL TENDENCY

In engineering and science, it is often desirable to characterize data by a single representative number referred to as a *descriptive measure*. These measures are numerical values that quantify the entire data set in a meaningful way and are easy to communicate to others. One of these measures is called *measure of central tendency*. As the name implies, a measure of central tendency is a number that represents the center of a data set. We will consider three measures of central tendency: *mean, median*, and *mode*.

9.3.1   Mean

You are familiar with the term *average* because this word is frequently used in our everyday language. You might hear someone say, "He has above average intelligence," or, "The temperature today is well below the seasonal average." In statistics, generally, we do not use the term *average*. Instead, we use the term *mean* or *arithmetic mean*. For a set of n numbers, **mean** is defined as *the sum of the numbers divided by n*. For example, suppose that we want to find the mean grade point average (GPA) of five students sitting in the front row of an engineering class. Their GPAs are 2.98, 3.50, 3.25, 3.74, and 3.18. The mean is

$$\text{mean} = \frac{x_1 + x_2 + x_3 + x_4 + x_5}{n} = \frac{2.98 + 3.50 + 3.25 + 3.74 + 3.18}{5} = 3.33$$

A more convenient mathematical shorthand notation for the sum of the numbers is

$$\sum_{i=1}^{n} x_i = x_1 + x_2 + x_3 + \cdots + x_n$$

where the symbol Σ denotes a sum, n is the number of data points, and i is a summation index that refers to the data point number, 1, 2, 3, etc. The sum is defined for all numbers in the data set, so the summation index i begins at 1 and ends at n, the number of data points. For brevity, we will hereafter drop the "$i = 1$" and "n" from the summation notation.

The mathematical notation used for the mean depends on whether the data set represents the population or a sample of the population. For a *population*, the notation used for mean is the Greek letter μ (pronounced *mew*). Thus,

$$\mu = \frac{\sum x_i}{n} \qquad (\text{population}, n > 30) \tag{9.3}$$

For a *sample*, the notation used for the mean is x with an overbar ($\bar{x}$). Therefore,

$$\bar{x} = \frac{\sum x_i}{n} \qquad (\text{sample}, n \leq 30) \tag{9.4}$$

Note that a population is defined as a data set with more than 30 numbers, and a sample is defined as a data set with 30 or fewer numbers. In Equation (9.3) the summation is over all numbers in the population, whereas in Equation (9.4) the summation is over only the numbers in the sample.

A mechanical analogy may be used to portray the mean. Imagine that the numbers of a data set are arranged in order and appropriately spaced along a massless beam supported by a fulcrum. This mechanical system is portrayed in Figure 9.3, where we plot the

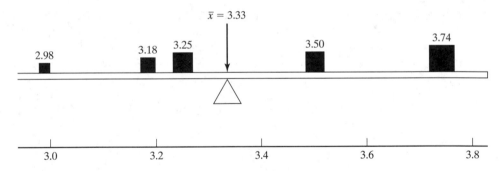

Figure 9.3. Mean is the "center of gravity" of the data.

GPA data we discussed earlier in Section 9.3. Now assume that the numbers represent "masses." For the beam to be in a state of balance, the fulcrum must be placed at precisely the mean of the data set. The mean may therefore be considered the "center of gravity" of the data.

The mean is a useful and popular measure of central tendency because it is easy to calculate, takes into account every number in the data set and can be used in other statistical calculations. Despite its advantages, the mean has a disadvantage of being susceptible to gross errors in the data set. For example, suppose that the fifth GPA in the previous example was mistakenly recorded as 3.81 instead of 3.18. The mean would be

$$\bar{x} = \frac{\sum x_i}{n} = \frac{2.98 + 3.50 + 3.25 + 3.74 + 3.81}{5} = 3.46$$

instead of the correct value of 3.33. Such errors can be minimized by using a different measure of central tendency, the median.

9.3.2   Median

The **median** is *the value of the number in the center of a data set arranged in ascending or descending order*. Arranging the numbers in our GPA example in ascending order, we have 2.98, 3.18, 3.25, 3.50, and 3.74. Thus, the median for this data set is 3.25, because it lies at the center of the data set. For data sets with an odd number of items, such as our GPA data set, the median is always the center number. For data sets with an even number of items, the median is defined as the mean of the two center numbers. For example, the median for the data set 2, 3, 6, 7, 12, 15 is $(6 + 7)/2 = 6.5$. Note that even though all the numbers in the data set are integers, the median is a decimal number.

Both mean and median describe the center of a data set, but they do so in different ways. The mean is the center of gravity of the data, and the median divides the data set into two halves. For a given data set, the mean and median may or may not be close to each other in value, and rarely do the mean and median coincide.

9.3.3   Mode

The **mode** is *one or more sets of numbers that occurs with the greatest frequency in a data set*. Unlike mean and median, which always exist, mode may not exist, since some data sets do not have a set of numbers that occur more often than the other numbers in the data set. To illustrate how to find mode, consider the following three data sets:

Data set 1   2, 2, 5, 7, 9, 9, 9, 10, 10, 11, 12, 18
Data set 2   2, 3, 4, 4, 4, 5, 5, 7, 7, 7, 9
Data set 3   3, 5, 8, 10, 12, 14, 17, 19, 22, 26

Data set 1 has a mode of 9 because the number 9 occurs with the greatest frequency in the data set. A data set with one mode is called *unimodal*. Data set 2 has two modes, 4 and 7, because *both* of these numbers occur with the greatest frequency in the data set. When two modes occur in a data set, it is called *bimodal*. Data set 3 has no mode because no number in the data set occurs with any greater frequency than any other number.

Modes are graphically exhibited in frequency distributions on histograms. The histograms in Figure 9.2(a), (b), and (c) show unimodal distributions, and the histogram in Figure 9.2(e) shows a bimodal distribution. The histograms in Figure 9.2(d) and (f) have no modes.

EXAMPLE 9.1

Professor Gauss has 125 students in his Engineering 101 class. After grading the final examination for this class, he randomly selects the following 30 scores for statistical analysis:

84	92	76	84	86	65
44	59	68	95	72	80
78	49	67	79	63	54
97	61	79	53	87	84
77	66	48	60	76	73

Using Professor Gauss's data, construct a histogram and find the mean, median, and mode.

SOLUTION

First, we sort the data into classes. We use the minimum recommended number of classes (six) and construct a table of the frequencies from which a histogram can be constructed. (See Table 9.3.) The histogram is shown in Figure 9.4. The calibrations below each bar denote the upper bound for each class.

TABLE 9.3 Classification of Examination Scores for Example 9.1.

Class	Score range	Frequency
1	41–50	3
2	51–60	4
3	61–70	6
4	71–80	9
5	81–90	5
6	91–100	3

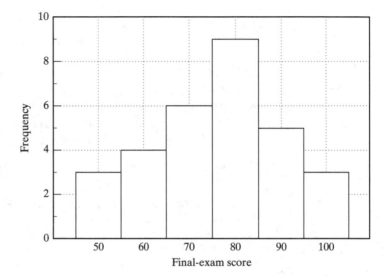

Figure 9.4. Histogram for Example 9.1.

The data set of 30 scores is a sample taken from the population of 125 students, so the mean is obtained from Equation (9.4):

$$\bar{x} = \frac{\sum x_i}{n} = \frac{2156}{30} = 71.9$$

To find the median, we arrange the test scores in ascending order:

44, 48, 49, 53, 54, 59, 60, 61, 63, 65, 66, 67, 68, 72, 73, 76, 76, 77, 78, 79, 79, 80, 84, 84, 84, 86, 87, 92, 95, 97

Note that we have an even number of scores. The two scores in the center of the data are 73 and 76, the 15th and 16th scores. Thus, the median is

$$\text{median} = \frac{(73 + 76)}{2} = 74.5$$

The mode is the number with the highest frequency in the data set. The score of 84 occurs three times—more than any other score—and is therefore the mode. The distribution is unimodal and somewhat resembles a normal distribution similar to that shown in Figure 9.2(a). Another name for normal distribution is *Gaussian* distribution.

Successful engineers are usually easy to spot. They are good communicators, good analysts, good designers, and good experimentalists. Another characteristic that makes a good engineer is the ability to make sound decisions. Decision making is a critical skill, particularly for engineering managers. Not all engineers are managers, but all engineers must be able to make decisions. A decision is a choice between alternatives and may be made by using analytical or nonanalytical techniques. Analytical techniques make up the main segment of engineering academics, emphasizing such topics as circuit analysis, structural analysis, energy analysis, and statistical analysis. Nonanalytical techniques are used to make choices in one's academic major, career, place of residence, spouse, and other matters.

While nonanalytical techniques are typically based on either judgement or intuition, analytical techniques are based on a more systematic approach, which may be broken down into the following steps:

- Recognize and define the decision issue
- Identify alternatives
- Evaluate and select alternative(s)
- Implement the selected alternative(s)
- Evaluate decision results
- Continue improvement

These steps are somewhat reminiscent of the general analysis procedure of problem statement, diagram, assumptions, governing equations, calculations, solution check, and discussion. In the broadest sense, engineering may be considered a decision-making process. Engineering sciences, physical sciences, and applied mathematics, including statistics, help engineers to make the best decisions possible.

9.4 MEASURES OF VARIATION

In the last section, we considered measures of central tendency. Another type of descriptive measure that is used to quantify a data set is referred to as a measure of variation or measure of dispersion. As the name implies, a **measure of variation** is *a number that indicates the extent to which data are spread out or bunched together around the mean*. To help us understand measure of variation, consider the following two data sets of grade point averages:

Data set 1   2.99   3.21   3.33   3.31   3.38   3.29   3.25   3.08

Data set 2   3.01   2.89   3.45   3.89   2.76   3.34   3.01   3.49

The mean of both data sets is 3.23, but the spread in the two data sets is clearly not the same. In the first data set, the GPAs are more tightly bunched around the mean than in the second data set. This can be shown in a simple way by subtracting the mean from the largest number in the data set:

Data set 1 $x_{max} - \bar{x} = 3.38 - 3.23 = 0.15$
Data set 2 $x_{max} - \bar{x} = 3.89 - 3.23 = 0.66$

We reach the same conclusion by subtracting the mean from the smallest number in the data set.

In order to characterize the variation of the entire data set, a formal mathematical definition must take all the data into account. We could expand our simple demonstration above by subtracting the mean from every number in the data set, add the results, and divide by the number of data points, n. For data set 1, this approach yields

$$\frac{\sum(x - \bar{x})}{n} = \frac{\begin{matrix}(2.99 - 3.23) + (3.21 - 3.23) + (3.33 - 3.23) + (3.31 - 3.23) \\ + (3.38 - 3.23) + (3.29 - 3.23) + (3.25 - 3.23) + (3.08 - 3.23)\end{matrix}}{8}$$
$$= 0$$

which is not a useful result. We would also obtain a value of zero for data set 2. The sum of the *deviations* from the mean is always zero. To avoid this difficulty, we square each deviation, sum the squares, divide the sum by the number of data points and take the square root. The result is called the **standard deviation**. For a population, the standard deviation σ is given by the formula

$$\sigma = \left[\frac{\sum (x_i - \mu)^2}{n} \right]^{1/2} \qquad \text{(population, } n > 30) \qquad (9.5)$$

where the Greek letter σ (sigma) denotes standard deviation for a population. Note that the mean for a population (μ) is used in Equation (9.5) rather than $\bar{x}$, which denotes the mean for a sample. If the standard deviation for a large sample is desired, Equation (9.5) may be used as long as the sample size is approximately 30 or larger. To distinguish standard deviation for samples and populations, we use a lowercase italic s for samples. Thus,

$$s = \left[\frac{\sum (x_i - \bar{x})^2}{n} \right]^{1/2} \qquad \text{(large sample, } n > 30) \qquad (9.6)$$

For small samples $(n \leq 30)$, statisticians have discovered that Equation (9.6) underestimates the standard deviation, and that if n is replaced with $n - 1$, the result is more accurate. Thus,

$$s = \left[\frac{\sum (x_i - \bar{x})^2}{n - 1} \right]^{1/2} \qquad \text{(small sample, } n \leq 30) \qquad (9.7)$$

A second measure of variation is called variance. **Variance** is *simply the square of the expressions* in Equations (9.5), (9.6), and (9.7). Thus, the variance for a population, large sample, and small sample are, respectively,

$$\sigma^2 = \frac{\sum (x_i - \mu)^2}{n} \qquad \text{(population, } n > 30) \qquad (9.8)$$

$$s^2 = \frac{\sum (x_i - \bar{x})^2}{n} \qquad \text{(large sample, } n > 30) \qquad (9.9)$$

$$s^2 = \frac{\sum (x_i - \bar{x})^2}{n - 1} \qquad \text{(small sample, } n \leq 30) \qquad (9.10)$$

EXAMPLE 9.2

The data sets discussed in Section 9.4 represent GPA samples from two different engineering classes. Find the standard deviation and variance for each class. The GPAs are repeated here:

Class 1   2.99   3.21   3.33   3.31   3.38   3.29   3.25   3.08

Class 2   3.01   2.89   3.45   3.89   2.76   3.34   3.01   3.49

SOLUTION

There are only eight numbers in each class, so we must use the equations for a small sample. Recalling that $\bar{x} = 3.23$ for both classes, the standard deviations are

$$\text{Class 1} \qquad s = \left[\frac{\sum(x_i - \bar{x})^2}{n - 1}\right]^{1/2} = \left[\frac{0.1234}{8 - 1}\right]^{1/2} = 0.133$$

and

$$\text{Class 2} \qquad s = \left[\frac{\sum(x_i - \bar{x})^2}{n - 1}\right]^{1/2} = \left[\frac{0.9970}{8 - 1}\right]^{1/2} = 0.377$$

Based on our earlier observations, these results are expected. In the first class, GPAs are tightly bunched around the mean, whereas GPAs in the second class have a large spread. Consequently, we have a smaller standard deviation in class 1 than in class 2. The variance is the square of the standard deviation,

$$\text{Class 1} \qquad s^2 = \frac{0.1234}{8 - 1} = 0.0176$$

$$\text{Class 2} \qquad s^2 = \frac{0.9970}{8 - 1} = 0.142$$

The widely used measures of central tendency and measures of variation are standard functions on scientific calculators, spreadsheets, and other computer-based tools. You are encouraged to become familiar with these tools and to learn how how to use them in your engineering courses.

9.5   NORMAL DISTRIBUTION

Earlier we examined the height distribution for a sample of residents in a hypothetical town. We began the study by sorting the height data into classes. From the sorted data, we constructed a histogram, a special type of graph that shows the frequency distribution of a measured quantity. To illustrate our next topic, let's consider the height distribution for a sample of residents in a different hypothetical town called Anyville, USA. After sorting the height data into classes, we construct the histogram shown in Figure 9.5. The calibrations on the horizontal axis represent the upper bounds for each data class. The histogram was constructed by dividing the data into 13 classes of 0.2 ft each from a sample size of 102.

Like all histograms, the histogram in Figure 9.5 is a graph of *discrete* quantities, since each bar represents the frequency for a range of distinct individual heights. The area of each bar represents the probability that the height of a person in Anyville will fall into a specific range. (Because the width of all bars is equal, we can say that the vertical length of each bar likewise represents these probabilities.) For example, the probability that a person will have a height between 5.0 ft and 5.2 ft is 16 out of 102, or 0.157. The probability that a person will have a height between 5.8 ft and 6.0 ft is 8 out of 102, or 0.0784, and so on. Histograms are not like graphs of *continuous* quantities such as length, flow rate, stress, or voltage. However, the discrete values on a histogram may be approximated as a continuous quantity by drawing a best-fit curve through the tops of the bars, as illustrated in Figure 9.6. In this manner, a continuous frequency distribution is derived from a discrete frequency distribution. A continuous frequency distribution can also be explained by visualizing an idealized situation in which the number of residents in our Anyville sample approaches infinity, thereby yielding infinitesimally narrow bars in the

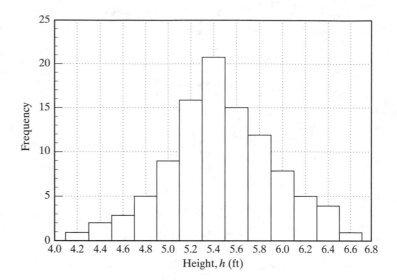

Figure 9.5. Histogram of heights for Anyville, USA.

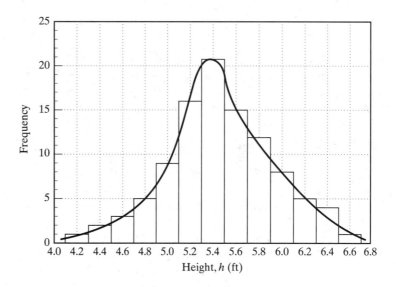

Figure 9.6. A continuous distribution approximates a discrete set of values.

histogram. Because each bar would essentially be reduced to a vertical line, a smooth curve drawn through the points at the top of each line would produce the desired continuous distribution. Using the continuous distribution, we find that the probability that a resident of Anyville will have a height in a given range is the area under the portion of the curve corresponding to that range, as illustrated in Figure 9.7. The probability that a person will have *any* height is the area under the *entire* curve and has a value of unity.

As one can see from the histogram in Figure 9.5 and the corresponding continuous distribution in Figure 9.7, the distribution of heights for the residents of Anyville is nearly symmetrical about the central peak. To perform a statistical analysis of data that approximates a symmetrical distribution, we use a special theoretical distribution called a *normal distribution* or *Gaussian distribution*, named after the German mathematician Carl Gauss

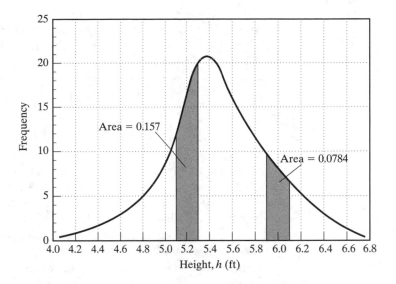

Figure 9.7. The area under a portion of a continuous frequency distribution represents a probability.

(1777–1855). A **normal distribution** is *a curve with a characteristic bell shape that is symmetrical about the mean and extends indefinitely in both directions*, as illustrated in Figure 9.8. The bell-shaped curve asymptotically approaches the horizontal axis on both

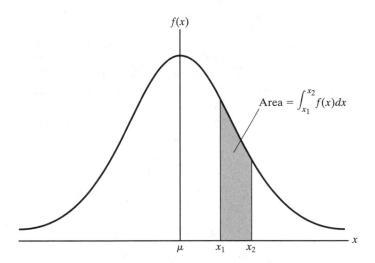

Figure 9.8. Normal distribution.

sides and is symmetrical about the mean. The location and shape of the normal distribution is specified by two quantities, the mean μ, which locates the center of the distribution, and the standard deviation σ, which describes the dispersion or spread of the data around the mean. The normal distribution is given by the mathematical function

$$f(x) = \frac{1}{\sigma\sqrt{2\pi}}e^{-\frac{1}{2}(x-\mu)^2/\sigma^2}$$

(9.11)

where x represents the continuous quantity being studied, which was height in our previous example. This equation can be used to find the probability that the quantity being studied will fall into a particular range of values. As we saw before, such a probability is represented by the area under the portion of the curve corresponding to that range. From calculus, the area under a curve is found by integrating over the interval of interest. Thus, the area (probability) for a specific range of x values is given by the relation

$$area = \int_{x_1}^{x_2} f(x)\, dx \tag{9.12}$$

where x_1 and x_2 are the lower and upper bounds for the range of interest, as shown in Figure 9.8, and the function $f(x)$ is given by Equation (9.11). This integration is cumbersome, so a special table has been developed that eliminates the need for performing the integration every time a new problem arises. The use of the table will be discussed later.

Equation (9.12) yields an area that reflects specific values of the mean μ and the standard deviation σ. This means that a separate table would have to be prepared for every different value of μ and σ, which would be extremely inconvenient. To avoid this difficulty, a transformation is applied to the normal distribution function given by Equation (9.11) such that a single table can be used. Applying the transformation

$$z = \frac{x - \mu}{\sigma} \tag{9.13}$$

normalizes the distribution to a *standard normal distribution* that has $\mu = 0$ and $\sigma = 1$. Hence, Equation (9.11) becomes

$$f(z) = \frac{1}{\sqrt{2\pi}} e^{-\frac{1}{2}z^2} \tag{9.14}$$

In order to find areas under the standard normal curve, we convert x values into z values by using Equation (9.13). The transformation results in a change of scale for the normal distribution shown in Figure 9.8. The x scale has a mean of μ and is graduated in terms of positive and negative values of σ from the mean, while the z scale has a mean of 0 and is graduated in terms of positive and negative numbers from the mean. For example, a data value that is 2 standard deviations from the mean (2σ from μ) has a z value of $z = (x - \mu)/\sigma = (2\sigma - 0)/\sigma = 2$. The standard normal distribution showing the x and z scales is illustrated in Figure 9.9. Consequently, the area under a specified portion of the standard normal distribution curve is given by the relation

$$area = \int_{z_1}^{z_2} f(z)\, dz = \frac{1}{\sqrt{2\pi}} \int_{z_1}^{z_2} e^{-\frac{1}{2}z^2} \tag{9.15}$$

where z is the transformed variable given by Equation (9.13) and z_1 and z_2 are the lower and upper bounds, respectively, for the interval of interest.

Earlier, we mentioned that the integration of the normal distribution function is cumbersome, necessitating the use of a special table. The transformation leading to Equation (9.15) does not make the integration any easier, but the transformation permits us to develop a *single* table that can be used for all values of μ and σ. As shown in Figure 9.9, the standard normal distribution is symmetrical about $z = 0$, so we need only to evaluate the integral in Equation (9.15) from $z = 0$ to $z = z_2$ to find any area of interest. The integration in Equation (9.15) has been evaluated over intervals from 0 to z, where z assumes a range of values from 0 to about 4. The results are presented in Table 9.4.

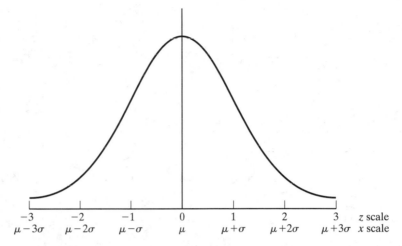

Figure 9.9. Standard normal distribution, showing the transformed scale.

Before using Table 9.4 to work some examples, let's explain how to read the table. The first column in the table contains z values from 0 to 3.9 in increments of 0.1. The numbers in the top row are used if the value of z has a nonzero hundredths digit. For example, the area under the curve from $z = 0$ to $z = 1.50$ is 0.4332. The area under the curve from $z = 0$ to $z = 1.57$ is 0.4418. Because the distribution is symmetrical about $z = 0$, we can deal with negative z values as well. For example, the area under the curve from $z = -1.46$ to $z = 0$ is 0.4279. The area under the curve from $z = -2.33$ to $z = 1.78$ is $(0.4901 + 0.4625) = 0.9526$. Note that the area from $z = 0$ to $z = 3.9$ is 0.5000, half the area under the entire curve. For z values greater than 3.9, the normal curve is so close to the horizontal axis that no significant additional area is obtained.

We should not lose sight of the physical significance of these "areas." Remember, the area under a specified region of a frequency distribution curve represents the probability that a data value will fall into that region or interval. For example, assuming that our data follows a normal distribution, the probability that a data value will fall into the interval $z = -1.32$ to $z = 0.87$ is $(0.4066 + 0.3078) = 0.7144$, or 71.44 percent. In many engineering applications, we consider data intervals centered on the mean at $z = 0$ that have spreads with integer values of the standard deviation σ. From Table 9.4, the probability that a value will lie within one standard deviation of the mean (i.e., within $\pm 1\sigma$ of μ) is

$$\frac{1}{\sqrt{2\pi}} \int_{-1}^{+1} e^{-\frac{1}{2}z^2}\, dz = 2(0.3413) = 0.6826$$

which means that $\pm 1\sigma$ about the mean encompasses 68.26 percent of the data. The probability that a value will lie within two standard deviations of the mean (i.e., within $\pm 2\sigma$ of μ) is

$$\frac{1}{\sqrt{2\pi}} \int_{-2}^{+2} e^{-\frac{1}{2}z^2} dz = 2(0.4772) = 0.9544$$

which means that $\pm 2\sigma$ about the mean encompasses 95.44 percent of the data. The probability that a value will lie within three standard deviations of the mean (i.e., within $\pm 3\sigma$ of μ) is

$$\frac{1}{\sqrt{2\pi}} \int_{-3}^{+3} e^{-\frac{1}{2}z^2}\, dz = 2(0.4987) = 0.9974$$

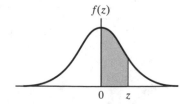

TABLE 9.4 Areas Under the Standard Normal Curve from 0 to z.

z	.00	.01	.02	.03	.04	.05	.06	.07	.08	.09
0.0	.0000	.0040	.0080	.0120	.0160	.0199	.0239	.0279	.0319	.0359
0.1	.0398	.0438	.0478	.0517	.0557	.0596	.0636	.0675	.0714	.0754
0.2	.0793	.0832	.0871	.0910	.0948	.0987	.1026	.1064	.1103	.1141
0.3	.1179	.1217	.1255	.1293	.1331	.1368	.1406	.1443	.1480	.1517
0.4	.1554	.1591	.1628	.1664	.1700	.1736	.1772	.1808	.1844	.1879
0.5	.1915	.1950	.1985	.2019	.2054	.2088	.2123	.2157	.2190	.2224
0.6	.2258	.2291	.2324	.2357	.2389	.2422	.2454	.2486	.2518	.2549
0.7	.2580	.2612	.2642	.2673	.2704	.2734	.2764	.2794	.2823	.2852
0.8	.2881	.2910	.2939	.2967	.2996	.3023	.3051	.3078	.3106	.3133
0.9	.3159	.3186	.3212	.3238	.3264	.3289	.3315	.3340	.3365	.3389
1.0	.3413	.3438	.3461	.3485	.3508	.3531	.3554	.3577	.3599	.3621
1.1	.3643	.3665	.3686	.3708	.3729	.3749	.3770	.3790	.3810	.3830
1.2	.3849	.3869	.3888	.3907	.3925	.3944	.3962	.3980	.3997	.4015
1.3	.4032	.4049	.4066	.4082	.4099	.4115	.4131	.4147	.4162	.4177
1.4	.4192	.4207	.4222	.4236	.4251	.4265	.4279	.4292	.4306	.4319
1.5	.4332	.4345	.4357	.4370	.4382	.4394	.4406	.4418	.4429	.4441
1.6	.4452	.4463	.4474	.4484	.4495	.4505	.4515	.4525	.4535	.4545
1.7	.4554	.4564	.4573	.4582	.4591	.4599	.4608	.4616	.4625	.4633
1.8	.4641	.4649	.4656	.4664	.4671	.4678	.4686	.4693	.4699	.4706
1.9	.4713	.4719	.4726	.4732	.4738	.4744	.4750	.4756	.4761	.4767
2.0	.4772	.4778	.4783	.4788	.4793	.4798	.4803	.4808	.4812	.4817
2.1	.4821	.4826	.4830	.4834	.4838	.4842	.4846	.4850	.4854	.4857
2.2	.4861	.4864	.4868	.4871	.4875	.4878	.4881	.4884	.4887	.4890
2.3	.4893	.4896	.4898	.4901	.4904	.4906	.4909	.4911	.4913	.4916
2.4	.4918	.4920	.4922	.4925	.4927	.4929	.4931	.4932	.4934	.4936
2.5	.4938	.4940	.4941	.4943	.4945	.4946	.4948	.4949	.4951	.4952
2.6	.4953	.4955	.4956	.4957	.4959	.4960	.4961	.4962	.4963	.4964
2.7	.4965	.4966	.4967	.4968	.4969	.4970	.4971	.4972	.4973	.4974
2.8	.4974	.4975	.4976	.4977	.4977	.4978	.4979	.4979	.4980	.4981
2.9	.4981	.4982	.4982	.4983	.4984	.4984	.4985	.4985	.4986	.4986
3.0	.4987	.4987	.4987	.4988	.4988	.4989	.4989	.4989	.4990	.4990
3.1	.4990	.4991	.4991	.4991	.4992	.4992	.4992	.4992	.4993	.4993
3.2	.4993	.4993	.4994	.4994	.4994	.4994	.4994	.4995	.4995	.4995
3.3	.4995	.4995	.4995	.4996	.4996	.4996	.4996	.4996	.4996	.4997
3.4	.4997	.4997	.4997	.4997	.4997	.4997	.4997	.4997	.4997	.4998
3.5	.4998	.4998	.4998	.4998	.4998	.4998	.4998	.4998	.4998	.4998
3.6	.4998	.4998	.4999	.4999	.4999	.4999	.4999	.4999	.4999	.4999
3.7	.4999	.4999	.4999	.4999	.4999	.4999	.4999	.4999	.4999	.4999
3.8	.4999	.4999	.4999	.4999	.4999	.4999	.4999	.4999	.4999	.4999
3.9	.5000	.5000	.5000	.5000	.5000	.5000	.5000	.5000	.5000	.5000

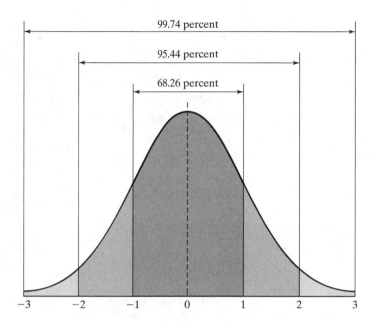

Figure 9.10.   Intervals of $\pm 1\sigma$, $\pm 2\sigma$, and $\pm 3\sigma$, centered on the mean.

which means that $\pm 3\sigma$ about the mean encompasses 99.74 percent of the data. These probabilities are illustrated in Figure 9.10. If we integrate from minus infinity to plus infinity, we obtain a probability of unity, or 100 percent.

In most statistical analyses of engineering data, we do not know the population parameters μ and σ, but we know the mean and standard deviation for a sample taken from a population. As long as the sample size is larger than $30(n > 30)$, we can substitute the sample parameters for the population parameters in the normal distribution.

EXAMPLE 9.3

In a production run of carbon-composition resistors, the mean resistance is $\mu = 100\ \Omega$ and the standard deviation is $\sigma = 4.7\ \Omega$. Assuming a normal distribution of resistances, what is the probability that a resistor will have a resistance R that lies in the range $95\ \Omega < R < 109\ \Omega$?

SOLUTION

In order to use the standard normal curve tabulated in Table 9.4, we must make a transformation to the z variable. We define our lower and upper limits as

$$x_1 = 95 \quad x_2 = 109$$

Noting that $\mu = 100$ and $\sigma = 4.7$, we obtain

$$z_1 = \frac{x_1 - \mu}{\sigma} = \frac{95 - 100}{4.7} = -1.06$$

and

$$z_2 = \frac{x_2 - \mu}{\sigma} = \frac{109 - 100}{4.7} = 1.91$$

The probability that a resistor will have a resistance in the interval $95\ \Omega < R < 100\ \Omega$ is the probability that z will lie in the interval $0 < z < 1.06$. Using Table 9.4, we find that this probability is 0.3554. Similarly, the probability that a resistor will have a resistance in the interval $100\ \Omega < R < 109\ \Omega$ is the probability that z will lie in the interval $0 < z < 1.91$.

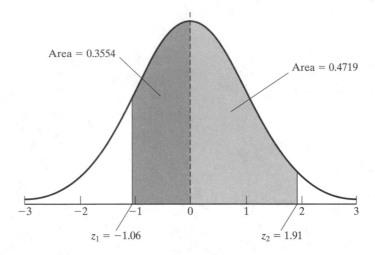

Figure 9.11. Probabilities for Example 9.3.

Using Table 9.4, we find that this probability is 0.4719. These two probabilities are represented as areas under the normal curve in Figure 9.11. The probability that a resistor will have a resistance in the range 95 Ω < R < 109 Ω is 0.3554 + 0.4719 = 0.8273. Hence, 82.73 percent of the resistors will have a resistance within this range. The remaining 17.27 percent of the resistors will have resistances that are lower than 95 Ω or higher than 109 Ω.

APPLICATION: USING THE NORMAL DISTRIBUTION TO EVALUATE LAMP LIFETIMES

One of the most widely used applications of statistics is the evaluation of manufactured products. Lamp lifetime is a crucial parameter in the lighting industry, because this number is typically printed on the product package for consumers to see. In conjunction with laboratory testing, lighting manufacturers use statistics to evaluate lifetimes of lamps.

A relatively new lighting product on the market is the compact flourescent lamp, which consumes less electrical power, produces more light for a given amount of electrical power supplied to it, and lasts about 10 times longer than standard incandescent lamps. The typical lifetime of a standard incandescent lamp is 1000 hours, whereas the typical lifetime of a compact flourescent lamp is 10,000 hours. Compact flourescent lamps cost more than standard incandescent lamps, but their long life is an attractive feature to many consumers because of the convenience afforded by a very long replacement schedule.

Based on customer complaints, the sales department for a major manufacturer of compact flourescent lamps claims that 11 percent of lamps sold are "burning out" after only 8700 hours of use. To address this claim, a quality engineer at the manufacturing facility pulls a sample of 100 lamps from the production line for testing.

Based on the tests, the engineer determines that the mean lifetime for the sample is 10,800 hours with a standard deviation of 1150 hours. Assuming that lamp lifetime follows a normal distribution, we have

$$z_1 = \frac{x_1 - \mu}{\sigma} = \frac{8700 - 10,800}{1150} = -1.83$$

Using Table 9.4, we find that the area corresponding to this z value is 0.4664, which means that the probability that a lamp will fail after only 8700 hours of use is

$$1. - (0.4664 + 0.5000) = 0.0336 \quad (3.36 \text{ percent})$$

This probability is depicted in Figure 9.12.

The claim made by the sales department that 11 percent of the lamps are failing after 8700 hours of use does not agree with the statistical analysis, which asserts that the percentage is much lower, about 3.4 percent. The discrepancy could be due to casual or sloppy data gathering by the sales department. However, the 11 percent failure rate could be accurate if it is based on a certain production lot with a manufacturing defect. The problem could be investigated further by doing a second statistical analysis on another sample or by recalling some of the defective lamps for testing.

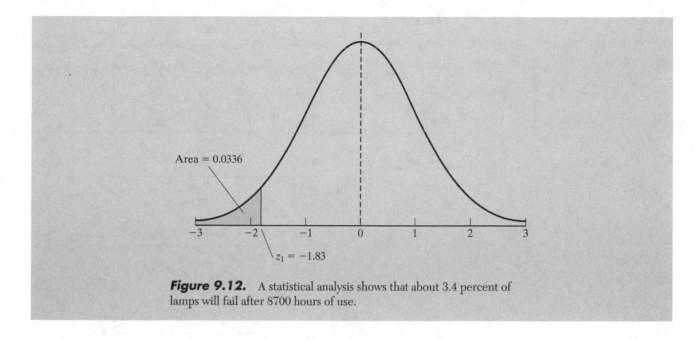

Figure 9.12. A statistical analysis shows that about 3.4 percent of lamps will fail after 8700 hours of use.

KEY TERMS

frequency distribution	median	sample
histogram	mode	standard deviation
mean	normal distribution	statistics
measure of variation	population	variance

REFERENCES

Ayyub, B.M. and R.H. McCuen, *Probability, Statistics, and Reliability for Engineers and Scientists*, 2d ed., Boca Raton, FL: Chapman & Hall/CRC, 2002.

Devore, J.L., *Probability and Statistics for Engineering and the Sciences*, 6th ed., Belmont, CA: Brooks/Cole, 2004.

Petruccelli, J.D., Nandram B., and M. Chen, *Applied Statistics for Engineers and Scientists*, Upper Saddle River, NJ: Prentice Hall, 1999.

Vining, G.G., *Statistical Methods for Engineers*, Pacific Grove, CA: Duxbury Press, 1998.

Vardeman, S.B., *Statistics for Engineering Problem Solving*, Boston, MA: PWS Publishing Company, 1994.

Problems

1. The grade point averages (GPAs) in a freshman level engineering class are given in Table P9.1. Subdivide the GPAs into at least six classes and construct a histogram. Also, find the mean, median, mode, and standard deviation.

2. The weights (in oz) of filled soup cans as they come off the production line are given in Table P9.2. Subdivide the weights into at least six classes and construct a histogram. Also, find the mean, median, mode, and standard deviation.

3. Using Table 9.4, find the area under the normal curve in each of the cases (a) through (g) in Figure P9.3.

TABLE P9.1

2.34	3.37	3.02	3.17	2.59	2.23	2.84	2.76
3.68	3.20	2.84	1.80	2.95	2.70	3.40	2.70
2.85	1.56	2.70	3.22	2.30	2.10	2.74	2.45
1.90	3.33	2.95	3.22	2.40	3.21	2.85	3.45
3.15	2.95	2.40	2.20	2.70	2.95	3.19	2.11
2.60	2.72	2.85	3.05	2.60	2.98	3.22	2.84

TABLE P9.2

15.73	16.25	16.10	16.69	16.05	15.92	16.10	16.30
15.30	15.02	15.85	16.23	16.80	16.40	15.91	15.42
15.70	16.10	16.23	16.33	16.66	15.70	15.85	16.20
16.41	16.54	16.37	15.80	16.19	16.33	15.81	16.18

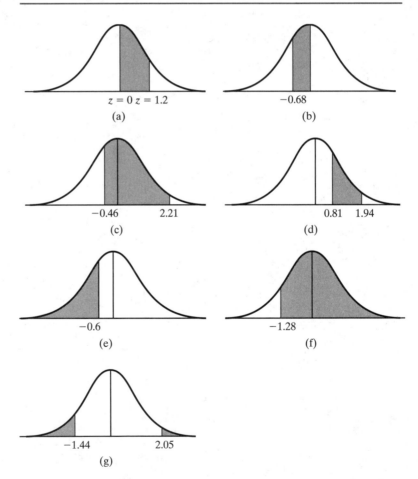

Figure P9.3.

4.   In a facility that manufactures electrical resistors, a sample of 35 1-kΩ resistors are randomly pulled from the production line, and their resistances are measured and recorded, as shown in Table P9.4. The desired resistance tolerance for the resistors is ± 10 percent, meaning that the acceptable range of resistance is 900 Ω to 1100 Ω.

TABLE P9.4

1005	1036	1082	940	972	1002	995
1060	900	1015	985	1055	1040	1010
955	1045	1090	1008	980	930	972
993	1020	1072	1045	928	1012	1032
1061	1018	978	952	1016	977	1019

a. Subdivide the resistances into at least six classes and construct a histogram.
b. Find the mean and standard deviation.
c. Assuming a normal distribution, how many standard deviations on either side of the mean does the ±10 percent tolerance represent?
d. Which resistors in the sample fall outside the ±1σ range? Which resistors fall outside the ±2σ range?

5. A sample of 40 microprocessor chips are randomly taken from a production run and tested to determine their processing speeds, as listed in Table P9.5. Chips

TABLE P9.5

3.05	3.30	2.80	3.90	2.26	3.20	2.85	3.65
3.02	3.15	3.45	2.52	3.60	3.33	2.70	3.40
2.78	3.55	3.35	2.70	2.72	3.12	3.19	3.28
2.79	2.27	3.02	3.08	3.54	2.92	2.80	3.03
3.31	3.45	3.10	3.36	2.82	3.21	3.05	3.37

that have speeds lower than a −2σ value are outside the specifications and are to be scrapped.
a. Subdivide the speeds into at least six classes and construct a histogram.
b. Find the mean and standard deviation.
c. Which chips in the sample are to be scrapped?

6. A quality engineer in a fastener-manufacturing plant extracts a random sample of 45 hex-head bolts from an assembly line to determine whether the bolts meet specifications which state that the length of the bolts must fall within ±2σ of a mean length of 2.000 in or be scrapped. Using the data in Table P9.6, do the following:

TABLE P9.6

2.003	1.999	1.998	2.007	1.996	1.992	2.002	1.991	2.011
2.005	2.000	1.988	1.995	1.993	2.000	2.007	2.003	2.000
1.996	1.998	2.007	2.004	1.995	2.003	2.000	1.997	2.012
2.000	2.004	1.994	1.991	2.004	2.001	1.995	2.000	2.003
2.007	1.995	2.003	2.000	1.994	2.000	2.003	1.995	1.999

a. Subdivide the bolt lengths into at least six classes and construct a histogram.
b. Find the mean and standard deviation.
c. Assuming a normal distribution, how many bolts are scrapped daily if 40,000 bolts are manufactured each day?
d. What length tolerance, measured in inches, does ±2σ represent?

7. Oxygen-free, high-conductivity (OFHC) copper has a minimum purity level of 99.99 percent copper. This type of copper is used in electrical and other applications where high-purity copper is required. In a materials-testing laboratory, elemental analyses of OFHC copper samples are performed to ascertain whether they have the minimum purity level. In a sample of 50, the mean purity level is 99.995 percent, and the standard deviation is 0.0045 percent. Assuming a normal distribution,

   a. How many standard deviations from the mean is 99.99 percent purity?
   b. If 10,000 parts are made from this OFHC copper daily, how many parts per day would not qualify as being made from OFHC copper?

8. The mean inside diameter of a sample of 200 washers manufactured by a machine is 0.502 in, and the standard deviation is 0.005 in. The application for the washers permits a tolerance in the diameter of 0.496 in to 0.508 in. If the diameter is outside this tolerance, the washers are deemed defective and are sold as scrap. Assuming a normal distribution of washer diameters,

   a. What percentage of washers are discarded?
   b. If 20,000 washers are manufactured each day, how many washers are discarded daily?
   c. Twenty-five washers have a combined mass of 1 lb_m. If the scrap dealer pays $0.85/$lb_m$, how much money is recovered per day from the sale of scrap washers?

9. The mean annual precipitation for Dilbertville is 44 in with a standard deviation of 6.5 in. Assuming the precipitation follows a normal distribution,

   a. Find the probability that the precipitation in any given year is greater than 55 in.
   b. Find the probability that the precipitation in any given year is less than 35 in.

10. The compressive strength of concrete specimens follows a normal distribution with a mean value of 2.75 ksi and a standard deviation of 0.30 ksi. If the applied stress is 2.5 ksi, what is the probability of failure?

11. The mean tensile yield strength of a large sample of structural steel specimens is 250 MPa. Fifteen percent of the specimens are observed to fail in a tension test that exerts a tensile stress of 225 MPa in the specimens. Assuming a normal distribution of yield strengths, find the number of standard deviations that corresponds to this failure rate.

12. The daily production of sulfuric acid is known to follow a normal distribution with a mean of 300 tons per day with a standard deviation of 75 tons per day.

   a. Find the probability that today's production will yield between 260 and 350 tons.
   b. Find the probability that today's production will yield less than 230 tons.

13. To evaluate the performance of a certain brand of alkaline battery, researchers at a consumer testing laboratory measure the lifetime of 160 1.5-volt batteries. Battery lifetime for this study is defined as the time it takes the voltage to drop to 1.0 V under a standard electrical load. The researchers determine that the mean lifetime is 48.3 hours with a standard deviation of 15.7 hours. Assuming a normal distribution of battery lifetimes,

a. Find the probability that a battery's voltage will drop below 1.0 V after 20 hours of use.

b. Find the probability that a battery's voltage will drop below 1.0 V after 70 hours of use.

c. If the daily production rate of 1.5-V batteries is 25,000, how many batteries per day will have lifetimes of 20 hours or less and 70 hours or more?

14. Journal bearings are ground to a mean diameter of 2.0002 in with a standard deviation of 0.0004 in. Assuming the diameters follow a normal distribution, what fraction of the bearings are within specifications if the allowable diameter is 2.0000 $\pm$ 0.0005 in?

15. A company manufactures aluminum rivets for use in the aircraft industry. From a sample of 1000 rivets, it is determined that the mean rivet diameter is 25.5 mm and the standard deviation is 0.8 mm. The company rejects rivets that do not meet the diameter specification of 25.2 $\pm$ 1.0 mm. If the cost of labor and materials is \$1.05/rivet, find the financial loss incurred per 1000 rivets manufactured by assuming a normal distribution on rivet diameter. What would the financial loss be if the specification was 25.2 $\pm$ 0.5 mm?

16. At cruising altitude, a commercial jet engine consumes an average of 850 gallons of fuel per hour with a standard deviation of 48 gallons per hour. Assuming that the fuel consumption follows a normal distribution at cruising altitude, find the probability that the hourly fuel consumption is

a. Between 700 and 950 gallons.

b. Less than 750 gallons.

c. More than 1000 gallons.

17. The cost of fuel for the commercial jet engine in Problem 16 is \$0.75/gallon. If the commercial aircraft cruises 225 hours per month, find the probability that the monthly cost of fuel exceeds \$150,000.

18. Big Brother Electronics, Inc., manufacturers compact-disc (CD) players. Their research and development (R&D) department determined that the mean life of the laser beam in their CD players is 4500 hours with a standard deviation of 400 hours. Big Brother Electronics wants to place a guarantee on the players so that no more than 5 percent fail during the guarantee period. Because the laser pickup is the part most likely to fail first, the guarantee period will be based on the laser beam device. Assuming a normal distribution, how many playing hours should the guarantee cover?

Appendix A
Mathematical Formulas

A.1.1   Quadratic equation

The quadratic equation

$$ax^2 + bx + c = 0 \quad (a \neq 0)$$

has the solution

$$x = \frac{-b \pm \sqrt{b^2 - 4ac}}{2a}$$

The two roots of the quadratic equation are either (a) both real or (b) complex conjugates.

A.1.2   Laws of exponents

$$x^m x^n = x^{m+n}$$
$$(x^m)^n = x^{mn}$$
$$(xy)^n = x^n y^n$$
$$(x/y)^n = x^n/y^n \qquad (y \neq 0)$$
$$(x^m/x^n) = x^{m-n} \qquad (x \neq 0)$$
$$x^m/x^n = 1 \qquad \text{if } m = n$$
$$x^0 = 1$$
$$x^{-n} = 1/x^n \qquad (x \neq 0)$$

A.1.3   Logarithms

In the relations that follow, the parameter a is called the *base*. These relations hold when $a > 0$ and $a \neq 1$. Typically, $a = 10$ (the common logarithm), or $a = e$ (the natural logarithm). The common logarithm is usually written as $\log (x)$, and the natural logarithm is usually written as $\ln(x)$:

$$v = \log_a u \qquad\qquad\qquad if\, a^v = u$$
$$x = \log_a a^x$$
$$\log_a(xy) = \log_a x + \log_a y$$

$$\log_a(x/y) = \log_a x - \log_a y$$
$$\log_a 1 = 0$$
$$\log_a(x^c) = c \log_a x$$
$$\log_a a = 1$$

A.1.4 Exponential Function

The same relations that apply to the laws of exponents apply to the exponential function $\exp(x) = e^x$:

$$e^m e^n = e^{m+n}$$
$$(e^m)^n = e^{mn}$$
$$(e^m/e^n) = e^{m-n}$$
$$e^m/e^n = 1 \qquad \text{if } m = n$$
$$e^0 = 1$$
$$e^{-n} = 1/e^n$$

The exponential function $\exp(x)$ and the natural logarithm $\ln(x)$ are inverse functions. Thus,

$$\ln(\exp(x)) = x$$
$$\exp(\ln(x)) = x$$

A.2 GEOMETRY

A.2.1 Areas

Rectangle $\qquad\qquad A = ab$

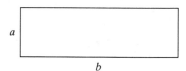

Parallelogram $\qquad\qquad A = bh$

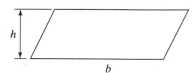

Trapezoid $\qquad\qquad A = \tfrac{1}{2} h(a + b)$

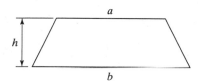

Triangle $\qquad\qquad A = \tfrac{1}{2} bh$

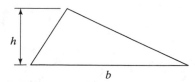

Circle $\qquad\qquad\qquad$ $A = \pi R^2 = \frac{1}{4}\pi D^2$

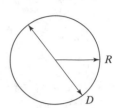

Circular sector $\qquad\qquad$ $A = \frac{1}{2}R^2\theta \;\; (\theta \text{ in radians})$

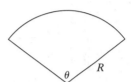

Circular segment $\qquad\quad$ $A = \frac{1}{2}R^2(\theta - \sin\theta)$
$\qquad\qquad\qquad\qquad\qquad$ $(\theta \text{ in radians})$

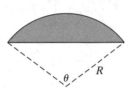

Regular polygon $\qquad\quad$ $A = nr^2\tan(180°/n)$
$\qquad\qquad\qquad\qquad\quad$ $= \frac{1}{2}nR^2\sin(360°/n)$
$\qquad\qquad\qquad\qquad$ $n = \text{number of sides}$

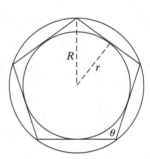

A.2.2 Solids

$A = $ Surface area
$V = $ Volume

Parallelpiped $\qquad\qquad$ $A = 2(ab + ac + bc)$
$\qquad\qquad\qquad\qquad$ $V = abc$

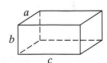

Cylinder $\qquad\qquad\qquad$ $A = 2\pi RL = \pi DL \text{ (ends excluded)}$
$\qquad\qquad\qquad\qquad$ $A = 2\pi R(L + R)(\text{total})$
$\qquad\qquad\qquad\qquad$ $V = \pi R^2 L$

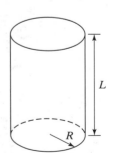

Sphere

$$A = 4\pi R^2 = \pi D^2$$

$$V = \frac{4\pi R^3}{3} = \frac{\pi D^3}{6}$$

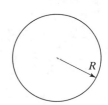

Cone

$$A = \pi R(R^2 + h^2)^{1/2} \text{ (base excluded)}$$

$$A = \pi R[R + (R^2 + h^2)^{1/2}] \text{ (total)}$$

$$V = \frac{\pi R^2 h}{3}$$

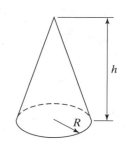

Torus

$$A = 4\pi^2 Rr$$

$$V = 2\pi^2 Rr^2$$

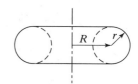

A.3   TRIGONOMETRY

A.3.1   Trigonometric Functions

$$\sin\theta = \frac{\text{opposite side}}{\text{hypotenuse}} = \frac{b}{c}$$

$$\cos\theta = \frac{\text{adjacent side}}{\text{hypotenuse}} = \frac{a}{c}$$

$$\tan\theta = \frac{\sin\theta}{\cos\theta} = \frac{\text{opposite side}}{\text{adjacent side}} = \frac{b}{a}$$

$$\cot\theta = \frac{1}{\tan\theta} = \frac{a}{b}$$

$$\sec\theta = \frac{1}{\cos\theta} = \frac{c}{a}$$

$$\csc\theta = \frac{1}{\sin\theta} = \frac{c}{b}$$

A.3.2   Identities and Relationships

$$\sin(-\theta) = -\sin(\theta)$$
$$\cos(-\theta) = \cos(\theta)$$
$$\tan(-\theta) = -\tan(\theta)$$
$$\sin^2\theta + \cos^2\theta = 1$$
$$1 + \tan^2\theta = \sec^2\theta$$

$$1 + \cot^2 \theta = \csc^2 \theta$$

$$\sin \theta = \cos(90° - \theta) = \sin(180° - \theta)$$

$$\cos \theta = \sin(90° - \theta) = -\cos(180° - \theta)$$

$$\tan \theta = \cot(90° - \theta) = -\tan(180° - \theta)$$

$$\sin(\theta \pm \alpha) = \sin \theta \cos \alpha \pm \cos \theta \sin \alpha$$

$$\cos(\theta \pm \alpha) = \cos \theta \cos \alpha \mp \sin \theta \sin \alpha$$

$$\tan(\theta \pm \alpha) = \frac{\tan \theta \pm \tan \alpha}{1 \mp \tan \theta \tan \alpha}$$

$$\sin 2\theta = 2 \sin \theta \cos \theta$$

$$\cos 2\theta = \cos^2 \theta - \sin^2 \theta = 2 \cos^2 \theta - 1 = 1 - 2 \sin^2 \theta$$

$$\tan 2\theta = \frac{2 \tan \theta}{1 - \tan^2 \theta}$$

$$\sin(\theta/2) = \pm\sqrt{\frac{1 - \cos \theta}{2}}$$

$$\cos(\theta/2) = \pm\sqrt{\frac{1 + \cos \theta}{2}}$$

$$\tan(\theta/2) = \frac{\sin \theta}{1 + \cos \theta} = \frac{1 - \cos \theta}{\sin \theta}$$

A.3.3   Laws of Sines and Cosines

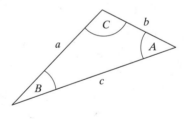

Law of sines
$$\frac{\sin A}{a} = \frac{\sin B}{b} = \frac{\sin C}{c}$$

Law of cosines
$$a^2 = b^2 + c^2 - 2bc \cos A$$
$$b^2 = a^2 + c^2 - 2ac \cos B$$
$$c^2 = a^2 + b^2 - 2ab \cos C$$

A.4   CALCULUS

In the following formulas, u and v represent functions of x, while a and n represent constants:

A.4.1   Derivatives

$$\frac{d(a)}{dx} = 0$$

$$\frac{d(x)}{dx} = 1$$

$$\frac{d(au)}{dx} = a\frac{du}{dx}$$

$$\frac{d(uv)}{dx} = u\frac{dv}{dx} + v\frac{du}{dx}$$

$$\frac{d(u^n)}{dx} = nu^{n-1}\frac{du}{dx}$$

$$\frac{d(\ln u)}{dx} = \frac{1}{u}\frac{du}{dx}$$

$$\frac{d(e^u)}{dx} = e^u\frac{du}{dx}$$

$$\frac{d(\sin u)}{dx} = \frac{du}{dx}(\cos u)$$

$$\frac{d(\cos u)}{dx} = -\frac{du}{dx}(\sin u)$$

A.4.2   Integrals

$$\int a\,dx = ax$$

$$\int a\,f(x)\,dx = a\int f(x)\,dx$$

$$\int x^n\,dx = \frac{x^{n+1}}{n+1} \quad (n \neq -1)$$

$$\int e^x\,dx = e^x$$

$$\int e^{ax}\,dx = \frac{e^{ax}}{a}$$

$$\int \ln(x)\,dx = x\ln(x) - x$$

Appendix B
Unit Conversions

Acceleration	1 m/s^2	$= 3.2808 \text{ ft/s}^2$
		$= 39.370 \text{ in/s}^2$
		$= 4.252 \times 10^7 \text{ ft/h}^2$
		$= 8053 \text{ mi/h}^2$
Area	1 m^2	$= 10^4 \text{ cm}^2 = 10^6 \text{ mm}^2$
		$= 10.7636 \text{ ft}^2$
		$= 1550 \text{ in}^2$
	1 acre	$= 43{,}560 \text{ } ft^2$
Density	1 kg/m^3	$= 1000 \text{ g/m}^3 = 0.001 \text{ g/cm}^3$
		$= 0.06243 \text{ lb}_\text{m}/\text{ft}^3$
		$= 3.6127 \times 10^{-5} \text{ lb}_\text{m}/\text{in}^3$
		$= 0.001940 \text{ slug/ft}^3$
Energy, work, heat	1055.06 J	= 1 Btu
	1.35582 J	$= 1 \text{ ft} \cdot \text{b}_\text{f}$
	4.1868 J	= 1 cal
	252 cal	= 1 Btu
	1 kWh	= 3412 Btu = 3600 kJ
Force	1 N	$= 10^5 \text{ dyne}$
		$= 0.22481 \text{ lb}_\text{f}$
	1 lb_f	$= 32.174 \text{ lb}_\text{m} \cdot \text{ft/s}^2$
Heat transfer, power	1 W	= 1 J/s
		= 3.6 kJ/h
		= 3.4121 Btu/h
	745.7 W	= 1 hp
		$= 550 \text{ lb}_\text{f} \cdot \text{ft/s}$
		= 2544.4 Btu/h
	1.3558 W	$= 1 \text{ lb}_\text{f} \cdot \text{ft/s}$

Length	1 m	$= 100$ cm $= 1000$ mm
		$= 3.2808$ ft
		$= 39.370$ in
		$= 3.0936$ yd
	2.54 cm	$= 1$ in
	1 ft	$= 12$ in
	5280 ft	$= 1$ mi
	1 km	$= 0.6214$ mi $= 0.5400$ nautical mi
Mass	1 kg	$= 1000$ g
		$= 2.20462$ lb_m
		$= 0.06852$ slug
	1 slug	$= 32.174$ lb_m
	1 short ton	$= 2000$ lb_m
	1 long ton	$= 2240$ lb_m
Mass flow rate	1 kg/s	$= 2.20462$ lb_m/s
		$= 7937$ lb_m/h
		$= 0.06852$ slug/s
		$= 246.68$ slug/h
Pressure	1 kN/m^2	$= 1$ kPa
		$= 20.8855$ lb_f/ft^2
		$= 0.14504$ lb_f/in^2 $= 0.14504$ psi
		$= 0.2953$ in Hg
		$= 4.0146$ in H_2O
	101.325 kPa	$= 1$ atm
		$= 14.6959$ lb_f/in^2 $= 14.6959$ psi
		$= 760$ mm Hg at 0°C
	1 bar	$= 10^5$ Pa
Specific heat	1 kJ/kg·°C	$= 1$ kJ/kg·K $= 1$ J/g·°C
		$= 0.2388$ Btu/lb_m·°F
		$= 0.2388$ Btu/lb_m·°R
Stress, modulus	1 kN/m^2	$= 1$ kPa
		$= 0.14504$ lb_f/in^2 $= 0.14504$ psi
	1 MN/m^2	$= 1$ MPa
		$= 1000$ kPa
		$= 145.04$ lb_f/in^2 $= 145.04$ psi
	1 GN/m^2	$= 1$ GPa
		$= 1000$ MPa
		$= 1.4504 \times 10^5$ lb_f/in^2
		$= 1.4504 \times 10^5$ psi
		$= 145$ ksi
Temperature	T(K)	$= $ T(°C) $+ 273.15$
		$= $ T(°R)/1.8
		$= $ [T(°F) $+ 459.67$]/1.8
	T(°F)	$= 1.8$ T(°C) $+ 32$

Temperature difference	$\Delta T(K)$	$= \Delta T(°C)$
		$= \Delta T(°F)/1.8$
		$= \Delta T(°R)/1.8$
Velocity	1 m/s	$= 3.2808 \text{ ft/s}$
		$= 11{,}811 \text{ ft/h}$
		$= 2.2369 \text{ mi/h}$
		$= 3.6000 \text{ km/h}$
		$= 0.5400 \text{ knot}$
Viscosity (dynamic)	$1 \text{ kg/m} \cdot \text{s}$	$= 1 \text{ Pa} \cdot \text{s} = 10 \text{ poise}$
		$= 0.6720 \text{ lb}_m/\text{ft} \cdot \text{s}$
		$= 2419 \text{ lb}_m/\text{ft} \cdot \text{h}$
Viscosity (kinematic)	$1 \text{ m}^2/\text{s}$	$= 10{,}000 \text{ stoke}$
		$= 10.7639 \text{ ft}^2/\text{s}$
		$= 38{,}750 \text{ ft}^2/\text{h}$
Volume	1 m^3	$= 1000 \text{ L}$
		$= 35.3134 \text{ ft}^3$
		$= 61{,}022 \text{ in}^3$
		$= 264.17 \text{ gal}$

Appendix C
Physical Properties of Materials

TABLE C.1. Physical Properties of Solids at 20°C

Property Definitions:

ρ = density
c_p = specific heat at constant pressure
E = modulus of elasticity
σ_y = yield stress, tension
σ_u = ultimate stress, tension[a]

Material	ρ (kg/m³)	c_p (J/kg · °C)	E (GPa)	σ_y (MPa)	σ_u (MPa)
Metals					
Aluminum (99.6% Al)	2710	921	70	100	110
Aluminum 2014-T6	2800	875	75	400	455
Aluminum 6061-T6	2710	963	70	240	260
Aluminum 7075-T6	2800	963	72	500	570
Copper					
Oxygen-free (99.9% Cu)	8940	385	120	70	220
Red brass, cold rolled	8710	385	120	435	585
Yellow brass, cold rolled	8470	377	105	410	510
Iron alloys					
Structural steel	7860	420	200	250	400
Cast iron, gray	7270	420	69		655[a]
AISI 1010 steel, cold rolled	7270	434	200	300	365
AISI 4130 steel, cold rolled	7840	460	200	760	850
AISI 302 stainless, cold rolled	8055	480	190	520	860
Magnesium (AZ31)	1770	1026	45	200	255
Monel 400 (67% Ni, 32% Cu) cold worked	8830	419	180	585	675
Titanium (6% Al, 4% V)	4420	610	115	830	900
Nonmetals					
Concrete, high strength	2320	900	30		40[a]
Glass	2190	750	65		50[a]
Plastic					
Acrylic	1180	1466	2.8	52	
Nylon 6/6	1140	1680	2.8	45	75

Material	ρ (kg/m³)	c_p (J/kg · °C)	E (GPa)	σ_y (MPa)	σ_u (MPa)
Polypropylene	905	1880	1.3	34	
Polystyrene	1030	1360	3.1	55	90
Polyvinylchloride (PVC)	1440	1170	3.1	45	70
Rock					
Granite	2770	775	70		240[a]
Sandstone	2300	745	40		85[a]
Wood					
Fir	470		13		50[a]
Oak	660		12		47[a]
Pine	415		9		36[a]

[a]Ultimate stress in compression.

TABLE C.2. Physical Properties of Fluids at 20°C

Property Definitions:

ρ = density

c_p = specific heat at constant pressure

μ = dynamic viscosity

ν = kinematic viscosity

Fluid	ρ (kg/m³)	c_p (J/kg · °C)	μ (kg/m · s)	ν (m²/s)
Liquids				
Ammonia	600	4825	1.31×10^{-4}	2.18×10^{-7}
Engine oil (SAE10W-30)	878	1800	0.191	2.17×10^{-4}
Ethyl alcohol	802	2457	1.05×10^{-3}	1.31×10^{-6}
Gasoline	751	2060	5.29×10^{-4}	7.04×10^{-7}
Glycerin	1260	2350	1.48	1.18×10^{-3}
Mercury	13,550	140	1.56×10^{-3}	1.15×10^{-7}
Water	998	4182	1.00×10^{-3}	1.00×10^{-6}
Water/ethylene glycol (50/50 mixture)	1073	3281	3.94×10^{-3}	3.67×10^{-6}
Gases (at 1 atm pressure)				
Air	1.194	1006	1.81×10^{-5}	1.52×10^{-5}
Carbon dioxide (CO_2)	1.818	844	1.46×10^{-5}	8.03×10^{-6}
Carbon monoxide (CO)	1.152	1043	1.72×10^{-5}	1.49×10^{-5}
Helium (He)	0.165	5193	1.95×10^{-5}	1.18×10^{-4}
Hydrogen (H)	0.0830	14,275	8.81×10^{-6}	1.06×10^{-4}
Nitrogen (N_2)	1.155	1041	1.75×10^{-5}	1.52×10^{-5}
Oxygen (O_2)	1.320	919	2.03×10^{-5}	1.54×10^{-5}
Water vapor, saturated (H_2O)	0.0173	1874	8.85×10^{-6}	5.12×10^{-4}

Appendix D

Areas Under the Standard Normal Curve from 0 to z

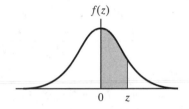

z	.00	.01	.02	.03	.04	.05	.06	.07	.08	.09
0.0	.0000	.0040	.0080	.0120	.0160	.0199	.0239	.0279	.0319	.0359
0.1	.0398	.0438	.0478	.0517	.0557	.0596	.0636	.0675	.0714	.0754
0.2	.0793	.0832	.0871	.0910	.0948	.0987	.1026	.1064	.1103	.1141
0.3	.1179	.1217	.1255	.1293	.1331	.1368	.1406	.1443	.1480	.1517
0.4	.1554	.1591	.1628	.1664	.1700	.1736	.1772	.1808	.1844	.1879
0.5	.1915	.1950	.1985	.2019	.2054	.2088	.2123	.2157	.2190	.2224
0.6	.2258	.2291	.2324	.2357	.2389	.2422	.2454	.2486	.2518	.2549
0.7	.2580	.2612	.2642	.2673	.2704	.2734	.2764	.2794	.2823	.2852
0.8	.2881	.2910	.2939	.2967	.2996	.3023	.3051	.3078	.3106	.3133
0.9	.3159	.3186	.3212	.3238	.3264	.3289	.3315	.3340	.3365	.3389
1.0	.3413	.3438	.3461	.3485	.3508	.3531	.3554	.3577	.3599	.3621
1.1	.3643	.3665	.3686	.3708	.3729	.3749	.3770	.3790	.3810	.3830
1.2	.3849	.3869	.3888	.3907	.3925	.3944	.3962	.3980	.3997	.4015
1.3	.4032	.4049	.4066	.4082	.4099	.4115	.4131	.4147	.4162	.4177
1.4	.4192	.4207	.4222	.4236	.4251	.4265	.4279	.4292	.4306	.4319
1.5	.4332	.4345	.4357	.4370	.4382	.4394	.4406	.4418	.4429	.4441
1.6	.4452	.4463	.4474	.4484	.4495	.4505	.4515	.4525	.4535	.4545
1.7	.4554	.4564	.4573	.4582	.4591	.4599	.4608	.4616	.4625	.4633
1.8	.4641	.4649	.4656	.4664	.4671	.4678	.4686	.4693	.4699	.4706
1.9	.4713	.4719	.4726	.4732	.4738	.4744	.4750	.4756	.4761	.4767
2.0	.4772	.4778	.4783	.4788	.4793	.4798	.4803	.4808	.4812	.4817
2.1	.4821	.4826	.4830	.4834	.4838	.4842	.4846	.4850	.4854	.4857
2.2	.4861	.4864	.4868	.4871	.4875	.4878	.4881	.4884	.4887	.4890
2.3	.4893	.4896	.4898	.4901	.4904	.4906	.4909	.4911	.4913	.4916
2.4	.4918	.4920	.4922	.4925	.4927	.4929	.4931	.4932	.4934	.4936
2.5	.4938	.4940	.4941	.4943	.4945	.4946	.4948	.4949	.4951	.4952
2.6	.4953	.4955	.4956	.4957	.4959	.4960	.4961	.4962	.4963	.4964
2.7	.4965	.4966	.4967	.4968	.4969	.4970	.4971	.4972	.4973	.4974
2.8	.4974	.4975	.4976	.4977	.4977	.4978	.4979	.4979	.4980	.4981
2.9	.4981	.4982	.4982	.4983	.4984	.4984	.4985	.4985	.4986	.4986

z	.00	.01	.02	.03	.04	.05	.06	.07	.08	.09
3.0	.4987	.4987	.4987	.4988	.4988	.4989	.4989	.4989	.4990	.4990
3.1	.4990	.4991	.4991	.4991	.4992	.4992	.4992	.4992	.4993	.4993
3.2	.4993	.4993	.4994	.4994	.4994	.4994	.4994	.4995	.4995	.4995
3.3	.4995	.4995	.4995	.4996	.4996	.4996	.4996	.4996	.4996	.4997
3.4	.4997	.4997	.4997	.4997	.4997	.4997	.4997	.4997	.4997	.4998
3.5	.4998	.4998	.4998	.4998	.4998	.4998	.4998	.4998	.4998	.4998
3.6	.4998	.4998	.4999	.4999	.4999	.4999	.4999	.4999	.4999	.4999
3.7	.4999	.4999	.4999	.4999	.4999	.4999	.4999	.4999	.4999	.4999
3.8	.4999	.4999	.4999	.4999	.4999	.4999	.4999	.4999	.4999	.4999
3.9	.5000	.5000	.5000	.5000	.5000	.5000	.5000	.5000	.5000	.5000

Appendix E
Greek Alphabet

Name of Letter	Uppercase Symbol	Lowercase Symbol	Name of Letter	Uppercase Symbol	Lowercase Symbol
Alpha	A	α	Nu	N	ν
Beta	B	β	Xi	Ξ	ξ
Gamma	Γ	γ	Omicron	O	o
Delta	Δ	δ	Pi	Π	π
Epsilon	E	ε	Rho	P	ρ
Zeta	Z	ζ	Sigma	Σ	σ
Eta	H	η	Tau	T	τ
Theta	Θ	θ	Upsilon	Υ	υ
Iota	I	ι	Phi	Φ	φ
Kappa	K	κ	Chi	X	χ
Lambda	Λ	λ	Psi	Ψ	ψ
Mu	M	μ	Omega	Ω	ω

Appendix F
Answers to Selected Problems

CHAPTER 1

1.2	y_{max} (mm)	h (mm)	b (mm)
	2.16	200	100
	2.70	200	80
	4.04	175	80
	1.11	250	100
	3.94	125	225
	1.85	175	175
	3.42	150	150

CHAPTER 2

2.2. The argument of any mathematical function must be dimensionless. Because the argument $[L][t]$ is not a dimensionless quantity, the equation is not dimensionally consistent.

2.4. Yes, because the argument $[M][M]^{-1}$ of the cosine function is dimensionless and $[T][N][T]$ appears on both sides of the equation.

2.6. $1 \text{ N} \cdot \text{m} = 1 \text{ J(joule)}$; work, energy, heat

2.8. $P = I^2R$

2.10. 20 kW

2.12. $3.66 \times 10^6 \text{kg}$, 36.0 MN

2.14. a) 4 lb$_f$, b) 1.51 lb$_f$

2.16. 2.30 kg, 22.6 N

2.18. 182

2.20. 15.0 mi/h, 6.71 m/s

2.22. $-40°$

2.24. 5.76 MJ

2.26.   11.0 lb_m/s, 1233 slug/h

2.28.   36.1°C, 36.1 K, 65°R

2.30. 3.162×10^7 s

2.32.   883 N, 198 lb_f

2.34.   3986 Pa, 0.578 psi

2.36. 4.83×10^6 Btu/h · ft^3

2.38. 3.15×10^{-4} m^3/s, 40.1 ft^3/h

2.40.   270 MJ, 2.56×10^5 Btu, 6.45×10^7 cal

CHAPTER 3

3.10.   b) <u>9.807</u>   c) 0.00<u>216</u>   d) <u>5</u>000   e) <u>7</u>000.   f) <u>12.00</u>   g) <u>2066</u>   h) <u>106.07</u>

        i) 0.0<u>2880</u>   j) <u>523.91</u>   k) <u>1</u>.<u>207</u> $\times 10^{-3}$

3.12. 2.5×10^3 N

3.14.   11 V

3.16.   36

3.18.   0.32 MN

3.20.   4

3.22.   170 kΩ, 0.587 mA

        20 kΩ: $V = 11.7$ V; 150 kΩ: $V = 88.1$ V; 250 Ω: $V = 0.147$ V

CHAPTER 4

4.2. $-7.01\,\mathbf{i} + 13.3\,\mathbf{j}$ N

4.4. $\mathbf{F}_R = 0\,\mathbf{i} - 38.9\,\mathbf{j}$ lb_f, $F_R = 38.9\,lb_f$, $\theta = -90°$

4.6.   No, because $\mathbf{F}_1 + \mathbf{F}_2 + \mathbf{F}_3 \neq \mathbf{0}$.

4.8. $F_1 = F_2 = 600$ N

4.10. $F = 1134$ N, $\theta = 41.4°$

4.12.   85.5 lb_f

4.14.   55.6 lb_f

4.16. $T_{AB} = T_{AC} = T_{BC} = 117$ N, $T_{BD} = T_{CE} = 219$ N

4.18.   19.9 cm

4.20. $\sigma_{AB} = 177$ MPa, $\sigma_{BC} = 78.6$ MPa, $\delta = 0.599$ mm

4.22.   horizontal: $\varepsilon = 1.67 \times 10^{-4}$, $\delta = 2.00 \times 10^{-5}$ m

        vertical: $\varepsilon = 1.25 \times 10^{-4}$, $\delta = 1.50 \times 10^{-5}$ m

4.24.   No. The factor of safety is acceptable at F.S. = 1.57, but the deformation is $\delta = 3.82$ cm, which exceeds the allowable deformation.

4.26.   1.57

CHAPTER 5

5.2. $t = 0.25$ s: $q = 0.25$ C

  $t = 0.75$ s: $q = 1.30$ C

5.4. 2.48, 5.70

5.6. 0.546 A; No, a large fraction of the electrical power is converted to heat.

5.10. (c)

5.12. 196, 148 Ω

5.14. 1933 Ω

5.16. 4 Ω

5.18. 1 Ω

5.22. 22 Ω: $V = 13.2$ V, $I = 0.601$ A

  75 Ω: $V = 36.8$ V, $I = 0.490$ A

  333 Ω: $V = 36.8$ V, $I = 0.111$ A

5.24. a) 0.172 A b) 77.6 V c) variable resistor: 13.4 W, fixed resistor: 3.86 W

5.26. 20 Ω: $V = 9.23$ V, $I = 0.426$ A, $P = 4.26$ W

  75 Ω: $V = 30.8$ V, $I = 0.410$ A

  100 Ω: $V = 5.12$ V, $I = 51.2$ mA, $P = 0.263$ W

  500 Ω: $V = 25.6$ V, $I = 51.2$ mA

5.28. 5 Ω: $V = 2.50$ V, $I = 0.5$ A

  10 Ω: $V = 1.67$ V, $I = 0.167$ A

  50 Ω: $V = 8.33$ V, $I = 0.167$ A

  25 Ω: $V = 8.33$ V, $I = 0.333$ A

  5 Ω: $V = 1.67$ V, $I = 0.333$ A

5.30. 7 Ω: $V = 1.40$ V, $I = 0.200$ A

  1 Ω: $V = 0.200$ V, $I = 0.200$ A

  25 Ω: $V = 1.15$ V, $I = 46.1$ mA

  5 Ω: $V = 0.231$ V, $I = 46.1$ mA

  10 Ω: $V = 0.461$ V, $I = 46.1$ mA

  3 Ω: $V = 0.107$ V, $I = 35.7$ mA

  40 Ω: $V = 1.43$ V, $I = 35.7$ mA

  2 Ω: $V = 0.308$ V, $I = 0.154$ A

  13 Ω: $V = 1.54$ V, $I = 0.118$ A

CHAPTER 6

6.2.	572 kPa
6.4.	6.2 psi
6.6.	558.3°R, 37.0°C, 310.2 K
6.8.	30.6°F, 30.6°R, 17 K
6.10.	$W_b = C\dfrac{(V_2^{1-n} - V_1^{1-n})}{1-n}, n \neq 1$
6.12.	12.1 N · m
6.14.	0.350 m
6.16.	5 air-conditioning units
6.18.	30 kJ
6.20.	280 MJ, 6.09 kg
6.22.	3.39 GJ, 1500 W/cm^2
6.24.	105°C
6.26.	60 kW
6.28.	0.556, 8 MW
6.30.	1.8 MJ
6.32.	3 GW, $\eta_{\text{carnot}} = 0.579$
6.36.	$\dot{Q}_{\text{in}} = 7.53$ MW, $\dot{Q}_{\text{out}} = 2.53$ MW
6.38.	$\dot{W}_{\text{max}} = 69.9$ kW, $\dot{W}_{\text{actual}} = 40$ kW

CHAPTER 7

7.2.	0.993 kg, 9.74 N
7.4.	5.43 kg, 53.3 N
7.6.	20.4 MPa
7.8.	0.375 m
7.10.	0.01 Pa
7.12.	110×10^3 kPa, 1.09×10^3 atm
7.14.	62.2 kPa
7.16.	7.27 MN
7.18.	2.49 kPa (assuming $h = 25.4$ cm and $\gamma = 9810$ N/m^3)
7.20.	$Q = 0.0393$ m^3/s, $\dot{m} = 39.3$ kg/s
7.22.	$\dot{m} = 1.24 \times 10^{-5}$ kg/s, $V = 3.57 \times 10^{-3}$ m/s, $t = 12.4$ h
7.24.	$V = 1.56$ m/s, $\dot{m} = 3.58$ kg/s

7.26. $\Delta V = 56.6$ m/s

7.28.   2 kg/s (enters the junction)

7.30.   small branch: $Q = 9.36$ m³/s, $\dot{m} = 11.2$ kg/s

   large branch: $Q = 18.4$ m³/s, $\dot{m} = 22.0$ kg/s

CHAPTER 8

8.4.   a. G   b. S, R   c. S   d. S   e. G   f. S   g. S, R   h. G

8.6.   Incomplete label on y-axis, missing label on x-axis, no legend, no minor graduations

8.10.   a. independent variable: pump speed, S
   dependent variable: pump output, Q

   b. $Q = 3.95 \times 10^{-4} \, S^{1.54}$ gal/min

   c. $S = 175$ rpm: $Q = 1.12$ gal/min

   $S = 350$ rpm: $Q = 3.27$ gal/min

   $S = 475$ rpm: $Q = 5.23$ gal/min

8.12.   a. independent variable: resistance, R
   dependent variable: current, I

   b. $I = 11.2 \, R^{-1.08}$ mA

   c. $R = 2.5$ kΩ: $I = 4.16$ mA

   $R = 4.7$ kΩ: $I = 2.11$ mA

   $R = 5.6$ kΩ: $I = 1.74$ mA

8.14.   a. independent variable: velocity, v
   dependent variable: drag force, F

   b. $F = 0.89 \, v^{1.97}$ N

   c. $v = 12$ m/s: $F = 119$ N

   $v = 45$ m/s: $F = 1608$ N

   $v = 60$ m/s: $F = 2834$ N

8.16.   a. independent variable: time, t
   dependent variable: temperature, T

   b. $T = 200 \, e^{-0.099t} {}^\circ$C

   c. $t = 7$ s: $T = 100°$C

   $t = 18$ s: $T = 33.7°$C

8.18.   a. independent variable: temperature, T
   dependent variable: voltage, V

   b. $V = -0.28 + 0.042 \, T$ mV

  c. $T = 150°C$: $V = 6.02$ mV

   $T = 575°C$: $V = 23.9$ mV

   $T = 850°C$: $V = 35.4$ mV

8.20. a. independent variable: temperature, T
   dependent variable: solubility, S

  b. $S = 10.0 + 0.023\ T$ kg

  c. $T = 302$ K: $S = 16.95$ kg

   $T = 330$ K: $S = 17.59$ kg

8.22. a. independent variable: diameter, d
   dependent variable: material removal rate, M

  b. $M = 9.95\ d^{2.00}$ in³/min

  c. $d = 0.875$ in: $M = 7.62$ in³/min

   $d = 1.25$ in: $M = 15.5$ in³/min

8.24. a. independent variable: speed, s
   dependent variable: power, P

  b. $P = 6.03 \times 10^{-5}\ s^{3.0}$ hp

  c. $s = 55$ mi/h: $P = 10.0$ hp

8.26. $T = 75°C$: $c = 4.199$ kJ/kg · °C

  $T = 120°C$: $c = 4.254$ kJ/kg · °C

  $T = 180°C$: $c = 4.422$ kJ/kg · °C

CHAPTER 9

9.2. $\bar{x} = 16.08$ oz, median = 16.14 oz, mode = 16.10 oz, s = 0.398 oz

9.4. b. 1005.7 Ω, 45.5 Ω

  c. left side: −2.32, right side: 2.07

  d. $\pm 1\ \sigma$: 940 Ω, 900 Ω, 955 Ω, 930 Ω, 928 Ω, 952 Ω;

    1082 Ω, 1060 Ω, 1055 Ω, 1090 Ω, 1072 Ω, 1061 Ω

   $\pm 2\ \sigma$: 900 Ω

9.6. b. 2.000 in, 0.00529 in

  c. 1824

  d. ±0.0106 in

9.8. a. 4.65 percent

  b. 912

  c. $31.01

9.10.   20.2 percent

9.12.   a. 45.1 percent

b. 17.5 percent

9.14.   73.3 percent

9.16.   a. 98.1 percent

b. 1.86 percent

9.18.   4550 h

Index